ROOFING
CONSTRUCTION & ESTIMATING

by Daniel Atcheson

Craftsman Book Company
6058 Corte del Cedro / P.O. Box 6500 / Carlsbad, CA 92018

Acknowledgements

The author wishes to thank the following individuals and companies for providing general assistance, information, and various illustrations used in the preparation of this book.

American Plywood Association (APA) 7011 South 19th Street, P.O. Box 11700, Tacoma, WA 98411-0700
ASC Pacific, Inc. 2110 Enterprise Boulevard, W. Sacramento, CA 95691-3493
Asphalt Roofing Manufacturers Association (ARMA) 6000 Executive Boulevard, Suite 201, Rockville, MD 20852-3803
Tim Atcheson, **Atcheson Building Systems** 3306 61st Street, Lubbock, TX 79413
James Edward Atcheson, 1906-1986
Tami Danielle Atcheson, Lubbock, TX
Mike Atcheson, AIA, Lubbock, TX
Cedar Shake & Shingle Bureau 515 116th Avenue N.E., Suite 275, Bellevue, WA 98004-5294
The Celotex Corporation 4010 Boy Scout Boulevard, Tampa, FL 33607-5750
Allan B. Jones, *Crowder Brothers Hardware* 1671 Venice ByPass, South Venice, FL 34293
Follansbee Steel General Offices, Follansbee, WV 26037
GAF Building Materials Corporation 1361 Alps Road, Wayne, NJ 07470-3689
Gladding McBean P.O. Box 97, 601 7th Street, Lincoln, CA 95648
Haag Engineering Co. P.O. Box 814245, Dallas, TX 75381-4245
Inland Buildings 175 N. Patrick Boulevard, P.O. Box 385, Brookfield, WI 53008-0385
Koppers Industries, Inc. 436 Seventh Avenue, Pittsburgh, PA 15219-1800
Randy Hooks, **Lydick-Hooks Roofing** P.O. Box 2605, Lubbock, TX 79408
Manville Sales Corporation, Roofing Systems Division P.O. Box 5108, Denver, CO 80217
Mayes Brothers Division of Great Neck Saw Manufacturers, Inc. 165 East Second Street, Mineola, NY 11501
MM Systems Corporation 4520 Elmdale Drive, Tucker, GA 30085-0326
Monier Roof Tile P.O. Box 19792, Irvine, CA 92713
Owens-Corning Owens-Corning Fiberglass Tower, Toledo, OH 43659
Petersen Aluminum Corporation 955 Estes Avenue, Elk Grove Village, IL 60007
Rubatex Corporation Adams Street, Bedford, VA 24523
Shakertown Cedar Siding, Shakertown Corporation 1200 Kerron Street/Box 400, Winlock, WA 98596
David Carlson, **Southwest Florida Roofing** 15491 S. Aron Circle, Port Charlotte, FL 33981
Stanley Tools Division of The Stanley Works New Britain, CT 06050
U.S. Department of Housing and Urban Development (HUD) 451 Seventh Street SW, Washington, DC 20410
Vermont Structural Slate Co. Box 98, 3 Prospect Street, Fair Haven, VT 05743
Zappone Manufacturing N. 2928 Pittsburg, Spokane, WA 99207

To my grandchildren Sophie & Cole Painter
And to my Lord and Savior, Jesus Christ, who gave me eyes to see, hands to type and a mind to think.

Looking for other construction reference manuals?
Craftsman has the books to fill your needs. **Call toll-free 1-800-829-8123**
or write to Craftsman Book Company, P.O. Box 6500, Carlsbad, CA 92018 for
a **FREE CATALOG** of over 100 books, including how-to manuals,
annual cost books, and estimating software.
Visit our Web site: http://www.craftsman-book.com

Library of Congress Cataloging-in-Publication Data

Atcheson, Daniel Benn.
  Roofing construction & estimating / by Daniel Atcheson.
    p. cm.
  Includes index.
  ISBN 978-1-57218-007-9
  1. Roofs--Design and construction. 2. Roofs--Estimates.
I. Title.
TH2401.A83 1995
695--dc20
                                    95-13394
                                      CIP

©1995 Craftsman Book Company
Seventh printing 2007

Contents

1 Measuring and Calculating Roofs   5
    Level Roofs. 7
    Sloped Roofs. 10
    How to Measure Roof Slope 12
    Perimeter of a Sloped Roof. 17
    Net Versus Gross Roof Area. 17
    Calculating Total Net Roof Area 18
    Roof Overhangs, Hips and Valleys. 21
    Length of Ridge (Hip Roofs). 22

2 Roof Sheathing, Decking and Loading   23
    Check the Framing 23
    Solid Roof Sheathing. 24
    Spaced Board Sheathing 29
    Roof Decking 32
    Loading the Roof 32
    Estimating Roof Sheathing 34

3 Underlayment on Sloping Roofs   35
    Saturated Felt Underlayment. 36
    Saturated Fiberglass Underlayment. 36
    Underlayment Requirements 37
    Drip Edge. 40
    Installing Underlayment 43
    Estimating Underlayment Quantities 49
    Interlayment (Lacing) 57
    Eaves Flashing (Ice Shield or Water Shield) . . . 61
    Valley Flashing 64

4 Asphalt Shingles   73
    UL Ratings for Shingles. 75
    Deck Requirements 76
    Asphalt Strip Shingles 78
    Flashing at Chimneys and
    Other Vertical Structures 96
    Fasteners 106
    Number of Shingles Required per Square. . . . 108
    Number of Shingle Courses 109
    Estimating Asphalt Strip Shingle Quantities . . . 113
    Ridge and Hip Units 114

    Estimating Ribbon-Course Quantities 124
    Individual Shingles 127
    Estimating Asphalt Shingle Roofing Costs 130

5 Mineral-Surfaced Roll Roofing   131
    Installing Mineral-Surfaced Roll Roofing 133
    Valley Flashing 134
    Estimating Mineral-Surfaced Roll Roofing 144
    Waste from Non-conforming Roof Layout 146
    Estimating Mineral-Surfaced Roll Roofing Costs 157

6 Wood Shingles and Shakes   159
    Installing Wood Shingles and Shakes. 164
    Covering Capacity of Shakes. 174
    Covering Capacity of Wood Shingles 174
    Estimating Wood Shingle and Shake Quantities . 176
    Staggered Patterns 185
    Sidewall Shakes and Wood Shingles 185
    Roof Junctures 192
    Estimating Wood Shingle Roofing Costs 196

7 Tile Roofing   197
    Underlayment Under Tile Roof Coverings 199
    Installing Roof Tiles 200
    The Starter Course 202
    Fastening Roofing Tiles 204
    Flashing at Vertical Walls. 217
    Replacing Broken Tiles 224
    Estimating Tile Quantities 225
    Estimating Total Tile Roofing Costs 229

8 Slate Roofing   231
    Slate Size, Color and Texture 231
    Felt Underlayment 234
    Installation on a Sloping Roof 234
    Fasteners 244
    Flashing 245
    Estimating Slate Quantities 249
    Estimating Slate Roofing Costs 254

9 Metal Roofing and Siding — 255
- Modern Metal Panel Systems — 256
- Installing Metal Roofing Panels — 257
- Job-Fabricated Seams — 263
- Estimating Metal Roofing and Siding — 269
- Steel Roofing and Siding Quantities — 270
- Ribbed Metal Panel Quantities — 271
- Miscellaneous Metal Roofing Quantities — 275

10 Built-Up Roofing — 291
- Roof Slopes — 292
- Substrate Design — 292
- Back Nailing — 297
- Base Sheets (Vapor Retarders) — 298
- Roofing Membranes — 300
- Hot Bitumens — 303
- Cold-applied Bitumens — 305
- Surface Aggregate — 306
- Smooth-surface Roofing — 308
- Cap Sheets — 308
- Aluminum Roof Coatings — 309
- Phasing — 310
- Cant Strips — 311
- Temporary Roofs — 313
- Roof Traffic Pads — 314
- Water-retaining Roofs — 315
- Flashing on Flat Roofs — 315
- Roof Expansion Joints — 319
- Estimating BUR Systems — 323
- Testing BUR Systems — 327
- Built-up Roofing Warranties — 327
- Built-up Roofing Repairs and Re-roofing — 329

11 Elastomeric Roofing — 333
- The Advantages of Elastomeric Systems — 334
- Liquid-applied Elastomers — 335
- Single-Ply Roofing Systems — 338
- EPDM Elastomeric Systems — 337
- CPE Elastomeric Roofing — 342
- CSPE Elastomeric Roofing — 342
- Hypalon Roofing — 343
- PVC Elastomeric Roofing — 343
- Composite Roofing Systems — 343
- Flashings for Elastomeric Roofs — 344
- Estimating Elastomeric Roofing — 345

12 Insulation, Vapor Retarders and Waterproofing — 347
- The Benefits of Insulation — 347
- Insulation Materials — 348
- Reducing Heat Loss — 355
- Insulation Values — 361
- Vapor Barriers — 362
- Weatherproofing Existing Homes — 364
- Caulking and Sealants — 364
- Wall Flashing — 370
- Waterproofing — 371
- Dampproofing — 377

13 Roofing Repair and Maintenance — 381
- Finding the Source of Leaks — 381
- Repairing Leaks — 384
- Roof Maintenance — 386
- Assessing Hail Damage — 388
- Roofing Demolition — 390
- Re-Roofing — 394
- Estimating Re-Roofing Quantities — 401
- Attic Ventilation — 402
- Gutters and Downspouts — 407

14 Estimating (and Maximizing) Production Rates — 411
- Labor Unit Prices — 411
- Estimating with Published Prices — 415
- Roofing Labor Tips — 420

Appendix A Roof-Slope Factors — 428

Appendix B Valley Length Factors — 429

Appendix C Equations Used in This Book — 430

Index — 436

1 Measuring and Calculating Roofs

▶ If you're like some roofing contractors, you estimate roofing quantities by calculating the area of a roof, then adding 10 percent for waste. That might be OK in a fat building market, but in a tight market you'll need a sharper pencil to compete successfully for the good jobs, and then make money on them. In this book, I'm going to show you how to make a quick and accurate takeoff for any kind of roof.

You'll also learn the latest and most acceptable roofing methods in an industry where installation practices are closely related to warranties. That's because material warranties may be invalid if you don't follow the manufacturer's recommendations for installation. Look here for general guidelines, but always follow the manufacturer's instructions to the letter.

New products come on the market every day to solve the complex roof covering requirements presented by modern building technology. Your job is to know as much as you can about those products. You also have to know how to install them so the job passes inspection and presents no future repair and maintenance problems. Callbacks are hard on your profit margin — and they don't do your reputation any good either. Know as much as you can about your roofing business, and you'll avoid them.

This book is more than an estimating book for roofing contractors. It develops a *system,* beginning with Chapter 1, for all types of roofing materials and installation methods. We'll cover the entire roofing trade, including how to manage your crews and keep them safe. So let's get started.

Before you can bid any job, you have to figure your costs. And before you can figure the costs, you have to know the size of the job. So you have to do two things: First, measure the roof and calculate the total area. Then find the lengths of the eaves, gables (or rakes), ridges, hips and valleys.

When you construct a roof on a new building, you can get these measurements from the plans. On repair or replacement jobs, you'll probably have to take your pencil, clipboard and tape measure, haul out your ladder, climb onto the roof, and start measuring.

To avoid mistakes, or a second trip to the job site, develop a system for taking measurements. Use a 100-foot flexible tape which has a ½-inch grout hook at the "stupid" end of the tape. Flexible tapes are made of metal, or fiberglass-reinforced nylon fabric. Find a tape that's marked with highlights at 5-inch intervals to match the exposure of most composition shingles.

There is no cardinal rule for the sequence you use to measure a roof, as long as you don't miss anything. Here's a system that works for me:

Start by measuring the length of the eaves. On a gable roof, you only have to measure in one direction. On a hip roof, you'll have to measure the eaves in two directions.

Next, measure the width of the roof. On a gable roof, hook the tape over one of the eaves, and run it over the ridge to the opposite eave. On a hip roof, measure the width the same way. To measure the length, hook the tape to the eaves at the ridge rafter (look ahead to Figure 1-16 on page 13 for an illustration of the parts of a roof), run the tape the length of the ridge and down the opposite ridge rafter. Measure the ridge at the same time.

Now, measure the hips and valleys by hooking the tape to a building corner and running the tape to the ridge. You use these measurements to calculate material requirements such as valley flashing and hip-covering material.

When you measure, some dimensions need to be more accurate than others. For instance, you could miss the length of ridge, hip or valley by a foot or more, and the error wouldn't affect your total bid price too much. But don't make a mistake in the length and width, because that error could be substantial. For example, assume you measure a roof at 100 feet by 200 feet, while the actual measurements are 100'6" by 200'6". The difference between the two measurements is 150 square feet, or 1½ squares of material.

Always make a sketch of the roof layout, including dimensions, roof slopes, location of penetrations and any unusual circumstances such as rotten deck areas, ventilation problems, or overhanging tree branches or other obstructions.

Once you have the measurements, you'll use them to calculate areas, slopes, angles, and allowance factors. Let's begin with an easy example.

Measuring and Calculating Roofs

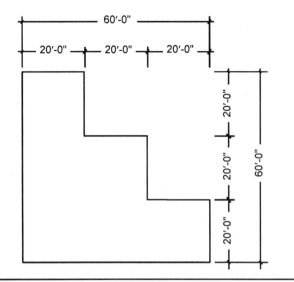

Figure 1-1 Roof plan of level roof

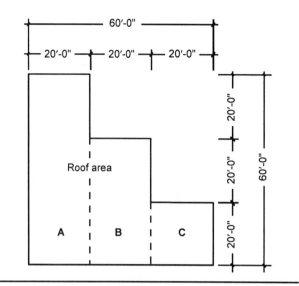

Figure 1-2 The positive method

Level Roofs

The dimensions on the plans give you the actual measurements for a level roof. To get the area of a rectangular roof, multiply its length by its width.

Area of a level rectangular roof = L x W | Equation 1-1 |

where L is the length and W is the width.

Of course, not every roof you work on will be a single rectangle. You may need to figure the area of a roof like the one in Figure 1-1. There are two ways to calculate this area:

1) The positive method

2) The negative method

In the *positive* method, you divide the roof into rectangular areas, then add the parts to get the total area. See Figure 1-2.

With the *negative* method, you extend the roof lines to form a single rectangle. Calculate the area of this rectangle, and subtract the areas of the rectangular spaces which lie outside the actual roof. Figure 1-3 illustrates this.

▼ **Example 1-1:** The Positive Method

Divide the roof into rectangles as shown in Figure 1-2. Calculate the area of each rectangle, then add them together:

Area A = 20 feet by 60 feet, or 1,200 square feet
Area B = 20 feet by 40 feet, or 800 square feet
Area C = 20 feet by 20 feet, or 400 square feet

Then, the total area = 1,200 SF + 800 SF + 400 SF, or 2,400 SF

▼ **Example 1-2:** The Negative Method

Extend the roof lines to form one rectangle, as in Figure 1-3. Calculate the total area of that rectangle, then subtract the areas of any rectangles which aren't in the actual roof:

Extended rectangle = 60' x 60' = 3,600 SF

Area A = 40 feet by 20 feet, or 800 square feet
Area B = 20 feet by 20 feet, or 400 square feet

Total area outside the roof layout is 800 SF + 400 SF, or 1,200 SF. Subtract that from the extended area to get the total area:

3,600 SF - 1,200 SF = 2,400 SF

You get the same answer both ways. So you might as well use the easiest method — the one that requires the fewest calculations. For example, in Figure 1-4 you'd have to calculate three areas, then add them together. But in Figure 1-5 you only have to calculate two areas, and then subtract one from the other.

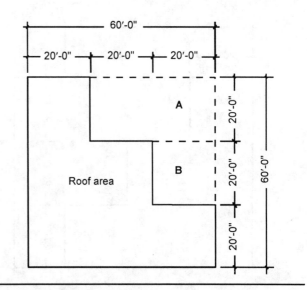

Figure 1-3 The negative method

Perimeter of a Level Roof

The *perimeter* (also called the perimetry or periphery) of a level roof is the total distance around the roof, measured from outside of roof to outside of roof. For example, in Figure 1-6, the perimeter is:

L + W + L + W + R + R, or 2L + 2W + 2R, or 2(L+W+R)   **Equation 1-2**

where L is the roof length, W is the width, and R is the depth of the recess.

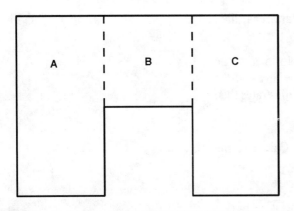

Figure 1-4 The positive method requiring three area calculations

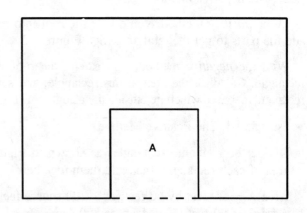

Figure 1-5 The negative method requiring two area calculations

If a building doesn't have any recesses, the equation is simply:

Perimeter = 2(L+W) | Equation 1-3 |

or 2 times the total of length plus width.

▼ **Example 1-3:** Find the perimeter of the level roof shown in Figure 1-7.

The perimeter of the roof = 2 x (40' + 45' + 12')
= 194 linear feet

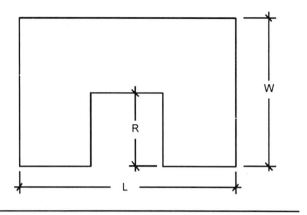

Figure 1-6   Roof perimeter

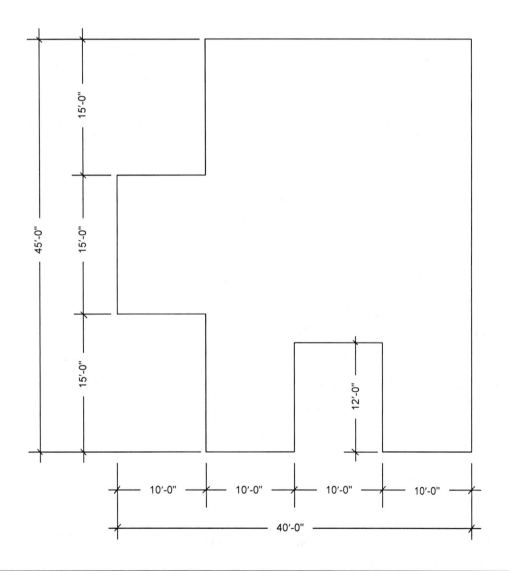

Figure 1-7   Roof perimeter example

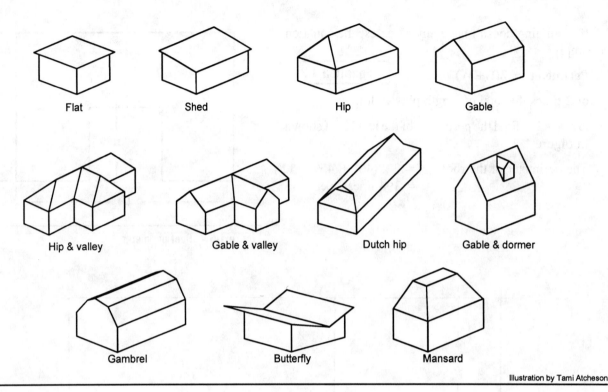

Figure 1-8   Roof types

Sloped Roofs

Figure 1-8 shows a few of the almost limitless types of sloped roofs. You define the *slope* of a roof in terms of rise (in inches) per 12 inches of run. For example, a "6 in 12 roof" is a roof that rises 6 inches for every 12 inches of horizontal run. That's illustrated in Figure 1-9.

You can determine the slope of any roof with the equation:

$$\text{Slope} = \frac{\text{Total Rise}}{\text{Total Run}} = \frac{\text{Rise (in inches)}}{12 \text{ (inches per foot of run)}}$$

Equation 1-4

▼ **Example 1-4:** Use the above equation to find the slope of the roof in Figure 1-10.

$$\text{Slope} = \frac{5 \text{ feet 5 inches (total rise)}}{13 \text{ feet (total run)}} = \frac{? \text{ inches}}{12}$$

Step 1: Convert feet and inches into feet and hundredths of a foot. To do that you divide 5 inches by 12 (inches). Notice that throughout the book, we usually round calculations to two decimal places.

$5 \div 12 = 0.4166 = 0.42'$ (rounded off)

Measuring and Calculating Roofs

Now the equation reads:

$$\text{Slope} = \frac{5.42' \text{ (total rise)}}{13' \text{ (total run)}} = \frac{? \text{ inches}}{12}$$

Step 2: To solve for "? inches," multiply both sides of the equation by 12:

$$\text{Slope} = \frac{5.42}{13} \times 12 = 0.417 \times 12 = 5.004$$

You'll round that answer down to 5, so the slope of the roof is 5 in 12.

You can change the original formula to find total rise if you already know the slope and the total run:

$$\text{Slope} = \frac{\text{Total Rise}}{\text{Total Run}}$$

$$\begin{aligned}\text{Total Rise} &= \text{Slope} \times \text{Total Run} \\ &= \frac{5}{12} \times 13 \\ &= 5.42' \text{ or } 5'5''\end{aligned}$$

You can also find the total run if you know the slope and the total rise:

Total Run = Total Rise ÷ Slope = 5.42 ÷ 0.417 = 13

(Remember, the 0.417 is 5 divided by 12.)

▼ **Example 1-5:** Calculate the total rise for a roof with a slope of 5 in 12 and a run of 20 feet.

$$\begin{aligned}\text{Total Rise} &= 0.417 \times 20 \\ &= 8.34' \text{ or } 8'4''\end{aligned}$$

Roof Pitch

The total *span* of a roof is the horizontal distance, from one eave to the other, as shown in Figure 1-11. You can use that information in a formula to find roof slope, if you know the roof *pitch,* by:

$$\text{Pitch} = \frac{\text{Total Rise}}{\text{Total Span}} \qquad \boxed{\text{Equation 1-5}}$$

$$\text{Slope} = 2 \times \text{Pitch} \qquad \boxed{\text{Equation 1-6}}$$

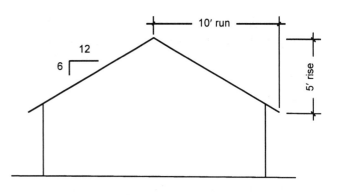

Figure 1-9   Roof rise and run

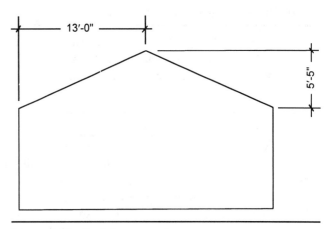

Figure 1-10   Roof slope example

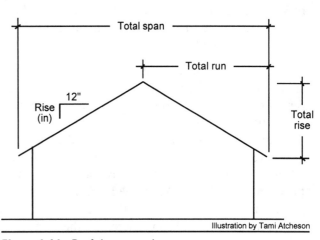

Figure 1-11   Roof rise, run and span

Occasionally a roof is described in terms of pitch, although that term means more to the framer than to a roofing estimator. But sometimes the pitch is the only information you have. Here's how to convert roof pitch to roof slope:

▼ **Example 1-6:** Convert a 1/3 pitch into terms of roof slope.

Slope = 2 x 1/3
      = 2/3

From Example 1-4, Step 2, you have:

Slope = $\frac{2}{3}$ x 12 = 8

Therefore, the roof slope is 8 in 12.

Roof Slope in Degrees of an Angle

Sometimes roof slope is described in terms of degrees of an angle. When it is, you can use Figure 1-12 to convert roof slope to degrees, and vice versa.

How to Measure Roof Slope

You can determine the slope of a roof with an adjustable device called a *Squangle®*. You simply place the Squangle® against an exposed rafter tail or a block placed over the fascia board, adjust the square so that it lines up with the slope of the roof and read the scale. Figure 1-13 shows a Squangle®.

You can also use a *sliding T-bevel* to size the angle between the fascia and roof deck. That's shown in Figure 1-14. Then transfer the angle to a board or sheet of paper and measure it with a Squangle® or protractor.

You can also place a bubble level and ruler over a straight board on the roof slope as shown in Figure 1-15. Since roof slopes are expressed in terms of rise per 12 inches of run, mark the level at 12 inches from one end. To determine slope, center the bubble in the level, place a ruler vertically so that its scale is lined up at your mark 12 inches from the up-slope end of the level, then read the distance to the bottom of the level. If, for instance, you read 4 inches on the ruler, the roof rises (or falls) 4 inches for each foot of run. Therefore, the roof slope is 4 in 12.

Roof slope (rise & run)	Roof slope (degrees)
1 in 12	4.76
2 in 12	9.46
3 in 12	14.04
4 in 12	18.43
5 in 12	22.62
6 in 12	26.57
7 in 12	30.26
8 in 12	33.69
9 in 12	36.87
10 in 12	39.81
11 in 12	42.51
12 in 12	45.00
13 in 12	47.29
14 in 12	49.40
15 in 12	51.34
16 in 12	53.13
17 in 12	54.78
18 in 12	56.31
19 in 12	57.72
20 in 12	59.04
21 in 12	60.26
22 in 12	61.39
23 in 12	62.45
24 in 12	63.43

Figure 1-12 Converting roof slope to degrees

Courtesy of Mayes Brothers Division of Great Neck Saw Manufacturers Inc.

Figure 1-13 Squangle®

Measuring and Calculating Roofs

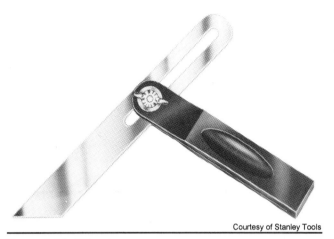

Figure 1-14 T-bevel

Figure 1-15 Determining roof slope

Rafters

Rafters are the inclined members of the roof frame. Figure 1-16 illustrates these rafter types:

- A rafter that extends perpendicularly from the top of an outside wall to the ridge board is called a *common rafter*.

- A common rafter that runs parallel to the ridge board is called a *ridge rafter*.

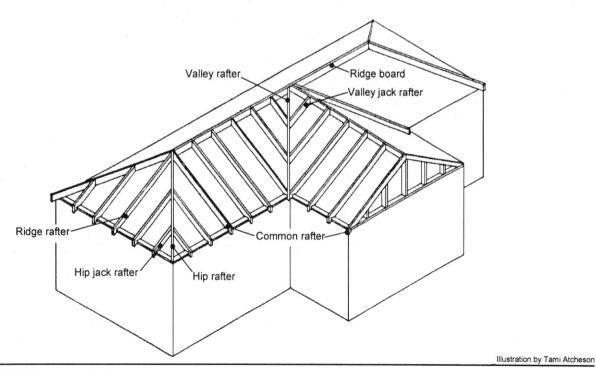

Figure 1-16 Various types of rafters

- A rafter that extends diagonally from an outside corner of a building to the ridge board is called a *hip rafter*.
- A rafter that extends diagonally from an inside corner of a building to the ridge board is called a *valley rafter*.
- A rafter that extends from an outside wall to a hip rafter is called a *hip jack rafter*.
- A rafter that extends from the ridge board to a valley rafter is called a *valley jack rafter*.

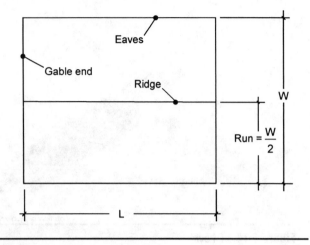

Figure 1-17 Plan view of gable roof

Rafter Length

Figure 1-17 is a plan view of a gable roof. The length (L on Figure 1-17) of the eaves edge (the roof dimension perpendicular to the run of the rafters) is horizontal. Therefore, you can read that dimension directly from the plans. But you can't see the exact size of the width (W in the figure) because the roof slopes. The *plan length* of a common rafter is called the *run* of the rafter. Figure 1-18 illustrates this for three kinds of rafters.

Sometimes you can scale the lengths of common rafters from an elevation or cross section drawing. But it's safer and more convenient to convert the plan dimensions. Figure 1-19 is a table which gives you the appropriate conversion factors. Column 2 of the table gives factors for common rafters, and columns 3 and 4 are for hip or valley rafters.

The values in column 2 are called *roof-slope factors*. The values in columns 3 and 4 are called *hip/valley-slope factors*. The conversion factors in Figure 1-19 assume that all hips and valleys are framed at an angle of 45 degrees with respect to the eaves line.

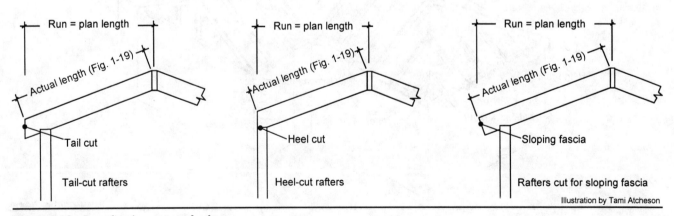

Figure 1-18 Run of various types of rafters

Measuring and Calculating Roofs

(1) Roof slope	(2) Common or jack rafters (factor x run = actual length)	(3) Hips or valleys (factor x run = actual length)	(4) Hips or valleys (factor x plan length = actual length)
1 in 12	1.004	1.417	1.002
2 in 12	1.014	1.424	1.007
3 in 12	1.031	1.436	1.015
4 in 12	1.054	1.453	1.027
5 in 12	1.083	1.474	1.042
6 in 12	1.118	1.500	1.061
7 in 12	1.158	1.530	1.082
8 in 12	1.202	1.564	1.106
9 in 12	1.250	1.601	1.132
10 in 12	1.302	1.642	1.161
11 in 12	1.357	1.685	1.191
12 in 12	1.414	1.732	1.225
13 in 12	1.474	1.782	1.260
14 in 12	1.537	1.833	1.296
15 in 12	1.601	1.888	1.335
16 in 12	1.667	1.944	1.375
17 in 12	1.734	2.002	1.416
18 in 12	1.803	2.062	1.458
19 in 12	1.873	2.123	1.501
20 in 12	1.944	2.186	1.546
21 in 12	2.016	2.250	1.591
22 in 12	2.088	2.315	1.637
23 in 12	2.162	2.382	1.684
24 in 12	2.236	2.450	1.732

Figure 1-19 Roof-slope factors for determining rafter lengths

Use roof-slope factors from column 2 of Figure 1-19 to determine the actual length of a common rafter or jack rafter.

You'll refer to this table again in later chapters. There's another copy, Appendix A, in the back of the book.

▼ **Example 1-7:** Look at the diagram in Figure 1-20. Assume a roof slope of 10 in 12, then find the actual length for the typical common rafters.

Actual Length (common rafter) = 10' x 1.302 (from column 2, Figure 1-19)
= 13.02 linear feet

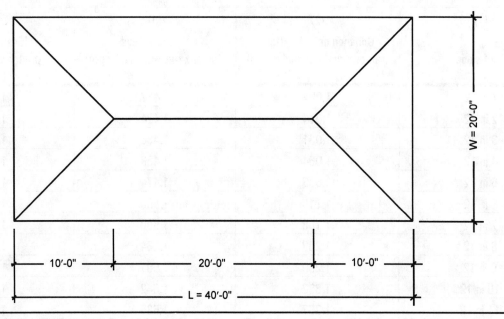

Figure 1-20 Hip roof example

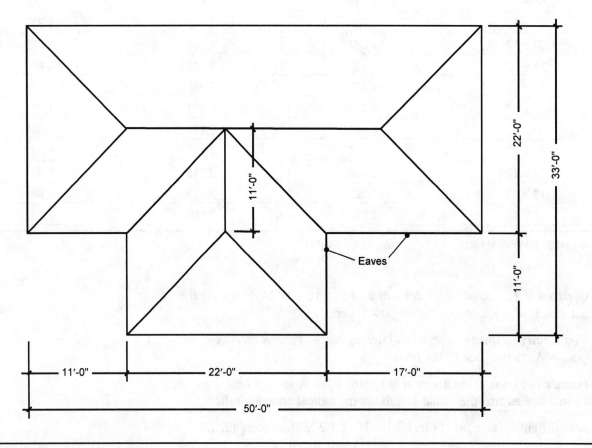

Figure 1-21 Hip-and-valley roof example

Measuring and Calculating Roofs

Perimeter of a Sloped Roof

The eaves of a hip roof (Figure 1-20) or a hip-and-valley roof (Figure 1-21) run horizontally all the way around the building, so you can determine the perimeter from the dimensions on the roof plan. The formula is the same as that for a level roof:

P = 2(L + W + R)   `Equation 1-7`

where L is the roof length, W is the roof width, and R is the depth of the recess.

If the building has no recess, the formula for the perimeter is simply:

P = 2(L + W)   `Equation 1-8`

▼ **Example 1-8:** Find the perimeter of the hip roof in Figure 1-20.

P = 2(40' + 20') = 120 linear feet

To find the perimeter of a gable roof like the one in Figure 1-17, the formula is:

Perimeter = 2(Length + Actual Width)   `Equation 1-9`

Actual Width = 2(Run x Roof-Slope Factor)   `Equation 1-10`

$$= 2\left(\frac{W}{2} \times \text{Roof-Slope Factor}\right)$$
$$= W \times \text{Roof-Slope Factor (from column 2 of Figure 1-19)}$$

Thus, the perimeter of a *gable roof* is:

Perimeter = 2[L + (W x Roof-Slope Factor)]   `Equation 1-11`

▼ **Example 1-9:** Find the perimeter of the gable roof in Figure 1-22 if the roof slope is 8 in 12.

P = 2(31' + [26' x 1.202])
  = 2(31' + 31.25')
  = 2 x 62.25'
  = 124.5 linear feet (For estimating purposes, round this to 125 feet.)

Net Versus Gross Roof Area

The *net* area of a roof is the area of roof sheathing that will be covered with roofing material. But you have to provide materials for an area much larger than the net roof area. You have to allow for such things as:

■ Additional felt underlayment at the ridge, hips, and valleys

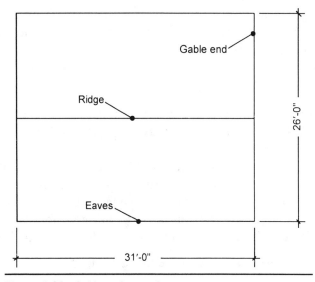

Figure 1-22 Gable roof example

- A starter course
- Hip and ridge units
- Cutting allowances at rakes, hips, and valleys (for shingles)

This larger roof area is called the *gross* area of the roof. For example, the net area of a roof might be 10 squares. However, you might have to provide additional material equal to a roof area requiring 12 squares. A roofing *square* is 100 square feet.

Allowance Factors

The simplest way to account for material required for overcutting and lapping is to use an *allowance factor*. An allowance factor is the ratio of the actual amount of material required to cover the roof (gross roof area) to the net area of roof deck covered:

$$\text{Allowance factor} = \frac{\text{Area Covered (including allowances)}}{\text{Net Roof Area}}$$

Equation 1-12

You can figure the percentage of material overrun by using the allowance factor. Here's an example:

▼ **Example 1-10:** Assume that it will require 12 squares of material (including allowance for waste and lapping) to cover 10 squares of roof deck. Calculate the allowance factor and the percentage of material allowance.

$$\text{Allowance factor} = \frac{12 \text{ squares}}{10 \text{ squares}}$$
$$= 1.20, \text{ or a 20 percent material allowance factor}$$

In later chapters, you'll see that allowance factors can be predicted, based on the roof type, roof size, roof slope, roofing material exposure, and type of roof construction.

Calculating Total Net Roof Area

Since the eaves and ridge of a roof run horizontally, their plan lengths are their actual lengths. And, as you've seen, you can find the actual length of any common rafter by multiplying its plan length by the roof-slope factor in Figure 1-19. So you can use a universal formula to calculate the actual (net) area of any roof that meets the following conditions:

- All roof planes have the same slope
- All hips and valleys are framed at 45 degrees with respect to the eaves

Here's the formula:

Actual (Net) Roof Area = Roof Plan Area x Roof-Slope Factor | Equation 1-13 |

where Roof Plan Area equals roof area as seen in plan view. The Roof-Slope Factor is from column 2 of Figure 1-19.

▼ **Example 1-11:** Assuming a roof slope of 5 in 12, find the net area of the roof shown in Figure 1-22.

Net Roof Area = 31' x 26' x 1.083
= 873 SF ÷ 100 SF/square
= 8.73 squares

▼ **Example 1-12:** Assume a roof slope of 4 in 12, then find the net area of the roof shown in Figure 1-21.

Total Roof Plan Area = (50' x 22') + (22' x 11')
= 1,342 SF

Net Roof Area = 1,342 SF x 1.054
= 1,415 SF ÷ 100
= 14.15 squares

When the slope of a roof changes from one section to another, you have to do a separate take-off for each area with a different slope. Here's an example:

▼ **Example 1-13:** Compute the net area of the roof in Figure 1-23.

Step 1: Section off the drawing, as shown in Figure 1-24, to isolate the two different slopes. Begin with the large 6 in 12 section. Notice you must deduct the triangle formed by the section of 4 in 12 roof (labeled ABC on the drawing). Multiply the length by the width and subtract the area of the triangle:

Roof Plan Area (6 in 12) = (100' x 48') - $\left(\frac{48' \times 8'}{2}\right)$
= 4,800 SF - 192 SF
= 4,608 SF

Step 2: Use the Roof-Slope factors from column 2 of Figure 1-19 to find the net roof area for the 6 in 12 roof:

Net Roof Area (6 in 12) = 4,608 SF x 1.118 = 5,152 SF

Step 3: Find the area of the 4 in 12 section. Notice that you *add* the area of the triangle to this section:

Roof Plan Area (4 in 12) = (48' x 16') + $\left(\frac{48' \times 8'}{2}\right)$
= 768 SF + 192 SF
= 960 SF

Roofing Construction & Estimating

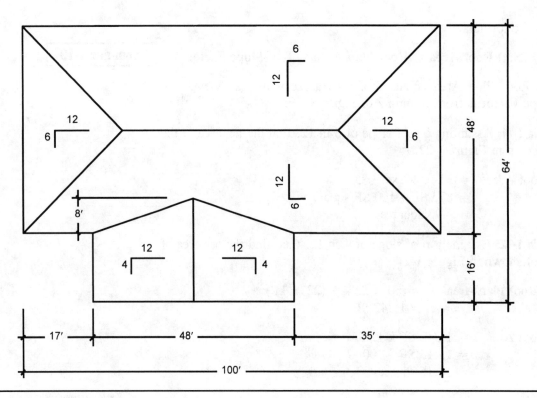

Figure 1-23 Roof with varied slopes

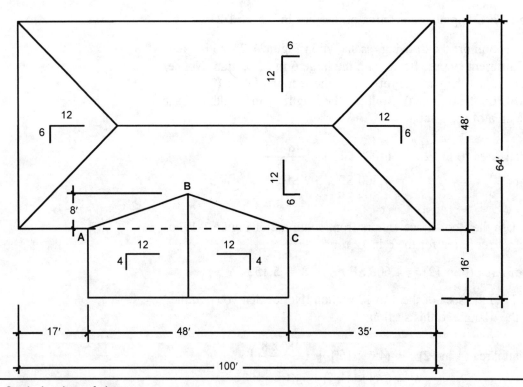

Figure 1-24 Sectioning the roof plan

Measuring and Calculating Roofs

Step 4: Repeat Step 2 for the 4 in 12 section:

Net Roof Area (4 in 12) = 960 SF x 1.054
                        = 1,012 SF

Step 5: Add together the two net areas:

Total Net Roof Area = 5,152 SF + 1,012 SF
                    = 6,164 SF

Roof Overhangs, Hips and Valleys

When you calculate the net area of a roof, be careful you don't omit the roof overhang that extends beyond the walls of the building. (See Figure 1-25.) Also watch for overhangs at interior gable end walls on multi-level roofs like the one in Figure 1-26, and on dormers.

You also need accurate measurements for hips and valleys, which require a variety of roofing materials. Again, refer to Figure 1-19 for conversion factors you can use to calculate the actual lengths for hip and valley rafters. Use the slope factors in column 3 if the hip or valley rafter dimensions are based on the run. Use column 4 if measurements are taken from the plan length. See Figure 1-27.

If the hips and valleys are framed conventionally at a 45-degree angle to the outside walls, you can calculate the plan length with this formula:

Plan Length (Hip or Valley) | Equation 1-14
= 1.414 x Run

▼ **Example 1-14:** Assume a roof slope of 10 in 12, then find the actual length of any hip rafter for the roof in Figure 1-20.

In this illustration, the run of the hip rafter is 10 feet. Refer to column 3 in Figure 1-19 and you see the conversion factor for a 10 in 12 slope is 1.642.

Rafter length = 10' x 1.642 = 16.42 linear feet

You can also use the formula above to calculate the plan length based on the run, then use the factor from column 4 in Figure 1-19 to get the actual length:

Plan length = 1.414 x 10' = 14.14 linear feet

Rafter length = 14.14 x 1.161 = 16.42 linear feet

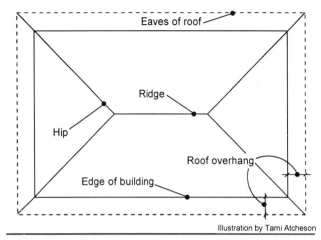

Figure 1-25 Roof overhangs

Figure 1-26 Multi-level roof overhang

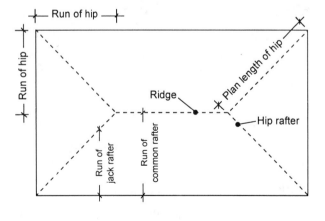

Figure 1-27 Lengths of hip rafters

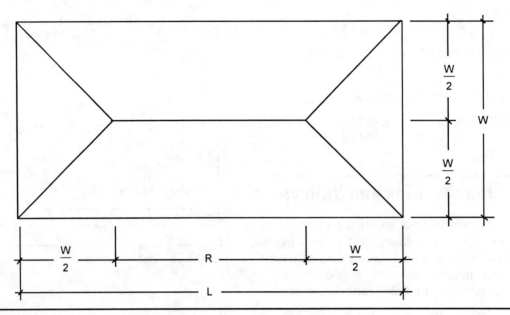

Figure 1-28   Ridge length

Now, what if a building is built with roofs of unequal slopes, such as the one shown in Figure 1-23? You can find the actual length of a valley where the roofs intersect by multiplying the run of the low-sloping roof by the appropriate factor from the table in Appendix B.

▼ **Example 1-15:** Determine the actual length of each valley of the roof diagrammed in Figure 1-23. The run of the low-sloping roof is 24 feet.

So the actual valley length (using the factor from Appendix B) is:

Valley length = 24' x 1.247 = 29.93 linear feet

Length of Ridge (Hip Roofs)

Refer to Figure 1-28. If you assume that the hips are conventionally framed, you find the ridge length on a hip roof with:

Ridge $= L - [2 \times (\frac{W}{2})] = L - W$ **Equation 1-15**

where L equals the length of the roof, and W is the width.

▼ **Example 1-16:** Determine the ridge length of the roof diagrammed in Figure 1-20.

Ridge = 40' - 20' = 20 linear feet

You can't do the example problems in later chapters if you don't know the formulas in this chapter. Don't go on until you're sure you know how these equations work and how to apply them.

2 Roof Sheathing, Decking and Loading

▶ In the last chapter, you learned how to measure a roof and figure out how much roofing material you need to cover it. The next step is to decide what type of roof deck or sheathing you'll need to support the roof covering. For a sloping roof, that depends on several things:

- The type of roofing material you'll install
- The roof slope
- The local climate
- The building code requirements in your area

Check the Framing

One problem you may face is faulty construction of the roof frame, especially at the overhang along the rakes. The correct method of framing this overhang is to construct a "ladder" of *lookout rafters*. That's shown in Figure 2-1. If the lookout rafters aren't there, the overhang will sag. When that happens, mineral granules can pop off the roof covering, leaving it discolored and deteriorated. If you run into this situation, tell the owner why the roof looks the way it does. Quote a price to repair the overhangs, assuming the rest of the roof frame is sound.

If you're replacing a relatively lightweight roof covering with a heavier one, or covering over an existing roof, make sure the roof frame will support the extra load. An overloaded roof can make the deck, rafters, and ridge sag. If the roof frame sags, so will the ceiling, since the mid-span of the rafters in a conventional roof frame is supported by struts and uprights fastened to the ceiling joists. See Figure 2-2.

A sagging deck will also stretch asphalt shingles, bending them over the rafters so the mineral granules pop off. This will leave light-colored lines over the rafters where the shingles are stressed and darker areas where the deck has sagged. And who does the customer blame for that? You guessed it — the roofing contractor. Most homeowners' insurance policies don't cover damage due to ignorance — whether it's the owner's or the contractor's. And if the roof fails, there may be injuries or even death. That's bad for your peace of mind and not real good for your insurance premiums either.

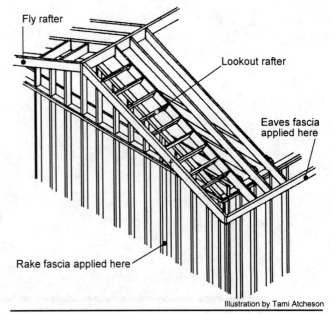

Figure 2-1 Lookout rafters

Solid Roof Sheathing

You always need solid roof sheathing, whether it's boards, plywood or waferboard, under asphalt, fiberglass and metal shingles, mineral-surfaced roll roofing, or built-up roofing. You must also use solid sheathing under tile, slate and metal roofing (except corrugated steel), or any other roof covering that requires continuous support. Building codes in some seismic zones require that you install solid roof sheathing under *all* roofs. Solid sheathing is also recommended (and sometimes required) in climates where wind-driven snows or rains occur.

I recommend that you install shakes or wood shingles over solid sheathing where:

- The outside design temperature (the average expected low temperature) is 0° F or colder. The map in Figure 2-3 shows those areas.

- The January mean temperature is 25° F or less.

- There's a possibility that ice will form along the eaves and cause leaks from a backup of water.

You can use 1 x 4, 1 x 6, 1 x 8, 1 x 10 or 1 x 12 boards for solid roof sheathing. But I don't recommend you use anything wider than a 1 x 6 because larger boards tend to swell and shrink excessively after they're installed.

Roof Sheathing, Decking and Loading

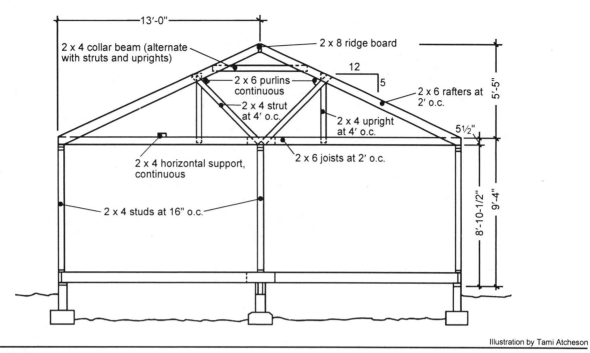

Figure 2-2   Roof supports

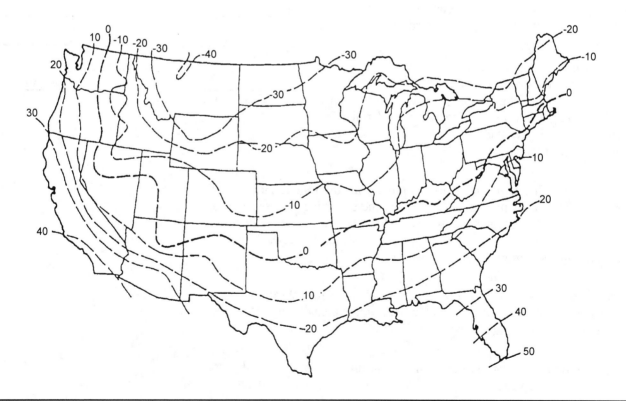

Figure 2-3   Outside design temperatures, U.S. Weather Bureau

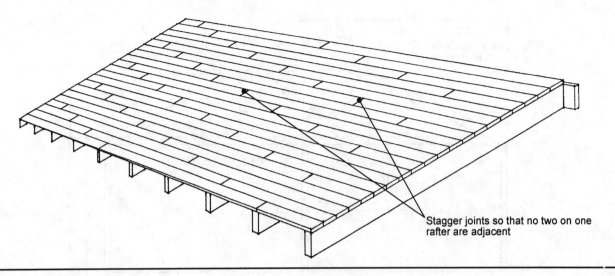

Figure 2-4 Solid board sheathing

Install sheathing boards with a fairly tight edge joint, but allow at least 1/8 inch of space at the end joints for expansion. If it's possible the sheathing will be rained on before you get the felts installed (which can cause warping), you might consider not drawing the edge joints up too close. Stagger the end joints of the boards and make sure they're centered over rafters. Figure 2-4 shows how.

You can use shiplap, tongue-and-groove or square-edged boards like those in Figure 2-5. But it's not a good idea to use square-edged boards on roof decks when you'll install a built-up or mineral-surfaced roll roof. That's because individual boards can warp, which may make the roll material split.

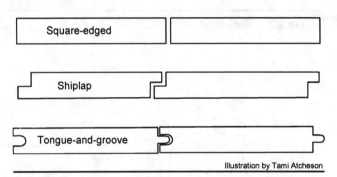

Figure 2-5 Sheathing boards

Always install tongue-and-groove lumber with the tongue laid toward the ridge. This helps prevent leaks through the sheathing in case the roof covering fails. Both tongue-and-groove and shiplap lumber are expensive, and so is installation labor. So when wood sheathing is specified under a flat, built-up roof, most people use plywood.

Never use "green" or poorly-seasoned lumber for roof sheathing material. Choose kiln-dried lumber. Wet lumber will eventually dry and warp, which can make the shingles buckle and wrinkle.

For the roof covering to be smooth, the sheathing must be smooth also. Never install scrap lumber, concrete form lumber, or boards of unequal thickness for sheathing material. Also, be sure all boards are securely nailed so the sheathing doesn't loosen and spring away from the rafters.

Don't use lumber with loose knots or areas where pitch runs over the board surface, because pitch (tree sap) reacts with asphalt, dissolving the asphalt. If you're covering an existing roof which has these defects, cover the affected area with sheet metal patches before you install the roof covering.

Plywood Sheathing

Most solid roof sheathing is made of plywood. The American Plywood Association recommends a minimum plywood grade of C-D Interior-and-Exterior Glue, or another equally durable exterior grade of plywood. And while the APA recommends it, most building codes *require* it.

Consider the rafter spacing and anticipated roof loads when you decide on plywood thickness. Figure 2-6 gives the American Plywood Association's specifications for plywood roof sheathing. Roofing material manufacturers recommend a minimum thickness of ½ inch for roof sheathing under a built-up roof. Be sure the index number of the plywood isn't less than 32/16 (unsanded) or Group 1 (sanded). The index number is a rating that says the plywood can span a minimum of 32 inches between rafters, or 16 inches between floor joists. This rating is based on plywood that's installed with the face grain of the plywood perpendicular to the supports and the sheets spanning at least two supports. That's shown in Figure 2-7.

Notice footnote "d" of Figure 2-6. You see that this data includes a 10-pound *dead load*. The dead load on a roof is the weight of the roof deck, underlayment, and shingles. Some roof coverings (clay tile for instance) weigh as much as 16 pounds per square foot. When you install that over a wood deck (3 pounds per square foot), 15-pound felt (0.15 pounds per square foot) and 90-pound roll roofing (0.9 pounds per square foot), the total dead load is more than 20 pounds.

If the dead load of the roof is more than 10 pounds per square foot (PSF), you must deduct the excess dead load from the allowable live loads. Live loads are loads caused by snow or workers — any temporary loads that may be added to the dead load. Keep in mind that building codes allow a minimum anticipated live load of 30 PSF for designing the required strength of a roof deck. For example, if you have a 20-pound dead load and a rafter spacing of 24 inches, the minimum permissible panel span rating is 24/16. That's because the allowable live load (40 PSF in Figure 2-6, Tables 21 and 22 of the American Plywood Association standards), less the excess dead load (10 PSF), is 30 PSF, which is the minimum anticipated roof live load allowed by the building code.

Table 21
Recommended uniform roof live loads for APA rated sheathing [c] and APA rated Sturd-I-Floor with long dimension perpendicular to supports [e]

Panel span rating	Minimum panel thickness (in.)	Maximum span (in.)		Allowable live loads (psf) [d]							
		With edge support [a]	Without edge support	Spacing of supports center-to-center (in.)							
				12	16	20	24	32	40	48	60
APA rated sheathing [c]											
12/0	5/16	12	12	30	----	----	----	----	----	----	----
16/0	5/16	16	16	70	30	----	----	----	----	----	----
20/0	5/16	20	20	120	50	30	----	----	----	----	----
24/0	3/8	24	20 [b]	190	100	60	30	----	----	----	----
24/16	7/16	24	24	190	100	65	40	----	----	----	----
32/16	15/32	32	28	325	180	120	70	30	----	----	----
40/20	19/32	40	32	----	305	205	130	60	30	----	----
48/24	23/32	48	36	----	----	280	175	95	45	35	----
60/32	7/8	60	48	----	----	----	305	165	100	70	35
APA rated Sturd-I-Floor [f]											
16 oc	19/32	24	24	185	100	65	40	----	----	----	----
20 oc	19/32	32	32	270	150	100	60	30	----	----	----
24 oc	23/32	40	36	----	240	160	100	50	30	25	----
32 oc	7/8	48	48	----	----	295	185	100	60	40	----
48 oc	1-3/32	60	48	----	----	----	290	160	100	65	40

[a] Tongue-and-groove edges, panel edge clips (one midway between each support, except two equally spaced between supports 48 inches on center), lumber blocking, or other. For low slope roofs, see Table 22.
[b] 24 inches for 15/32-inch and 1/2-inch panels.
[c] Includes APA RATED SHEATHING/CEILING DECK.
[d] 10 psf dead load assumed.
[e] Applies to panels 24 inches or wider.
[f] Also applies to C-C Plugged grade plywood.

Table 22
Recommended maximum spans for APA panel roof decks for low slope roofs [a]
(long panel dimension perpendicular to supports and continuous over two or more spans)

Grade	Minimum nominal panel thickness (in.)	Minimum span rating	Maximum span (in.)	Panel clips per span [b] (number)
APA rated sheathing	15/32	32/16	24	1
	19/32	40/20	32	1
	23/32	48/24	48	2
	7/8	60/32	60	2

[a] Low slope roofs are applicable to built-up, single-ply and modified bitumen roofing systems. For guaranteed or warranted roofs contact membrane manufacturer for acceptable deck.
[b] Edge support may also be provided by tongue-and-groove edges or solid blocking.

Courtesy of American Plywood Association

Figure 2-6 Recommended roof live loads and maximum spans

Support for Panel Sheathing

For a built-up roof with 1/2-inch plywood sheathing, roofing material manufacturers recommend a minimum rafter spacing of 24 inches. Unless the plywood is tongue-and-groove or you use edge clips, install blocking at the unsupported edges of the panels. Use 2 x 4s (or larger) installed edgewise between rafters for blocking. Space the edges of the plywood panels at least 1/16 inch between end joints and 1/8 inch between side joints to allow for expansion. That's shown in Figure 2-7. Double this spacing in climates where the relative humidity is high. (See Figure 6-1 on page 159.) Nail plywood panels to a wood structure with 8d annular-threaded or ring-shank nails spaced at 6-inch centers along the edges and at 12-inch centers for the rest of the roof.

You can also use waferboard for roof sheathing. Figure 2-8 shows the specifications. If you install OSB, use a minimum 15/32 inch (nominal) panel.

Install any wood sheet roofing material with staggered end joints and be sure that all end joints lie on the tops of rafters. (Again, see Figure 2-7.) Whenever you install solid sheathing, make sure the attic is well-ventilated to reduce the possibility of shingle and deck deterioration. Here's why:

Today's better-built homes are fairly airtight, so water vapor is trapped in the house. Eventually, that moisture passes through the ceiling and accumulates in the attic. In cold weather, that warm, moist air will condense on the underside of the sheathing in a poorly-ventilated attic. When the weather is hot, the moist air will try to escape through gaps in the roof covering. The end result is a buckled, rotten deck, deteriorated underlayment and buckling shingles. With proper ventilation, air circulates freely in the attic so the water vapor dissipates before it can condense and cause trouble.

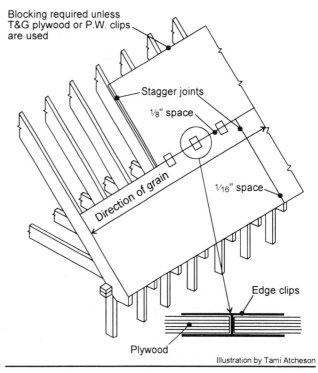

Figure 2-7 Plywood installation

Roof sheathing (1)	
Thickness (in.)	Maximum span (in.)
3/8	16
7/16	24
1/2	24

(1) Refer to appropriate research reports for specific installation information. There are some differences among the code agencies and the applicable requirements are to be followed.

Figure 2-8 Waferboard roof sheathing

Spaced Board Sheathing

In milder climates, you can install spaced sheathing (skip sheathing) under wood shingles or shakes. In fact, some roofing contractors recommend that you always use spaced sheathing under wood shingles or shakes to permit air circulation. You can also use spaced sheathing under corrugated metal panel roofs, if the local building code allows it.

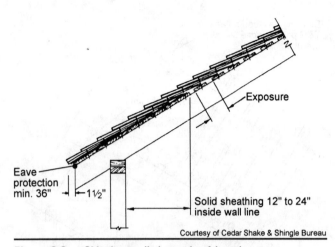

Figure 2-9 Shingles applied over 1 x 4 boards

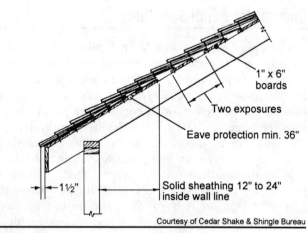

Figure 2-10 Shingles applied over 1 x 6 boards

Spaced sheathing is usually square-edged 1 x 4s and wider for wood shingles, or 1 x 4s, 1 x 6s and wider for shakes.

There are two acceptable ways to install spaced sheathing under wood shingles. With one, you install 1 x 4s spaced on centers equal to the shingle exposure. If you lay the shingles with 5½ inches to the weather, space the sheathing boards at 5½-inch centers. Nail each shingle at the center of each board, as shown in Figure 2-9. Here's something to watch for. Some building codes specify that the board spacing can't exceed the nominal width of the board.

In the second application method, you install 1 x 6s, and nail two courses of shingles to each 1 x 6, up to and including 5½ inches weather exposure. That's shown in Figure 2-10. With 7½-inch exposures, the center-to-center spacing of the sheathing boards is equal to the weather exposure.

If you're installing spaced sheathing under shakes, you'll normally use 1 x 6s. Make the spacing between the shakes equal to their exposure to the weather, as shown in Figure 2-11. Don't space the sheathing more than 2½ inches in order to prevent interlayment from sagging into the attic.

When you use spaced sheathing, install solid sheathing from the eaves of the roof up to 12 to 24 inches beyond the interior face of the exterior wall line. That's also shown in Figures 2-9 through 2-11. Check the local building code for the required distance. In some locations, the code may specify 36 inches on slopes that are less than 4 in 12.

If your roof has a gable overhang with an open cornice (an overhang with the rafters exposed from beneath), you also need to install solid sheathing over the exposed overhang. I recommend that you solidly cover the roof at least 18 inches on each side of the ridges as well. That gives you backing to adjust the exposure of the last few shingles, as shown in Figure 2-12.

Some building codes allow spaced sheathing under corrugated metal roofs. However, most sheet metal roofs require a solid deck for support.

Roof Sheathing, Decking and Loading

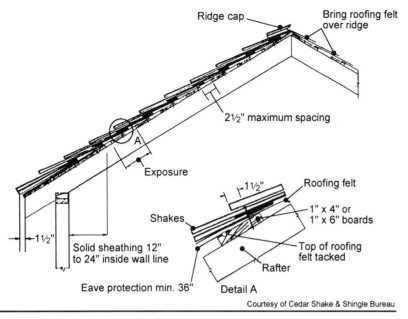

Figure 2-11 Shakes applied over 1 x 6 boards

Spaced Sheathing over Solid Sheathing

The minimum roof slope recommended for wood shingles is 4 in 12, and for shakes, 4 in 12. However, you can install wood shingles and shakes on lower slopes if you install 2 x 4 spacers over a built-up roof covering applied to the sheathing. Then install 1 x 4 or 1 x 6 nailing strips as shown in Figure 2-13.

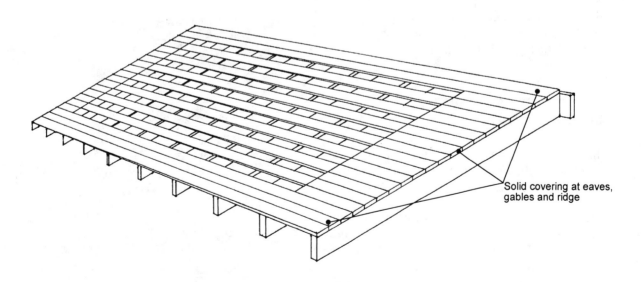

Figure 2-12 Spaced sheathing

Roofing Construction & Estimating

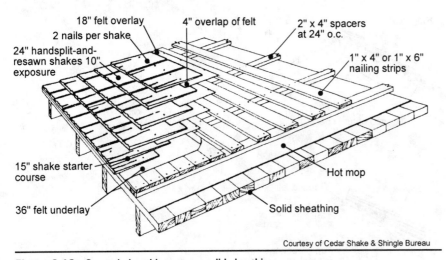

Figure 2-13 Spaced sheathing over a solid sheathing

Roof Decking

Roof decking serves both as a solid roof deck and wood ceiling. Use it on projects which call for an attractive exposed wood ceiling. You can get roof decking in solid tongue-and-groove planks from 1 to 5 inches (nominal) thick and 4 to 12 inches (nominal) wide. You can also get laminated tongue-and-groove roof decking that's 2 to 5 inches thick and 6 and 8 inches wide. See Figure 2-14. Roof decking is sold in random lengths of 6 to 20 feet. Decking is manufactured in commercial grades (where appearance and strength aren't important) and select grades. Many patterns are available for the ceiling side of the decking. Laminated decking is available with the same, or different, lumber species for the top, intermediate, and bottom layers.

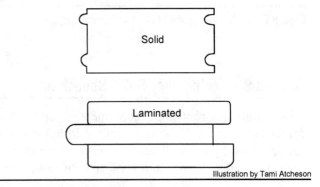

Figure 2-14 Roof decking

Loading the Roof

The roof deck is an integral supporting part of the roof structure. Here's the proper way to load roofing materials onto the deck:

Load the roof by stacking bundles of shingles or roofing felt along the ridge. Distribute the weight as evenly as possible on both sides of the ridge. This keeps the roofing materials out of the way until you need them, and eliminates uneven and concentrated loads. Figure 2-15 shows an even load.

Figure 2-15 Loading the roof along the ridge

Roof Sheathing, Decking and Loading

Figure 2-16 Distributing the load throughout the roof

Distribute roofing tiles throughout the roof and take special care not to overload the roof at the mid-span of rafters. This is the weakest area of the roof framing system. See Figure 2-16.

Load roofing tiles on a gable roof in stacks of eight along every fourth course, except along the ridge where you load in stacks of four. Figure 2-17 is a diagram for tile loading on a gable roof.

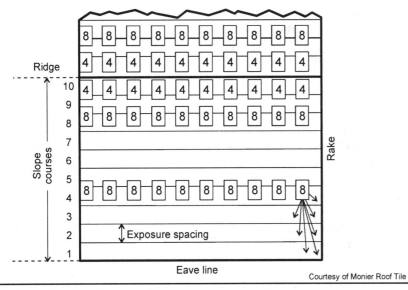

Figure 2-17 Loading tiles on a gable roof

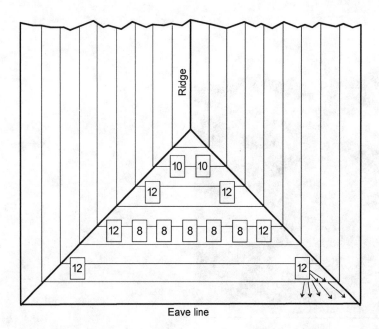

Figure 2-18 Loading tiles on a hip roof

Load roofing tiles on a hip roof as shown in Figure 2-18. On a gable or hip roof, allow a horizontal space of about 1 foot between stacks. Also, pick tiles from different pallets as you load them to avoid color patterning. On a re-roof, load shingles before starting the job to allow the building to adjust to the weight.

Estimating Roof Sheathing

In most cases, the carpenters will have installed the roof sheathing before you arrive on the site. But it won't hurt for you to know how to estimate this part of the job. Assume you're required to install 3,000 square feet of roof deck (including waste) using 1 x 8 tongue-and-groove boards which cost $0.80 per square foot.

3000 x $0.80 = $2,400

According to the *National Construction Estimator* (see Chapter 14), a crew of one carpenter and one laborer can install board sheathing at the rate of 1 square foot per 0.026 manhour. At an average cost of $17.50 per manhour for this operation, the labor cost is:

3000 SF x 0.026 x $17.50 = $1,365

The total cost for the job is $3,765 (the total of materials and labor). The manhour rate to install plywood sheathing (for the same two-man crew) is 0.013 manhours per square foot.

In the next chapter, we'll discuss underlayment and flashings before we go on to specific finish roofing materials.

3 Underlayment on Sloping Roofs

▶ In this chapter, we'll cover underlayment and metal items such as drip edge and valley flashing. Other flashings, such as those for chimneys, vents and parapet walls, appear in chapters where they apply to particular kinds of roofing. By *underlayment,* I mean any roll-type waterproofing material you install under the finished roof covering on a sloping roof. (We'll cover underlayment and drainage for level roofs in Chapter 10 on built-up roofing.) When the code says you need additional weather protection, underlayment also includes fortified eaves flashing. Installing underlayment is often called *capping in, drying in* or *pre-felting.*

Underlayment serves several purposes:

1) It protects roof sheathing before the roof covering is applied.

2) It acts as added weather protection for the building. In fact, in the case of shake and tile roofs, underlayment *is* the weather protection.

3) It separates asphalt shingles from areas where surface resins in wood sheathing may damage the shingle.

4) It provides additional insulation, especially over roofs with nailing strips that create air space between the sheathing and roof covering.

5) It cushions heavy roofing units such as tile or slate. While it doesn't absorb a lot of shock, underlayment does help to prevent breakage of these brittle materials under foot traffic.

Type	Length (ft.)	Width (ft.)	Factory squares per roll[1]	Weight (lb/square) (single coverage)
#15	144	3	4	15
#20	108	3	3	20
#30	72	3	2	30

[1] A factory square is 108 square feet

Figure 3-1　Saturated felt

6) When low-slope roofs are mopped with hot asphalt, underlayment prevents the asphalt from leaking into the building through joints in the roof deck.

Underlayment must be water-resistant, but not vapor-resistant. Water-resistant materials allow moisture or frost to accumulate between the roof sheathing and the underlayment. Then, unless the attic is well-ventilated, the deck will deteriorate.

Saturated Felt Underlayment

One type of underlayment is made of asphalt- or tar-saturated felt. The felt is primarily rag, shredded wood, cellulose and animal fibers. A saturated, organic felt is saturated with bitumen. An *asphalt felt* is saturated with an asphalt flux. A *tarred felt* is saturated with coal tar. These saturated materials allow vapor to pass through them. *Coated felts* differ from saturated felts in that they are actually coated with a glossy, vapor-resistant finish.

Your choice of underlayment depends on the roof slope, local temperatures, and type of material being applied over it. (See Figure 3-3 on page 38.) There's more information about asphalt and tar saturants in Chapter 10.

Asphalt-saturated felts are sold in quantities shown in Figure 3-1.

Saturated Fiberglass Underlayment

This type of underlayment is made of asphalt-saturated fiberglass. The glass fiber mats are made of continuous or random glass fibers bonded with plastic binders reinforced with chopped, continuous-random, or parallel glass fiber strands. Fiberglass underlayment comes in three types:

1) Base sheets (the first membrane in a built-up roofing system)

Type	Length (ft.)	Width (ft.)	Factory squares per roll	Weight (lb/square) (single coverage)
Base sheet	108	3	3	25
Ply-sheet	180	3	5	12
(Type IV) cap sheet	36	3	1	75

Figure 3-2 Saturated fiberglass

2) Ply-sheets (intermediate layers in a built-up system, or underlayment under shingle, slate or tile roofs)

3) Cap sheets (a heavy membrane, sometimes used instead of aggregate in a built-up roof system)

Asphalt-saturated fiberglass felts are sold in quantities as shown in Figure 3-2.

Organic felts can deteriorate due to oxidation and wicking (absorption of water by capillary action). Fiberglass felts don't rot or absorb water, so they last longer than organic felts, and they're fire resistant. But the down side is they cost about 65 percent more (for the same weight material) than organic felts.

Fiberglass felts don't buckle as much as organic felts, but they're less resistant to tearing, so they're more prone to wind damage. Use fiberglass felts on roofs with a long life expectancy, such as slate. Lower-cost organic felts are adequate under composition shingles.

Underlayment Requirements

You must use underlayment with all shingle roofs — with two exceptions. First, you don't use it under a wood shingle roof because it'll keep air from circulating and let moisture collect. That in turn reduces the life of the roof covering and sheathing. Second, you don't have to install underlayment between the two layers of roof coverings when you re-roof over existing shingles.

The type and amount of underlayment required depends on the roof slope and the type of roof covering you install. See Figure 3-3 for those requirements. In that table, solid sheathing is assumed, unless otherwise noted. These underlayment specs are the *minimums* required. Local building codes may have stricter requirements, so be sure to check with your inspector.

Roof type	Description	Min. slope	On slopes	Underlayment recommended[1][2][13]
Asphalt shingles	2- or 3-tab strip (self sealing)	2 in 12	2/12	Double coverage with 15 lb felt (nailed)[3]
			3/12 and steeper	Single coverage with 15 lb felt (nailed)
	3-tab strip (standard)	2 in 12	2/12-3/12	Double coverage with 15 lb felt (nailed)[3]
			4/12 and steeper	Single coverage with 15 lb felt (nailed)
	Individual Dutch lap or American	4 in 12	4/12 and steeper	Single coverage with 15 lb felt (nailed)
	2- or 3-tab hexagonal	4 in 12	4/12 and steeper	Single coverage with 15 lb felt (nailed)
Fiberglass shingles	Light weight (to 260 lbs)	2 in 12	2/12-3/12	Double coverage with 15 lb felt (nailed)[3]
			4/12 and steeper	Double coverage with 15 lb felt (nailed)
	Medium weight (260-280 lbs)	2 in 12	2/12-3/12	Double coverage with 30 lb felt (nailed)[3][4]
			4/12 and steeper	Double coverage with 30 lb felt (nailed)
	Heavy weight (280 lbs and more)	2 in 12	2/12-3/12	Double coverage, 1st ply is 15 lb felt (nailed), 2nd ply is Type IV fiberglass (nailed)
			4/12 and steeper	Single coverage with 30 lb felt (nailed)[4]
Mineral-surfaced roll roofing	Single coverage	1 in 12[8]	1/12-3/12	Double coverage with 15 lb felt (nailed)[3][10]
			4/12 and steeper	Single coverage with 15 lb felt (nailed)[10]
	Double coverage	1 in 12	1/12 and steeper	None required [6][10]
	Pattern edge	4 in 12	4/12 and steeper	None required [6][10]
Clay tile	All types	3 in 12	3/12 and steeper	Triple coverage. 30 lb felt (nailed) followed by two plies of Type IV fiberglass (mopped)[4]
Cement tile	All types	2.5 in 12	2.5/12 and steeper	Double coverage. 30 lb felt (nailed) followed by one ply of 90 lb mineral-surfaced roll roofing (nailed)
Slate	Standard	4 in 12	4/12 and steeper	Single coverage with 15 lb felt (nailed)[9]
	Textural	4 in 12	4/12 and steeper	Single coverage with 30 lb felt (nailed)[9]
	Graduated	4 in 12	4/12 and steeper	Single coverage with 45 lb mineral-surfaced roll roofing (nailed)[9][11]
Aluminum	All types	4 in 12	4/12 and steeper	Single coverage with 30 lb felt (nailed)
Porcelain	All types	3 in 12	3/12 and steeper	30 lb felt underlayment (nailed) with 30 lb felt interlayment

Figure 3-3   Underlayment recommendations

Roof type	Description	Min. slope	On slopes	Underlayment recommended[1][2][13]
Wood	Shakes, spaced sheathing	4 in 12	4/12 and steeper	Single coverage. 30 lb felt underlayment starter course with 30 lb felt interlayment
	Shakes, solid sheathing	3 in 12	3/12-4/12	Double coverage. 30 lb felt underlayment (nailed) with 30 lb felt interlayment
			4/12 and steeper	Single coverage. 30 lb felt underlayment starter course with 30 lb felt interlayment
	Shingles, solid or spaced sheathing	3 in 12	3/12 and steeper	None required[6][7]
Copper	Standing seam, pan or roll method	2.5 in 12	2.5/12 and steeper	Single coverage with 15 lb felt and rosin paper
	Batten seam	3 in 12	3/12 and steeper	Single coverage with 15 lb felt and rosin paper
	Flat seam	0.25 in 12	0.25/12 and steeper	Single coverage with 15 lb felt and rosin paper
Copper-bearing steel	Standing seam, roll method	2 in 12	2/12 and steeper	None required[6]
	Pressed standing seam	0.25 in 12	0.25/12 and steeper	None required[6]
Lead	All types	0.25 in 12	0.25/12 and steeper	Single coverage with 30 lb felt and rosin paper
Copper-zinc alloy	All types	3 in 12	3/12 and steeper	Single coverage with 15 lb felt and rosin paper
Stainless steel	All types	0.25 in 12	0.25/12 and steeper	None required[6]
Terne plate	Batten seam or standing	2.5 in 12	2.5/12 and steeper	Rosin paper[12]
	Flat seam	0.25 in 12	0.25/12 and steeper	Rosin paper[12]

Footnotes:

[1] Apply underlayment horizontally. Exception: Hot-mopped mineral-surfaced roll roofing underlayment can be installed parallel to the slope.

[2] Check your local building code. It may require heavier underlayment than those given.

[3] To help keep water from leaking through the nail holes on roof slopes less than 3 in 12, first nail on a 15-pound felt, and follow that by mopping on a 15-pound felt (or Type IV fiberglass ply-sheet).

[4] Use heavy, durable underlayment for shingles with long life spans (such as slate or tile) or the shingle will outlast the underlayment.

[5] As an alternate to interlayment over solid decks (see Figure 3-36)
　a) Roof slopes less than 6 in 12: Nail one 30-pound felt followed by mopping two 15-pound felts (or two Type IV fiberglass ply-sheets).
　b) Roof slopes equal to 6 in 12: Nail two 30-pound felts.
　c) Roof slopes greater than 6 in 12: Nail one 30-pound felt.

[6] Underlayment might be desirable for protection of the sheathing.

[7] Some roofing contractors recommend:
　a) Roof slopes less than 6 in 12: Nail one 30-pound felt, followed by nailing one Type IV fiberglass ply-sheet.
　b) Roof slopes of 6 in 12 and greater: Nail one 30-pound felt.

[8] Minimum slope varies, depending on the installation method. (See Figure 5-3).

[9] Some roofing contractors recommend the underlayment as specified for tile roofs.

[10] Some roofing contractors recommend that you never install underlayment under mineral-surfaced roll roofing.

[11] On graduated roofs, you usually use 30-pound felt under slates that are ¾ inches or less thick, and 50-pound underlayment under slates 1 inch thick and thicker. In some areas, it's common to install a 30-pound felt followed by a 15-pound felt under any graduated roof.

[12] Terne-coated stainless steel can be underlaid with a single 15-pound felt, with a slip sheet of rosin paper.

[13] Solid sheathing is assumed, unless otherwise noted

Figure 3-3 (cont.) Underlayment recommendations

Keep in mind that some roofing materials (like heavy laminated fiberglass shingles) have a life expectancy of 20 to 30 years, while other roof coverings (like tile or slate) should last 50 years or more. When a long-life roof fails, it's usually because the flashing or underlayment failed, and not the roof covering. For that reason, I recommend a heavier underlayment than that recommended in Figure 3-3 whenever you install a long-life roof covering.

Many roofing contractors don't recommend the use of underlayment under asphalt or fiberglass shingles on roof slopes steeper than 4 in 12. They install it only because the building code requires it. They argue that it does more harm than good because the underlayment can buckle, causing the shingles to buckle. If underlayment is required, these contractors install heavy fiberglass felts. I've personally never seen shingles buckle due to warped underlayment. Since roofers sometimes disagree on the best installation methods, I've included options. Follow the guidelines in this book, and use your own judgment and experience where application methods aren't specified by codes or manufacturer's warranty requirements.

Figure 3-4 Installing the drip edge along rake

Drip Edge

Install a drip edge (nosing or roof edging) at the eaves and rake to help shed water and prevent runoff water from getting to the underlayment and sheathing where it can lead to deterioration. In Figure 3-4 the drip edge is installed under the felt along the eaves and over the underlayment along the rakes.

You install drip edge *under* the felt at the eaves. Otherwise, water might get under the shingles and seep under the drip edge. You don't have that problem at the rakes because of rapid runoff. In that case, put the drip edge *over* the felt to help secure the felt. If the drip edge is installed under the felt along the rakes, wind-driven water can penetrate between the felt and the drip edge.

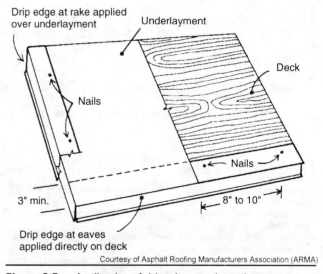

Figure 3-5 Application of drip edge at rake and eaves

Make drip edges from 28-gauge (minimum) galvanized metal or other non-corrosive, stain-resistant material such as copper or terne metal. Extend the drip edge back over the roof deck no less than 3 inches and nail it on 8- to 10-inch centers as shown in Figure 3-5. In high-wind areas, nail the drip edge on 4-inch centers. If you need more than one piece to cover the eaves, lap the joint at least 3 inches.

Underlayment on Sloping Roofs

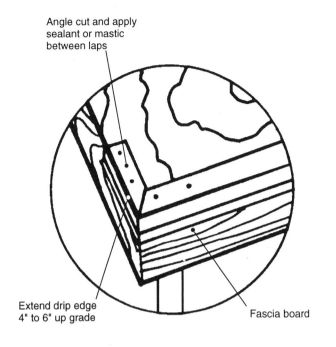

Figure 3-6 Application of drip edge at corners

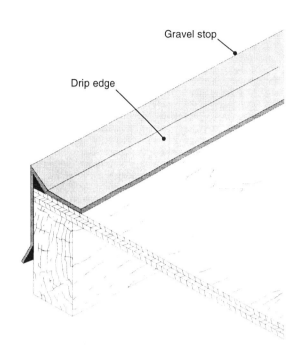

Figure 3-7 Drip edge and gravel stop

Run the drip edge down the eaves 4 to 6 inches beyond the corner of the roof, then make an angle cut in the metal and extend the drip edge up the rake. Place roofing cement between the laps where you made the angle cut. That's shown in Figure 3-6.

At gable ends, install the drip edge beginning at the lower end of the rakes. If you need more than one piece to cover the rake, lap the upper unit over the lower by at least 3 inches.

You can install a sheet metal product called a *gravel stop* at the edge of aggregate-surfaced roofs to help prevent loose gravel from falling off the edge of the roof. Gravel stop is similar to the drip edge in Figure 3-7, except the gravel stop has a more pronounced lip at the edge of the roof to retain the gravel.

After you've installed the drip edge, place a bundle of shingles at the eaves to support the upper end of your ladder. This also keeps the ladder from slipping, and from denting the drip edge. That's shown in Figure 3-8.

Figure 3-8 Protecting the drip edge

41

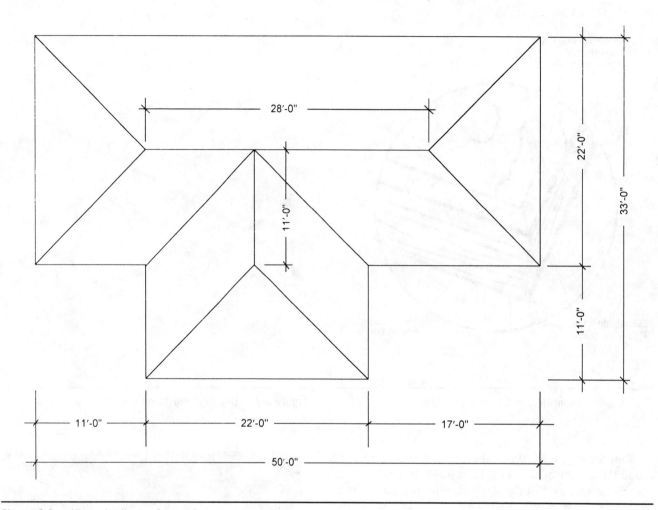

Figure 3-9   Hip-and-valley roof example

Calculating Drip Edge

Here's how to calculate how much drip edge you need:

▼ **Example 3-1:** Assume a 10 percent material cutting allowance and find the total linear feet of drip edge required for the roof of the building in Figure 3-9.

First, you remember from Chapter 1 that the formula for the perimeter of a building is 2(L+W). The perimeter of this building is 2 x (50' + 33'), or 166 linear feet. Multiply that by 1.10 to add the 10 percent allowance factor, and round your answer up to 183 linear feet.

Use 166 linear feet to figure manhours at the rate of 1.4 manhours per 100 linear feet.

Installing Underlayment

Underlayment is applied single-coverage (Figure 3-10) or double-coverage (Figure 3-11). In either case, install felts *over* the drip edge at the eaves and *under* the drip edge at the rakes. Roll out the felt, nail down one end, and stretch the felt. Roll out only as much felt as you can install in about an hour. Otherwise, the loose felt may buckle because heat causes the felt to relax. The hotter the weather, the more the felt will buckle. Wind is also a factor.

Moisture on the deck can also cause buckling. (This problem is worst in early morning.) If you can't pull a wrinkle out, cut it, lap the edges, and nail the felt flat. Never install felt over a wet deck.

Install felt underlayment (sweat sheet) in the valleys before you install felts over the main roof deck. Nail the valley felt on 24-inch centers in rows along the edges of

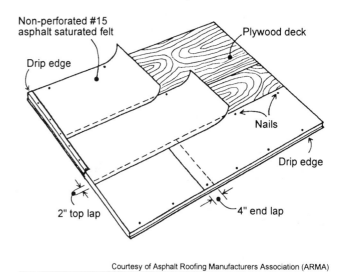

Figure 3-10 Application of single-coverage underlayment

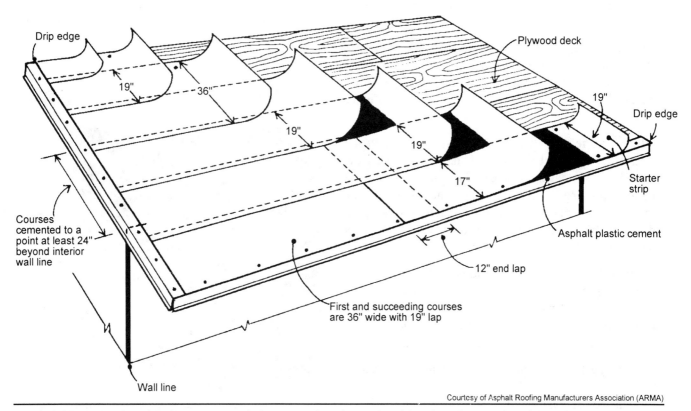

Figure 3-11 Application of double-coverage underlayment on low slopes where icing along the eaves is anticipated

Figure 3-12 Installing felt in a valley

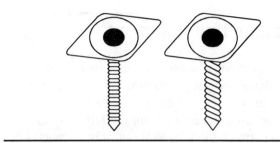

Figure 3-13 Simplex nails

Figure 3-14 Extended valley underlayment beyond the ridge

the felts as shown in Figure 3-12. Drive the nails through tin caps, or use Simplex nails. If you use tin caps (tin tags), they should be 32-gauge caps that are at least 1⅝ inches in diameter. Use galvanized nails at least ¾-inch long, or long enough to go all the way through the sheathing. Simplex nails are shown in Figure 3-13.

On any roof with a valley, extend the valley underlayment at least 6 inches beyond the ridge and cut it so that it conforms to the surface of the sheathing with no bulges. Then nail it down. It should look like Figure 3-14.

Install all felt parallel to the eaves, beginning at the eaves. Be sure to let the edge of the felt overhang the drip edge at the eaves by at least ⅜ inch. Better yet, let the felt overhang the drip edge the same distance as the final roof covering. That prevents water from seeping under the drip edge where it can damage the deck or rot the fascia.

Align the first felt with the lower left corner of the roof and secure it with one nail. Roll out the felt, straighten it, and nail it randomly as you go, on 3- to 4-foot centers. See Figure 3-15. Don't nail just at opposite ends or you'll get humps in the felt.

Then, go back and nail in rows along the edges on 12- to 18-inch centers, and along the center on 18- to 24-inch centers. Nail end laps on 6- to 8-inch centers as shown in Figure 3-16.

Figure 3-15 Nail the felt as you roll it out

Underlayment on Sloping Roofs

Figure 3-16 Running felts into a valley

Figure 3-17 Lapping the felt over the ridge

Continue up the roof the same way. When you reach the ridge, fold the last felt over the ridge at least 6 inches and nail it on the other side. Make sure the felt covers at least 6 inches on both sides of the ridge. See Figure 3-17.

You must also roll and lap the felt at least 6 inches beyond the centerlines on both sides of hips as shown in Figure 3-18. Overlap the 36-inch-wide felt in the valley by 6 inches from both sides, as shown in Figure 3-16. Extend the felt 3 to 4 inches up any vertical surface such as a chimney or wall. That's illustrated in Chapter 4, as Figure 4-62.

Use a 2-inch top lap (allowing a 34-inch exposure) for single underlayment coverage, as shown in Figure 3-10. For double coverage, use a 19-inch top lap and a 17-inch exposure, as in Figure 3-11. You'll find guidelines for various top laps printed on the rolls. End laps vary, depending on the type of underlayment, the slope of the roof, the severity of the climate, and local code requirements. Figure 3-19 shows end-lap requirements for various underlayment materials. Be sure to stagger end laps.

To work around a vent pipe, temporarily roll the felt out so that the edge of the roll just touches the pipe. Mark the location on the edge of the roll and cut a slit to the location where the pipe will penetrate the roll. You can

Figure 3-18 Lapping the felt over a hip

Material type	End lap (inches)
15, 20 or 30# saturated felt (single coverage)	4
15, 20 or 30# saturated felt (double coverage)	6[1]
Fiberglass base sheet	6
Fiberglass ply-sheet	12
Fiberglass cap sheet	6
Mineral-surfaced roll	6
Mineral-surfaced roll (double coverage)	6
Coated roll	6

[1] 12 inches is recommended by some manufacturers on low-slope roofs in colder climates.

Figure 3-19 End-lap requirements

see that in Figure 3-15. Slide the felt into final position and cut the felt around the vent pipe for a smooth fit. Then drive nails along each side of the slit.

Underlayment for Tile Roof Coverings

When you choose which type of underlayment to use under a tile roof, you should consider your confidence in the roofing crew to install a watertight roof covering, as well as the:

- Local climate
- Method of fastening the tiles
- Roof slope
- Weatherproofing integrity of the tile
- Local building code

Except in highly humid areas, you can install a non-sealed underlayment system as in Figure 3-20 when you direct-nail tiles to the sheathing or battens on roof slopes of 4 in 12 and steeper. This system is a single 43-pound coated base sheet nailed to the deck. Install sheets with a 2-inch top lap and a 6-inch end lap. Drive nails along the eaves at 24-inch centers and along the top laps at 12-inch centers.

When you direct-nail tiles on slopes of 4 in 12 and steeper in more severe climates, install the underlayment as described above, but seal the laps with roofing cement on slopes up to 6 in 12, as shown in Figure 3-21. This method is called a *sealed underlayment system*. You can omit the roofing cement on slopes of 6 in 12 and steeper.

Underlayment on Sloping Roofs

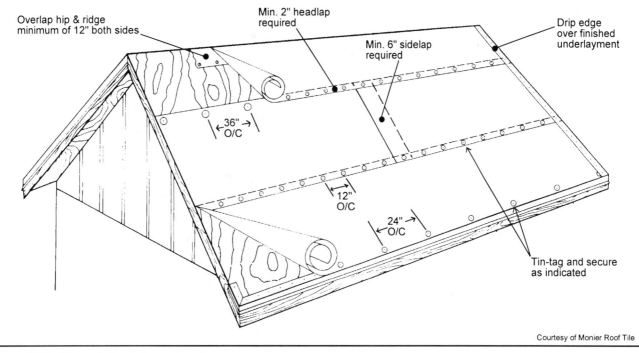

Figure 3-20 Non-sealed underlayment system

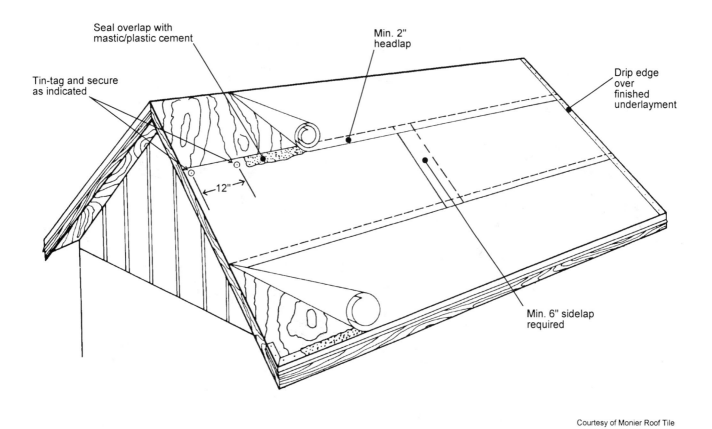

Figure 3-21 Sealed underlayment system

In high-wind and hurricane areas, nail down a 30-pound felt followed by 90-pound mineral-surfaced roll roofing embedded in a solid hot mopping of Type IV (steep) asphalt. That's shown in Figure 3-22. You can also use a flood coat of cold-process roofing cement (Figure 3-23). These methods are called the *30/90 hot-mop* and *2-ply sealed systems*, respectively.

Use the 30/90 hot-mop method whenever you'll be installing the tiles with mortar (mortar-set or mud-on system). You can also use it on roof slopes between 2½ and 4 in 12 if you're going to use nails to install tiles. Use hot asphalt on roofs 5 in 12 to 6 in 12. You can use cold-applied roofing cement instead of hot asphalt (2-ply sealed system) on roof slopes steeper than 6 in 12.

When you hot-mop the 90-pound material along the eaves, nail the top of the sheet and pull back the loose edge. Mop the back of the sheet (Figure 3-24) and flip it over to its original position. Then walk along the sheet to help stick it to the underlying felt. That's called *walking in*. Mopping the back of the sheet as opposed to

Figure 3-22 Installing mineral-surfaced roll roofing over hot asphalt

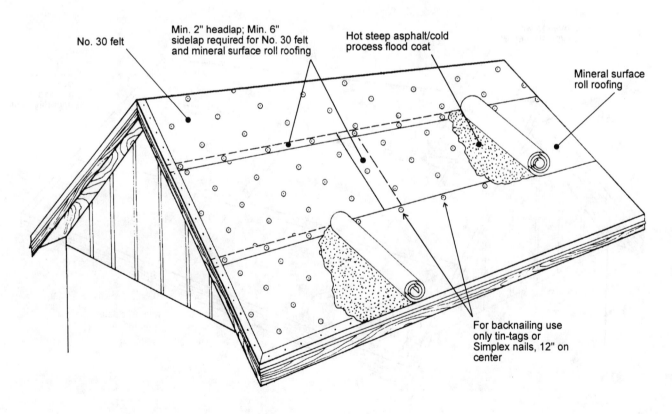

Courtesy of Monier Roof Tile

Figure 3-23 Two-ply sealed system

Underlayment on Sloping Roofs

mopping the roof surface and laying the sheet on top helps prevent hot bitumen from running off the edge of the roof.

You can use a similar method to mop the 90-pound material into a valley by mopping half of the valley at a time. That's shown in Figure 3-25. With this type of installation you can install the 90-pound material down slope, as well as perpendicular to the slope (Figure 3-26).

When you come to a roof penetration, cut the 90-pound material so it slips over the penetration, as in Figure 3-27. Then seal the edges with roofing cement, as in Figure 3-28.

Estimating Underlayment Quantities

Before we get into takeoffs and estimating, let's discuss *rounding off*. In the estimating calculations in this book, I generally round up if the decimal is 0.5 or greater, and down otherwise. The theory is that you'll round up about as often as you round down, and the final answer is close enough even if it's not accurate to four decimal places. But there are exceptions, depending on variations in material cost and manhours.

Figure 3-24 Mopping the back of the first sheet

Figure 3-25 Mopping mineral-surfaced roll roofing into a valley

Figure 3-26 Installing mineral-surfaced roll roofing down slope

If valley flashing metal was pure gold, you'd want to be very precise in your takeoff. And you'll want to figure slate more closely than the felt which goes under it. But you don't want to waste time figuring materials down to the nearest inch if that won't make more than a couple of dollars' difference on a $20,000 job. You're better off spending your time negotiating with subcontractors, finding the most economical material source, or training your crews to be more efficient. So use your judgment and common sense before you spend needless time figuring everything down to the last nickel.

Roofing felt is relatively inexpensive and it's eventually concealed, so roofing contractors don't worry too much about waste or beauty when they install it. As you can see in Figure 3-29, it may be hard to figure out how much extra to allow for. You certainly can't estimate it with any precision.

You begin estimating underlayment by calculating square feet of coverage. But this material is sold by the roll, so you have to convert square feet to rolls.

Figure 3-27 Rolling over a roof penetration

Figure 3-28 Sealing a penetration

Figure 3-29 It's hard to predict underlayment waste

No. of plies	Coverage (factory squares per square)	Coverage (square feet per factory square)
1	1	100
2	2	50
3	3	33
4	4	25

Figure 3-30 Covering capacity of underlayment

- After you deduct the material lost due to a 2-inch top lap and a 4-inch end lap, one factory square (108 square feet) will actually cover only 101 square feet (single coverage).

- After you deduct material lost due to a 19-inch top lap and a 6-inch end lap, one factory square will cover only 49.5 square feet (double coverage).

To make estimating easy (but still accurate enough), you can figure that one factory square of underlayment will cover 100 square feet (single coverage) or 50 square feet (double coverage) of roof area as shown in Figure 3-30.

The quantities in Figure 3-30 don't allow for waste due to crew error or the material you need for overlap at the ridge, hips, and valleys.

Lap Allowance

Remember, you have to lap underlayment at least 6 inches over each side of the centerlines of ridges and hips. You can see that in Figures 3-17 and 3-18. Also, you must install underlayment running into valleys with two 6-inch laps as shown in Figure 3-16. So, add 1 square foot to the roof area to be covered for each linear foot of ridge, hip, and valley.

Lap Area (ridge, hip and valley) = Total Length x 1 SF/LF   **Equation 3-1**

For added protection, I recommend that you apply a 6-inch-wide strip of 30-pound felt over the hips and ridges before you install the hip and ridge units. That's relatively inexpensive leak insurance.

Overcut Allowance

On a gable roof, cut the underlayment about 4 inches beyond the rake edge as shown in Figure 3-31. Trim it flush with the edge later. If you're going to cover the gable roof with tiles and trim it with rake tiles, it's customary to wrap the underlayment down over the rake fascia and nail it to the barge board on 6-inch centers. The rule for this is:

Overcut Area (gables) = LF of Rake x 0.34 SF/LF   Equation 3-2

In this case, overlaps are covered using the information in Figure 3-30.

A Shortcut Allowance Calculation

As a rule of thumb, some contractors add 10 percent to the net underlayment quantities on small gable roofs and 15 percent on small hip roofs. This additional material allowance is for lapping at the ridges, hips and valleys, overcutting at rakes, and crew error. They allow a larger percentage for more complex roofs and a smaller percentage for large, simple roofs.

I don't like that method for figuring allowances. Without considering crew errors or job conditions, actual allowances depend on the roof type, the ratio of building length to building width, and the roof slope. Use Figures 3-32 and 3-33 to get a ballpark percentage for underlayment allowances for single or double coverage. Crew-error waste isn't included in the figures. Figure 3-32 shows the percentage to add to the net roof area because of the overlap at the ridge and the 4-inch overcut at the rakes. Figure 3-33 shows the percentage you add to the net roof area for overlaps at the hips and ridge.

The following examples demonstrate the reliability of Figures 3-32 and 3-33. You can use them to find underlayment quantities for any roof size and slope.

▼ **Example 3-2:** Find the quantity of 15 pound single-coverage underlayment required for the roof shown in Figure 3-34. Assume a roof slope of 5 in 12 and a 4-inch overcut at each rake.

Use the equation for net roof area from Chapter 1, with the roof-slope factor from the slope factor conversion chart, Appendix A.

Figure 3-31 Underlayment waste along the rake

Underlayment on Sloping Roofs

Building dimensions	Roof slope		
(L x W)	3/12	6/12	12/12
30 x 20	7	7	6
40 x 30	5	5	4
45 x 30	5	5	4
50 x 30	4	4	3
60 x 30	4	4	3
70 x 30	4	4	3
80 x 30	4	4	3
50 x 40	3	3	3
60 x 40	3	3	3
70 x 40	3	3	3
80 x 40	3	3	3
90 x 40	3	3	3
60 x 50	3	3	2
70 x 50	3	3	2
80 x 50	3	3	2
90 x 50	3	3	2

Figure 3-32 Underlayment material overrun percentage (gable roof)

Building dimensions	Roof slope		
(L x W)	3/12	6/12	12/12
30 x 20	10	10	9
40 x 30	8	8	7
45 x 30	7	7	6
50 x 30	7	7	6
60 x 30	6	6	5
70 x 30	6	6	5
80 x 30	6	5	5
50 x 40	6	6	5
60 x 40	6	5	5
70 x 40	5	5	4
80 x 40	5	5	4
90 x 40	5	4	4
60 x 50	5	5	4
70 x 50	5	4	4
80 x 50	4	4	4
90 x 50	4	4	3

Figure 3-33 Underlayment material overrun percentage (hip roof)

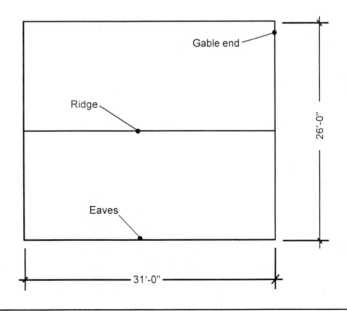

Figure 3-34 Gable roof example

Actual (Net) Roof Area = Roof Plan Area x Roof-Slope Factor
Net Roof Area = 31' x 26' x 1.083
                 = 873 square feet ÷100
                 = 8.73 squares (Use this quantity
                     to estimate labor costs.)

The ridge length is 31 feet, so you add an additional 31 square feet (from Equation 3-1):

Lap Area (Ridge) = 31 feet x 1 SF/LF
                   = 31 square feet

The total length of roof on the plan view is 52 feet (4 x 13 feet). To convert that to actual length, use the equation from Chapter 1 and Column 2 of Appendix A for common rafter length:

Actual Length = Plan Length x Roof-Slope Factor
                  = 52 feet x 1.083
                  = 56 linear feet

Now, use Equation 3-2 to find how much additional underlayment to order for the rakes:

Overcut Area (Gables) = LF of Rake x 0.34 SF/LF
                           = 56 feet x 0.34 SF/LF
                           = 19 square feet

Now, add the three answers together to find the total underlayment required:

Gross Roof Area = 873 SF + 31 SF + 19 SF
                     = 923 square feet ÷ 100
                     = 9.23 squares

Next, convert that figure to rolls. Look back at Figure 3-1. There are 4 factory squares (FS) per roll of 15-pound material.

$$\text{Rolls (\#15)} = \frac{9.23 \text{ FS}}{4 \text{ FS/Roll}} = 2.3 \text{ rolls}$$

The allowance factor is the area covered (including cutting allowance) divided by the net roof area.

$$\text{Allowance Factor} = \frac{9.23 \text{ squares}}{8.73 \text{ squares}} = 1.06$$

That's equal to 6 percent material lap and overcut allowance. Now, look back at Figure 3-32. A 30- x 20-foot roof with a slope of 6 in 12 is about like the one in this example, and that table shows a material overrun of 7 percent. Our calculations agree with that. Now, try another one:

▼ **Example 3-3:** Assume single coverage and a roof slope of 6 in 12, then find how much 15-pound underlayment is required for the roof of the building in Figure 3-35.

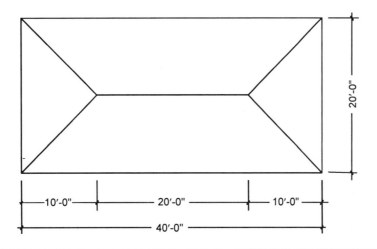

Figure 3-35 Hip roof example

First, use the formula for net roof area from the last example:

Net Roof Area = 40' x 20' x 1.118
= 894 square feet ÷ 100
= 8.94 squares (Use this quantity to estimate labor costs.)

You can see from Figure 3-35 the total ridge length is 20 feet. Also, using the run of any hip (10 feet), and the slope factor from Column 3, Appendix A (1.5), each hip is 15 feet long. The total for the four hips is 60 feet. That makes the total length of ridge and hips 80 feet. So, you add 80 SF (1 SF per linear foot of ridge and hip) to the net roof area for lapping at the ridge and hips:

Gross Roof Area = 894 SF + 80 SF
= 974 square feet ÷ 100
= 9.74 squares

Now, use Figure 3-1 to see how much felt to order:

$$\text{Rolls (\#15)} = \frac{9.74 \text{ FS}}{4 \text{ FS/Roll}} = 2.4 \text{ rolls}$$

The allowance factor is:

$$\text{Allowance Factor} = \frac{9.74 \text{ squares}}{8.94 \text{ squares}} = 1.09$$

That's in line with Figure 3-33, where a roof of similar size and slope requires from 8 to 10 percent allowance.

We'll do one more example, this time for a hip-and-valley roof.

▼ **Example 3-4:** Assuming single coverage, a crew-error waste of 3 percent, and a slope of 6 in 12, calculate the amount of 15-pound underlayment required for the roof of the building in Figure 3-9.

First, calculate the roof plan area:

Roof Plan Area = (50' x 22') + (22' x 11') = 1,342 square feet

The actual net roof area is the plan area times the roof-slope factor from Appendix A:

Net Roof Area = 1,342 SF x 1.118
= 1,500 square feet ÷ 100
= 15 squares

Use this quantity to estimate labor cost at 0.36 manhours per square.

From Figure 3-9, the total ridge length is 28 + 11, or 39 linear feet.

Each hip or valley is 16.5 feet long (11 x 1.5, the slope factor from Appendix A).

There are six hips and two valleys for a total length of 132 linear feet (8 x 16.5 feet).

The total length of ridge, hips and valleys is 171 linear feet (39 + 132). At 1 square foot per linear foot of ridge, hip, and valley, the amount to add to the roof plan area for lap allowance is 171 square feet. Add that to the gross roof area to see how much material to order:

Gross Roof Area = 1,500 SF + 171 SF
= 1,671 square feet ÷ 100 = 16.71 squares

Multiply that by 1.03 for the 3 percent crew waste allowance:

16.71 x 1.03 = 17.21 squares, which you convert to rolls using Figure 3-1:

$$\text{Rolls (\#15)} = \frac{17.21 \text{ FS}}{4 \text{ FS/Roll}} = 4.3 \text{ rolls}$$

The allowance factor is:

$$\text{Allowance Factor} = \frac{17.21 \text{ squares}}{15 \text{ squares}} = 1.15, \text{ or 15 percent allowance.}$$

If you want to use double coverage for the previous example, refer back to Figure 3-30. The material quantity required is:

FS = 17.21 squares x 2 FS/Square = 34.42 FS

Use Figure 3-1 to convert this to rolls:

$$\text{Rolls (\#15)} = \frac{34.42 \text{ FS}}{4 \text{ FS/Roll}} = 8.6 \text{ rolls}$$

Underlayment on Sloping Roofs

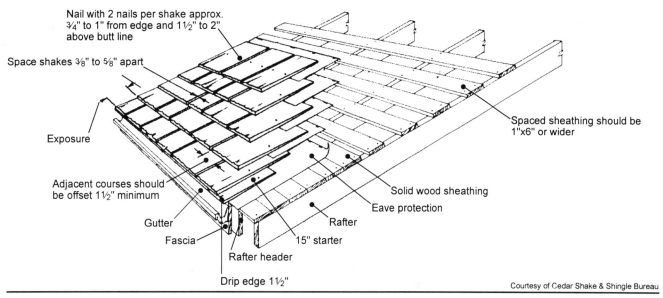

Figure 3-36 Interlayment

Interlayment (Lacing)

Whether you use solid or spaced sheathing for shake roofs, you're normally required to put felt *interlayment* between courses (see Figure 3-36). The interlayment acts as a baffle to keep wind-driven snow or other foreign material from getting into the attic cavity during extreme weather conditions. It also increases the roof's insulating value.

To be an effective baffle, the top edge of the interlayment felt must rest on the sheathing. Space open (spaced) sheathing carefully so that you can properly attach the top of the felt to the sheathing board.

Code permitting, you can omit felt interlayment when you apply straight-split or taper-split shakes in snow-free areas at weather exposures less than one-third the total shake length (3-ply roof).

Usually you apply shakes over a 30-pound, 36-inch-wide underlayment starter course installed at the eaves (eaves protection). Follow that with 30-pound, 18-inch-wide interlayment courses. Some building codes allow 15-pound felt. Figure 3-36 shows interlayment between shake courses.

Install the underlayment course under the double starter-course shakes at the eaves with the lower edge aligned with the butts of the shakes.

Install the first 18-inch interlayment course so the bottom edge of the felt is twice the exposure distance above the butts of the starter course. For example, for 24-inch shakes laid with a 10-inch exposure, apply the first interlayment felt so the bottom edge is 20 inches above the butts of the starter-course shingles, including the overhang. The felt will cover the top

Exposure (in.)	FS per square	Coverage (SF/FS)
5½	3.04	32.9
7½	2.23	44.8
8	2.09	47.8
8½	1.97	50.8
10	1.67	59.8
11½	1.45	68.8
12	1.39	71.8
14	1.20	83.7
16	1.05	95.7
18	0.93	107.6

Figure 3-37 Interlayment coverage (30-pound felt)

4 inches of the shakes and extend 14 inches onto the sheathing. In heavy snow areas, I recommend that you install two strips of felt for the first course of interlayment.

Install subsequent interlayment courses so the distance between their bottom edges is equal to the weather exposure. Nail the top edges of the felts to the sheathing at about 1- to 2-foot centers.

▼ **Example 3-5:** Assume an exposure of 7½ inches, and a 1½-inch overhang at the eaves. To find the distance from the eaves to the bottom edge of the first course of interlayment, multiply the exposure by 2, then subtract the overhang from the total:

(2 x 7½") - 1½" = 13½"

Then, the distance between the bottom edges of the succeeding courses of underlayment is 7½ inches (equal to the exposure).

Estimating Interlayment Quantities

Use Figure 3-37 to find interlayment quantities. Data in Figure 3-37 doesn't include:

- The 36-inch-wide underlayment starter-course felt installed at the eaves

- The additional felt required due to lapping at ridges, hips and valleys. Take off the starter-course felt separately, then add 1 square foot per linear foot of ridge, hip and valley to be covered.

- The cutting waste, which depends on roof complexity and the crew's experience.

To take off the 36-inch-wide underlayment starter course, assume that the length of underlayment you need is equal to the total eaves length. On a hip roof, this estimate is a bit long on the underlayment starter-course quantities. So, don't add material for lap or cutting allowance involved with the underlayment starter course to your estimate.

▼ **Example 3-6:** Assume a 5 in 12 roof slope, 30-pound interlayment laid at a 10-inch exposure and a 30-pound underlayment starter course at the eaves. Find the interlayment and underlayment quantities required for the roof of the building in Figure 3-35. Assume a 4 percent cutting allowance for the interlayment and no allowance for the underlayment starter course.

Use the formula:

Net Roof Area = Length x Width x Slope Factor

The net roof area of this building is 8.66 squares (40' x 20' x 1.083). That's also the area you use to estimate labor.

The area requiring interlayment begins at a point above the eaves equal to twice the shingle exposure (2 x 10" = 20"). You deduct 20 inches from each edge of the roof, so the interlayment area is:

Interlayment Area = (40' - 40") x (20' - 40") x 1.083

Remember to convert inches to hundredths of a foot, then proceed:

Interlayment Area = (40' - 3.33') x (20' - 3.33') x 1.083
= 36.67 x 16.67 x 1.083 = 662 SF

From Figure 3-35 you can see the ridge length is 20 feet.

You calculate the hip length using Column 2 of Appendix A (plan length times slope factor)

Length (Hip) = 10' x 1.474 = 14.74 linear feet

There are four hips, so the total length is 4 x 14.74 = 58.96 or 59 linear feet.

The total length of ridge and hips is 20 + 59 = 79 linear feet.

At 1 square foot per linear foot, the total allowance for lapping is 79 SF.

Including material required for lapping at the ridge and hips, the gross interlayment area is the same as the interlayment area plus the lap allowance:

Gross Interlayment Area = 662 SF + 79 SF
= 741 SF ÷ 100
= 7.41 squares

Now, from Figure 3-37, the interlayment quantity is:

FS (30-pound interlayment) = 7.41 squares x 1.67 FS/Square
= 12.37 FS

Now, multiply by 1.04 for a 4 percent cutting allowance:

12.37 x 1.04 = 12.9 factory squares

The roof perimeter (twice the total of length plus width) taken at the eaves, is:

Perimeter = 2 x (40' + 20') = 120 linear feet

The area covered by the underlayment starter course is the perimeter times 3 feet:

Area of Underlayment = 120' x 3'
= 360 square feet ÷ 100
= 3.6 squares

The total quantity of 30-pound felt required for all interlayment and underlayment is 16.5 factory squares (12.9 FS + 3.6 FS).

Look back at Figure 3-1 to find that 30-pound felt yields 2 FS per roll, so you need to order 8.3 rolls of felt:

$$\text{Rolls (30-pound felt)} = \frac{16.5 \text{ FS}}{2 \text{ FS/Roll}} = 8.3 \text{ rolls}$$

Figure 3-37 gives interlayment coverage data for common shake exposures. But remember, it doesn't allow for eaves underlayment, laps at ridges, hips and valleys, or crew waste. Use the following equations to find the interlayment coverage based on *any* exposure:

$$\text{Coverage (SF/FS)} = \frac{(\text{Roll Length} - \text{End Lap}) \times \text{Exposure}}{\text{Number of FS per Roll}} \quad \boxed{\text{Equation 3-3}}$$

where:

$$\text{Exposure} = \frac{\text{Roll Width} - \text{Top Lap}}{\text{Number of Plies}} \quad \boxed{\text{Equation 3-4}}$$

The coverage, in factory squares installed per square (100 SF) of roof surface covered, is:

$$\text{FS/Sq.} = \frac{100 \text{ SF/Square}}{\text{Coverage /FS}} \quad \boxed{\text{Equation 3-5}}$$

▼ **Example 3-7:** Assume an exposure of 6 inches and a 6-inch end lap, then find the coverage of 30-pound felt interlayment, using Equation 3-3 above.

You see in Figure 3-1 that 30-pound felt is normally sold in 72-foot-long rolls. You split each roll in half for 18-inch-wide strips, so the total roll length is 2 x 72, or 144 linear feet. Use Equation 3-3 to find the coverage in terms of square feet covered per factory square:

$$\text{Coverage (SF/FS)} = \frac{(144' - 0.5') \times 0.5'}{2} = 35.9 \text{ SF/FS}$$

Now, use Equation 3-5 to find the number of factory squares required per square of roof surface:

$$\text{FS/Square} = \frac{100 \text{ FS/Square}}{35.9 \text{ SF/FS}} = 2.79 \text{ factory squares per square}$$

Eaves Flashing (Ice Shield or Water Shield)

In snow-belt areas, there's sometimes enough heat loss from inside a building to melt snow on the roof. When that happens, water can leak into the walls and ceiling areas due to the freeze-thaw cycle of the snow. That cycle causes ice dams (freezebacks) to form on the roof. Water backs up under the shingles and down through the roof sheathing. This is shown in Figure 3-38. From there, the water drips down onto the ceiling or enters through the top of the walls where it finally ends up on the floor under the carpet.

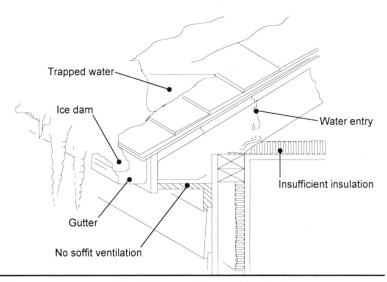

Figure 3-38 Ice dam caused by insufficient insulation and ventilation

This particular interior heat loss is caused by insufficient ceiling insulation. Ice dams can also occur over windows with large headers where there isn't enough wall insulation over the windows. Heat escapes up the wall and into the attic immediately above. This causes ice on the shingles to melt from the bottom up. Then, the weight of the ice remaining above presses down, forcing the water up the slope and under the shingles.

The action you take to prevent leaks caused by ice dams will depend on the roof slope and the local building code.

On slopes of 4 in 12 and steeper, in addition to underlayment, install an ice shield made of one ply of 50-pound smooth-coated roll roofing parallel to the eaves. Provide a ¼- to ⅜-inch overhang beyond the drip edge. Install the flashing 12 inches (some building codes require 24 inches) beyond the interior face of the exterior wall line as shown in Figure 3-39. If it takes more than one width of flashing to reach that point, locate the lap outside the exterior wall face. Overlap the flashing at least 2 inches and cement the entire length of the lap. If you have to use more than one roll over the length of the eaves, lap the ends at least 12 inches and cement the end lap.

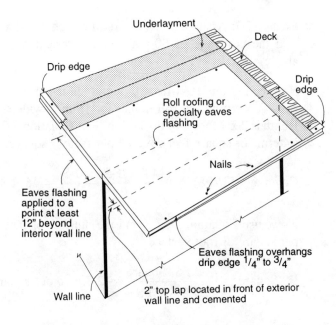

Courtesy of Asphalt Roofing Manufacturers Association (ARMA)

Figure 3-39 Application of eaves flashing

On slopes less than 4 in 12, you can use double-coverage underlayment for the eaves flashing. However, you have to install the flashing 24 inches (some building codes require 36 inches) beyond the interior face of the exterior wall line. Also, you must embed each course of the flashing material into roofing cement, which you've applied at the rate of 2 gallons per 100 square feet. That's shown back in Figure 3-11.

I believe a heavyweight roll-roofing ice shield under asphalt shingles does more harm than good. The heavy material can expand and buckle during hot weather, forcing shingles up and making them more susceptible to hail damage and foot traffic. You only need two layers of 15-pound felt under double-coverage asphalt shingles. If a freezeback does force moisture under the shingles, no leak will occur, provided the felts are smooth and watertight. But remember that you have to follow your local building code.

W. R. Grace Co. manufactures a rubberized product you can use as ice-shield material. It comes in 3-foot-wide rolls with an adhesive back covered with release paper. Remove the release paper and stick the material down to the sheathing, underlayment, or existing asphalt shingles. This material won't buckle, even in hot weather. You can also stick this material to flanges on flashing to help waterproof the flashing.

To cut down on ice dams, check the design of the building. The roof should be well ventilated, as shown in Figure 3-40. There should be a constant flow of cold air between the ceiling insulation and the roofing material. The vent space must let air flow freely from eaves to roof top. The steeper the roof, the better the venting system works.

Discourage wide overhangs at the eaves. They make cold areas where snow and ice can build up. Warn against putting doors and large windows with large headers and no wall insulation at the bottom of a roof slope. Instead, locate them beneath the gable ends of the roof. Be sure insulation doesn't block the soffit vents.

Ice dams don't form just at the eaves. They can occur anywhere heat escapes to the underside of the roof. It's best if chimneys are located at the ridge or gable ends. Recommend that plumbers stack out plumbing vent pipes so they go through the roof near the ridge, and use cast iron or galvanized iron vent pipes because sliding snow can move plastic ones. Make sure all vent pipes are securely anchored to the structure.

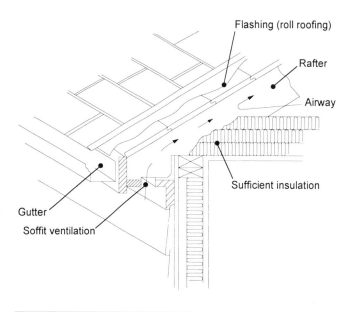

Figure 3-40 Good ventilation, insulation and roof flashing

Ice dams can also occur at the lower end of a valley the same way they do at the eaves. To minimize or prevent this problem, building owners can install an electric thermal wire in the valley in a zigzag pattern, and turn on the current when it snows.

▼ **Example 3-8:** Assume a potential for ice dams, a roof slope of 6 in 12, and that the interior face of the exterior wall line is 30 inches (measured horizontally) from the eaves line. Find the amount of eaves flashing required for the hip and valley roof in Figure 3-9.

Since the roof slope is 6 in 12, you need to put a 50-pound coated felt flashing along the eaves 24 inches beyond the inside face of the outside wall line. The actual flashing width is:

Flashing Width = (30" + 24") x 1.118
 (the slope factor from Column 2, Appendix A)
 = (2.5' + 2') x 1.118 = 5.59' = 6 feet (rounded up)

The formula (from Chapter 1) for the building perimeter is:

Perimeter = 2(L+W)

The total length of flashing required (using the formula for building perimeter) is:

Flashing Length = 2 x (50' + 33') = 166 linear feet

The total area of flashing required is:

Flashing Area = 166' x 6'
 = 996 square feet ÷ 100
 = 9.96 or 10 squares

(Use this figure to estimate manhours at the rate of 0.738 manhours per square.)

Since the length of flashing required is equal to the total eaves length, you don't need to add for cutting allowance.

Valley Flashing

If a roof leaks, it's most likely to leak at a valley. So you need to flash valleys for added weather protection. The type of valley flashing you use depends on the type of valley you'll construct. That in turn depends on the type of roof covering you'll install.

Figure 3-41 W-type metal valley on an asphalt shingle roof

Closed and Open Valleys

In a closed valley, you lap the roof covering from both sides of the valley to cover the valley flashing. In an open valley, there's space between the roof covering on adjacent roofs. Shingles don't extend across the valley, so the valley flashing is exposed as in Figure 3-41. You can use open valleys with all types of asphalt or fiberglass shingle products. In fact, this type of valley is recommended for T-locks and mineral-surfaced roll roofing.

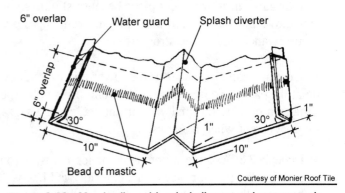

Figure 3-42 Metal valley with splash diverter and water guards

Open valleys tend to leak more than other types of valleys because heavy rain or debris collected in the valley can force water up under shingles adjacent to the valley.

If you have a choice, avoid installing open valleys where the two roofs forming the valley have considerably different slopes or areas which let very different amounts of water flow into the valley. For example, at a dormer, there would be a much higher volume of water flowing into the valley off the main roof than from the dormer. The heavier flow might back up under the shingles on the opposite side of the valley.

On strip-shingle roofs with varying slopes, you can reduce the problem by installing a closed-cut valley, and shingling the lower-sloped roof surface first. This is the slope that will receive most of the problem water. When you have to install an open valley on this type of roof, be sure to use metal valley flashing with a splash diverter and water guards as shown in Figure 3-42 to break the force of the water from the steeper or longer slope to keep it from being driven up under the shingles on the opposite side.

You are required to use open valleys on tile, wood shingle, or shake roofs. You can use a closed valley on a slate roof, but the open valley is more common. Chapter 8 has more information about that. You can use the closed-cut (half-lace) valley (Figure 3-46) with all types of strip shingles. The woven (full-lace) closed valley is recommended only with 3-tab shingles.

Valley Flashing Materials

Use mineral-surfaced roll roofing material to flash valleys on roofs covered with 3-tab, T-lock and laminated shingles, and mineral-surfaced roll roofing. You can also use metal flashing in these valleys, but this method (shown in Figure 3-41) isn't common because the metal expands and contracts when the temperature changes.

Use metal flashing on open valleys of roofs covered with slate, wood shingles, and shakes. Use metal, or a combination of metal overlaid with mineral-surfaced roll roofing, to flash open valleys on tile or metal roofs.

Figure 3-43 Roll metal valley flashing

Here are some of the many types of metal valley flashing materials you can get:

- Galvanized steel
- Aluminum
- Copper
- Tin (terne plate)
- Lead
- Zinc

Galvanized steel and aluminum are the least expensive, so they're used the most. You'll usually find the other (more expensive) metals only on certain commercial or long-lasting roofs such as tile or slate.

Galvanized steel valley flashing (roll valley metal) is made of 26- or 29-gauge galvanized steel or other corrosion-resistant metal. It comes in 50-foot rolls 12, 14, 16, 18 and 20 inches wide (Figure 3-43). I recommend you use (and some building codes require) 26-gauge metal because rainwater from a roof carries erosive roof dust and grit that'll quickly undermine lesser metal gauges. For this reason, use at least 16-ounce copper (16 ounces per square foot), 12-ounce tin, 3-pound hard lead and 11-gauge (0.024-inch-thick) zinc whenever those materials are specified for any type of flashing.

Some roofing manufacturers recommend that you paint galvanized steel on both sides with metal paint or bituminous paint. Paint tin valley metal on the underside. If you have to bend the flashing at a sharp angle, apply the paint after you bend it. The paint, especially on the underside, adds another layer to help protect the metal from corrosion.

To help keep water from getting under the shingles adjacent to the valley, install metal with a W-shaped crimp (splash diverter rib) in the center and water guards along the edge. See Figure 3-42. In valleys likely to accumulate debris, install flashing with a double crimp like the one in Figure 3-44. If you fabricate the flashing with water guards, you'll need to anchor the flashing to the sheathing with metal clips, as shown in Figure 3-45, on 12-inch centers. Some manufacturers recommend

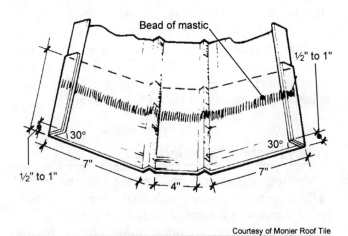

Figure 3-44 Metal valley with a double crimp

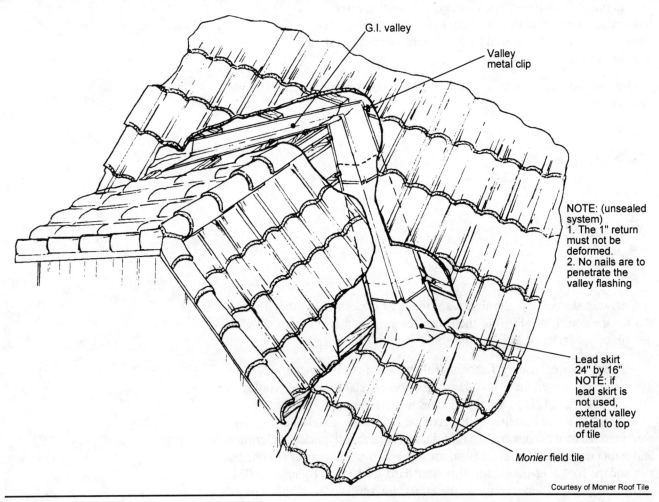

Figure 3-45 Metal valley fastened with metal clips

Underlayment on Sloping Roofs

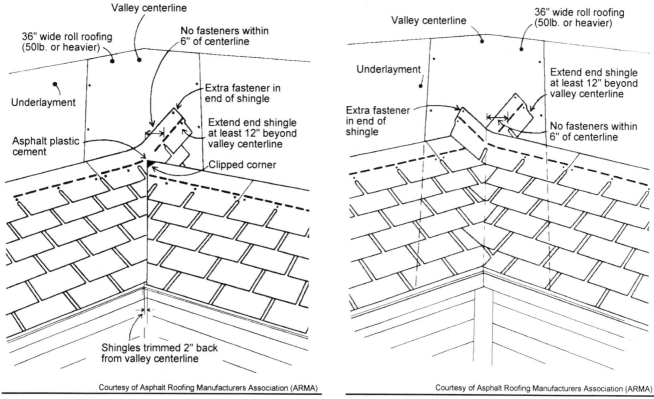

Figure 3-46 Flashing a closed-cut valley

Figure 3-47 Flashing a woven valley

using metal clips rather than driving nails into the valley metal because the clips can expand and contract when the temperature changes and they pivot to move with the valley metal.

You can leave out the splash diverter ribs and water guards where a valley takes a sharp turn, because those features make it hard to change direction. Instead, install metal from a smooth stock metal roll, or use an open mineral-surfaced roll-roofing valley.

Installing Non-Metal Valley Flashing

In a closed valley, in addition to the required felt underlayment described above, flash the valley with a 36-inch-wide layer of No. 15 or No. 30 asphalt-saturated felt. Many building codes allow this; however, I recommend using mineral- or smooth-surfaced roll roofing (50 pounds or heavier). That's shown in Figures 3-46 and 3-47. If you need more than one strip of roll roofing over the length of the valley, lap the upper strip over the lower by at least 12 inches and bond the lap with roofing cement. Don't apply too much cement because it could cause blistering. Don't drive nails through the lap within the valley. Nail the flashing strip to the deck along a line 1 inch from the edge of the flashing, using only enough

Figure 3-48 Extend the valley flashing beyond the ridge

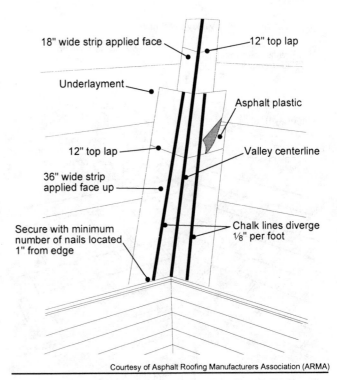

Figure 3-49 Application of 90-pound roll roofing as flashing for an open valley

nails to hold the strip in place. Extend the flashing beyond the ridge and cut it so it lays close to the surface of the sheathing with no bulges. Then nail it down. See Figure 3-48.

In an open valley, in addition to the felt underlayment previously discussed, flash the valley with two layers of mineral-surfaced roll roofing (90 pounds). Cement the layers together with the bottom 18-inch-wide layer mineral surface down, and the top 36-inch wide layer with the mineral surface facing up. See Figure 3-49. If you need to splice the roll roofing over the length of the valley, lap the upper strip over the lower by at least 12 inches and bond the lap with roofing cement. Don't drive nails through the lap within the valley. Nail both strips to the deck along a line 1 inch from the edge of the flashing, driving only enough nails to hold each strip in place. Install a top layer of roll roofing that's the same color as the shingles, or a neutral color.

Installing Metal Valley Flashing

To install light-gauge roll metal valley flashing, roll the flashing down the valley. Step into the center of the sheet so that it conforms to the valley, nailing as you walk up the valley. Drive the nails on 12-inch centers along rows located 1 inch from the edges of the flashing. Use metal clips if the flashing has water guards.

Underlayment on Sloping Roofs

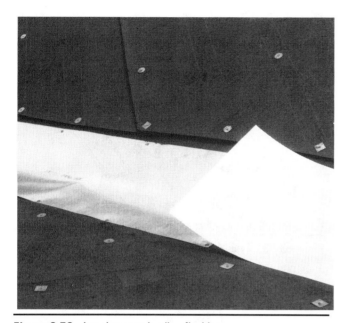

Figure 3-50 Lapping metal valley flashing

Figure 3-51 Extend the metal valley flashing beyond the ridge

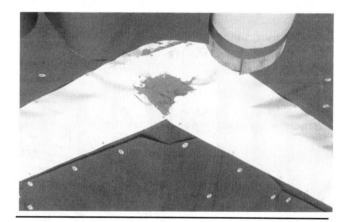

Figure 3-52 Metal valley flashing in two intersecting valleys

Don't fabricate heavy-gauge metal flashing pieces longer than 10 feet. If you need more than one piece or roll of flashing over the length of a valley, lap the upper layer over the lower by at least 6 inches, as shown in Figure 3-50 and bond the lap with roofing cement. Don't drive nails through the lap within the valley.

Extend roll metal flashing beyond the ridge and cut it so that it conforms with the surface of the sheathing. Then nail it down as shown in Figure 3-51. Lap the metal where two valleys intersect and seal the joint with roofing cement, as in Figure 3-52. For added leak protection, install a lead saddle at the juncture of two metal valleys. That's shown in Figure 3-53.

On wood-shingle roofs, extend metal valley flashing at least 10 inches beyond each side of the centerline of the valley on roofs with slopes up to 6 in 12, and 7 inches beyond the centerline on roofs with slopes 6 in 12 and steeper. Snap a chalk line as a shingling guide 2 to 4 inches on either side of the centerline, depending on the anticipated water volume. See Figure 3-54.

You don't need underlayment beneath the metal valley on a wood-shingle roof. But you should use a 36-inch-wide strip of 15-pound felt, especially in cold-weather areas where there are wind-driven snows. The underlayment helps to prevent leaks. It also helps to prevent condensation that would eventually cause corrosion on the underside of the metal. That's also shown in Figure 3-54.

On shake roofs, extend the metal valley flashing at least 10 inches beyond each side of the centerline of the valley. Install a 36-inch-wide roll of 15-pound felt (minimum) beneath the metal flashing. Code permitting, you can omit the valley felt when you install the shakes over spaced sheathing.

Some roofing manufacturers don't recommend using felt beneath any valley metal except copper or lead since the felt keeps condensed water beneath the metal from running off or evaporating. When you use felt to head off wind-driven precipitation, it's very important to paint the underside of a metal valley.

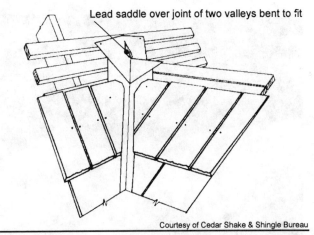

Figure 3-53 Typical saddle flashing

Estimating Valley Flashing Material

For estimating, you take off roll roofing flashing material by the square foot, then convert to squares. Take off metal flashing material by the linear foot. You have to account for end laps and overcuts for valley ends at inside roof corners. Figure 3-55 shows that the length of additional material you need is one-half the width of the flashing material. In most cases, add 1 foot to the valley length to allow for this additional material. Some roofing contractors cut the material off at the lower end, forming a small triangular extension that helps shed runoff water away from the building. That's shown in Figure 3-56.

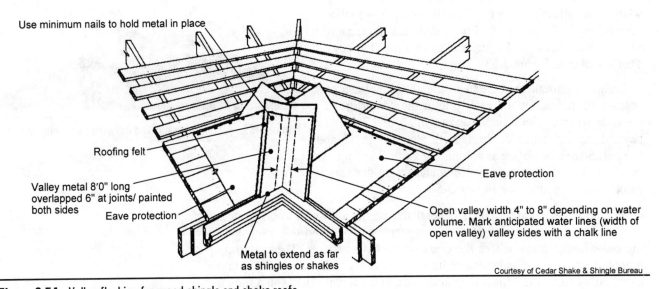

Figure 3-54 Valley flashing for wood shingle and shake roofs

Underlayment on Sloping Roofs

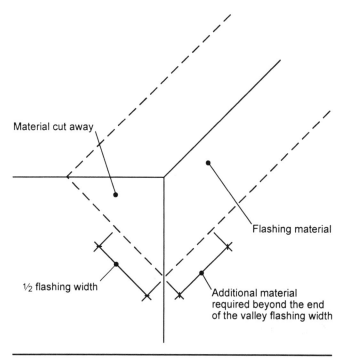

Figure 3-55 Additional flashing material required at the lower end of a valley

Figure 3-56 Valley flashing extended beyond the eaves to help shed water out of the valley

Here's how to estimate valley flashing quantities:

▼ **Example 3-9:** Assume you're applying 3-tab asphalt shingles on a 6 in 12 roof. Find the quantity of valley flashing required for the roof in Figure 3-9, for both a closed and open valley. Assume the flashing ends where the valleys meet.

Using the slope factor conversion table from Column 3 of Appendix A, the actual length of material required at either valley (including 1 foot of additional material required beyond the lower end) is:

Length (Valley) = 12' x 1.5 = 18 linear feet

Multiply by 2 for the total length of both valleys, 36 linear feet.

For a closed valley, the total area to be covered with 50-pound roll roofing is:

Flashing Area = 36' x 3'
              = 108 square feet ÷ 100
              = 1.08 squares

For an open valley, the total area to be covered with 90-pound mineral-surfaced roll roofing is:

Flashing Area = 36' x (3' + 1.5')
= 162 square feet ÷ 100
= 1.62 squares

The manhour rate for the above example is 1.03 manhours per square.

When you're estimating roll roofing for flashings, be alert to unusual colors. If the roof is a color you seldom install, you might never be able to use leftover roll roofing material. In that case, figure the entire roll into your estimate, including any excess you won't use on the current job.

▼ **Example 3-10:** Assuming a roof slope of 6 in 12, find the total square feet of 18-inch-wide metal valley flashing material required for the roof of the building in Figure 3-9. Assume 10-foot lengths of metal installed with a 6-inch end lap.

The actual length of either valley (using the slope factor from Column 3 of Appendix A) is:

Length (Valley) = 11' x 1.5 = 16.5 linear feet

Multiply by 2 for the total length of both valleys, 33 linear feet

Also, add 0.5 foot for material at each lap. Since there are two joints, add 2 x 0.5 foot, or 1 foot.

The metal must extend 9 inches beyond the end of the valley, so add 0.75 foot for each valley, rounded up to 2 feet total.

The total is:

Total Valley Metal = 33 feet + 1 foot + 2 feet = 36 feet

Now that we've covered underlayment and sheathing, let's go on to roof coverings. In the next chapter, we'll begin with asphalt shingles.

4 Asphalt Shingles

▶ Asphalt roofing materials have been manufactured since the early 1890s. Today, asphalt shingles cover about 70 to 80 percent of all roofs in the United States. Those roofs are attractive, versatile, and fire- and wind-resistant. Asphalt shingle roofs are relatively inexpensive, easy to install, and require little maintenance. The normal life expectancy of an organic asphalt shingle roof is 15 to 20 years. Heavyweight laminated fiberglass shingles will last 20 to 30 years.

Organic Shingles

The base mat of organic-based asphalt shingles (organic shingles) was originally composed of cellulose fibers made from recycled paper or wood chips, and cotton or wool fibers made from rags. Now, it's made of a tough, asphalt-saturated roofing felt, coated on both sides with asphalt, as shown in Figure 4-1.

The base mat gives shingles their strength. The base material is saturated and covered with a high-melting-point flexible asphalt called a *saturant*. The saturant is reinforced with mineral stabilizers such as ground limestone, slate, trap rock (weathered volcanic rock) or other inert materials such as ceramic-coated rock granules. Coarse mineral granules are pressed into the asphalt coating on the exposed face. This gives the shingle its color and helps it resist weather and fire.

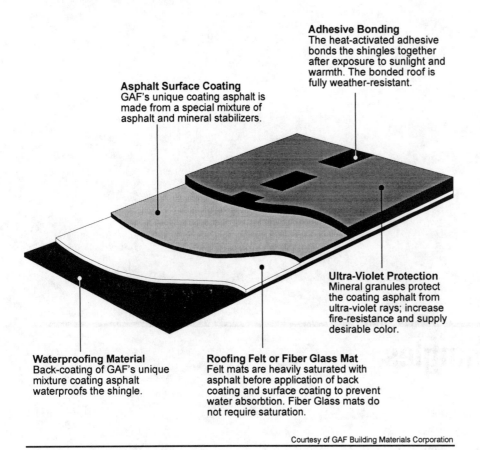

Figure 4-1 Asphalt shingle components

The materials most often used for coarse mineral surfacing are natural-colored slate, natural-colored rock granules, or ceramic-coated rock granules. The back of each shingle is covered with talc, sand or mica to prevent shingles from sticking together in the bundle.

Fiberglass Shingles

Fiberglass shingles first appeared in the late 1950s. By the late 1970s, they had improved so much they were as good as traditional asphalt shingles. Fiberglass shingles have a fiberglass base mat saturated and covered with flexible asphalt and surfaced with mineral granules. The weight and thickness of a fiberglass mat is usually much less than a cellulose-fiber mat. Fiberglass shingles contain more asphalt than organic-based asphalt shingles.

Nowadays, organic shingles aren't used very often. They soak up water from underneath, which makes the corners at the bottom of the tabs curl up.

Throughout this chapter, the term "asphalt shingle" means an organic- or fiberglass-based shingle saturated with asphalt. I'll distinguish between the two products only when it's necessary for the sake of accuracy.

UL Ratings for Shingles

The Underwriters' Laboratory (UL) is a non-profit organization founded in 1894 under the sponsorship of the National Board of Fire Underwriters. The UL has the most widely-accepted standards for fire resistance of building materials. The UL classifies a fire-resistant shingle as A, B, or C. Class A shingles withstand severe fire exposure. Class B shingles withstand moderate fire exposure and Class C, light fire exposure. With all three ratings, "exposure" means exposure to fire that comes from sources outside the building. To qualify for any UL classification, a shingle must not:

Figure 4-2 Underwriters' Laboratory label

a) disintegrate and fall off the roof as glowing brands (airborne embers)

b) break, slide, warp, or crack, exposing the deck

c) allow the roof deck to fall away as glowing particles

d) allow continued flaming beneath the roof deck.

To bear a UL "wind-resistant" label, a shingle must withstand winds up to 63 miles per hour for two hours without a single tab being uplifted. Figure 4-2 shows a UL shingle label. Look for this label on each bundle of shingles, and be sure to install them according to the manufacturer's instructions.

To increase their wind resistance, many asphalt shingles come with a self-sealing thermoplastic adhesive strip (tar strip) above the cutouts on the face of the shingle. That's shown in Figure 4-3. Heat from the sun makes the strip sticky and helps to bond each shingle to the one above it. The adhesive strip takes longer to bond in cold weather or when the roof is shaded, has a low slope, or faces north or east.

Figure 4-3 Shingle with adhesive strip

The best temperature range for installing asphalt shingles is between 40° F and 85° F. Before you install asphalt shingles during cold weather, store the shingles in a warm location or lay them in the sun until they soften up.

If you will be storing shingles, keep them in a cool dry area in stacks no more than 4 feet high. Your local roofing materials supplier can advise you about how long it's safe to stockpile asphalt shingles. Rotate the bundles so the shingles stored the longest will be the first ones you'll use. That's so the shingles at the bottom of the stacks don't become discolored. Light-colored shingles may darken because oils in the asphalt move. Dark-colored shingles may show light smudges when backing materials (such as talc, which helps keep shingles from sticking together in the bundle) transfer to adjacent shingles.

If you store shingles outdoors, place them on a raised platform so they don't touch the ground. Cover the shingles to protect them from wet weather. Don't store shingles in the hot sun because heat makes them stick together.

Deck Requirements

Asphalt shingles require a solid roof deck. As a general rule, you can install asphalt shingles on roof slopes ranging from 4 in 12 through 21 in 12 using standard application methods. You can also install asphalt shingles on slopes as flat as 2 in 12, or steeper than 21 in 12, but you'll have to follow special application procedures. We'll cover this later in this chapter (see Figure 4-27 and related text). Figure 4-4 gives minimum roof slope requirements for various asphalt roofing materials.

Shingle Colors

The color of the shingles you use can dramatically affect the appearance of a building. For example, a light-colored roof directs the eye upward and gives the illusion of spaciousness. Dark colors create the opposite effect. In the case of a large, steep roof, you can use that illusion to scale down the roof structure and make the building look more proportional and attractive. Use Figure 4-5 as a guide for choosing shingle colors that go with various colors of siding, trim, shutters and doors.

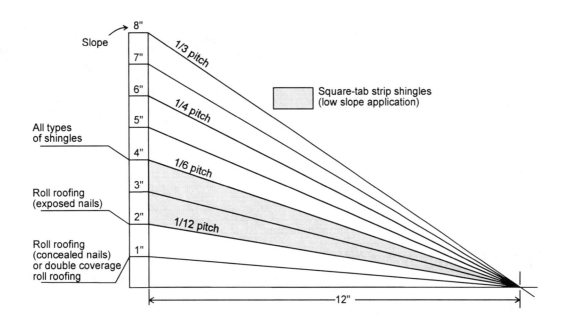

Figure 4-4 Minimum pitch and slope requirements for various asphalt roofing products

Roof Shingles	Siding	Trim	Shutters and Doors	Roof Shingles	Siding	Trim	Shutters and Doors
White	White	White	Deep Gold, Maroon	**Brown**	White	White	Dark Brown, Terra Cotta
	White	Gray	Charcoal		Green	White	Dark Brown, Dark Green
	Green	White	Dark Brown, Dark Green		Yellow	White	Dark Brown, White
Black	White	White	Black, Maroon	**Green**	White	White	Dark Green, Black
	Yellow	White	Black, Deep Olive Green		Yellow	White	Dark Green
	Gold	White	Black, Deep Olive Green		Lt. Green	White	Dark Green, Terra Cotta
Gray	Red	White	Black, White	**Blue**	White	White	Blue
	Yellow	White	Gray, Charcoal, Green		Yellow	White	White
	Coral Pink	Lt. Gray	Charcoal		Lt. Blue	White	Dark Blue, White
Red	White	Gray	Charcoal				
	White	White	Red				
	Beige	White	Dark Brown				

Courtesy of Asphalt Roofing Manufacturers Association (ARMA)

Figure 4-5 Asphalt shingle color guide

Roofing Construction & Estimating

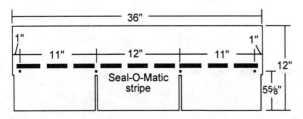

Figure 4-6 Three-tab Fire-Glass III fiberglass shingle

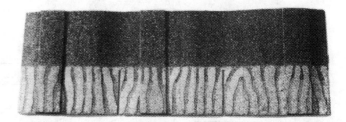

Figure 4-7 Laminated shingle

Asphalt Strip Shingles

The most widely-used type of asphalt shingle is the 3-tab (triple-tab) strip shingle like the one in Figure 4-6. This is also called a square-butt or thick-butt shingle because some manufacturers make a shingle that's thicker at the butt edge. Cutouts (also called keys, water lines, bond lines, tab notches, and water jackets) make a roof look like it's finished with many smaller units.

You can also find 2-tab (twin-tab) strip shingles, strip shingles with no cutouts, and shingles with as many as five tabs (random-tab strip shingles). Some shingles have staggered butt lines. Shingles whose tabs are all the same size are called square-tab shingles. Strip shingles with more than one layer of tabs are called laminated, dimensional, or three-dimensional shingles. These shingles create extra thickness and give a three-dimensional effect. See Figure 4-7. Other strip shingles include 2- and 3-tab hexagonal shingles. You can see those at the end of the chapter in Figure 4-99.

Asphalt strip shingles weigh from 135 to 390 pounds per square depending on:

- the shape of the shingle
- the thickness of the base mat
- the amount of asphalt absorbed by the base mat
- the thickness of the asphalt coating
- the amount of surface material pressed into the exposed face.

You can use Figure 4-8 to estimate approximate asphalt shingle weights.

The chart in Figure 4-9 is a quick reference to the specs and coverage for several typical asphalt shingles. Asphalt shingles are from 35 5/16 to 42 inches long, and from 12 to 14 1/8 inches wide. They have recommended

Shingle type	Approximate weight per square (pounds)
Two-tab strip	300
Three-tab strip	235
Two- or three-tab hexagonal	200
Individual Dutch lap [1]	165
Individual American [1]	330

[1] See Figures 4-97 and 4-98

Figure 4-8 Approximate asphalt shingle weights

PRODUCT	Configuration	Per Square			Size			ASTM* fire and wind ratings
		Approximate Shipping Weight	Shingles	Bundles	Width	Length	Exposure	
Self-sealing random-tab strip shingle 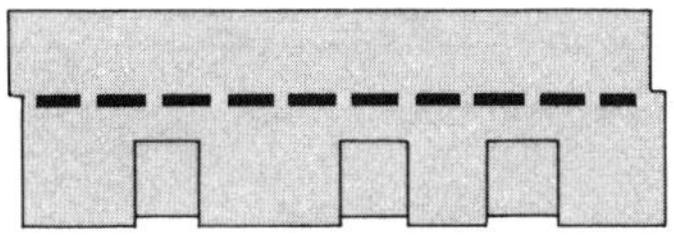 Multi-thickness	Various edge, surface texture and application treatments	240# to 360#	64 to 90	3, 4 or 5	11½" to 14"	36" to 40"	4" to 6"	A or C - Many wind resistant
Self-sealing random-tab strip shingle 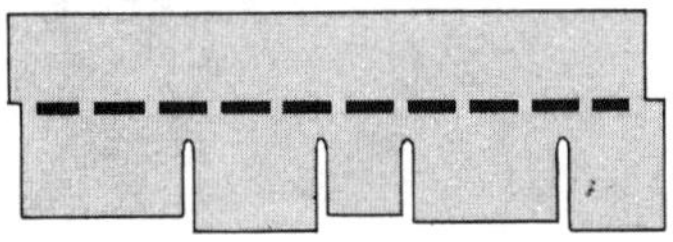 Single-thickness	Various edge, surface texture and application treatments	240# to 300#	65 to 80	3 or 4	12" to 13¼"	36" to 40"	4" to 5⅝"	A or C - Many wind resistant
Self-sealing square-tab strip shingle 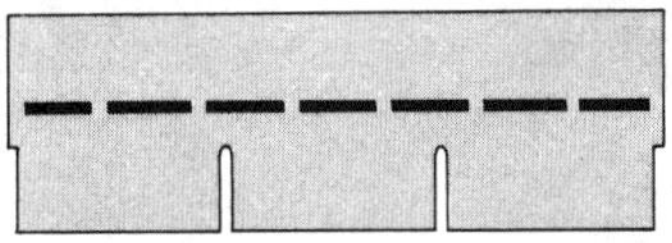 Three-tab	Three-tab or Four-tab	200# to 300#	65 to 80	3 or 4	12" to 13¼"	36" to 40"	5" to 5⅝"	A or C - All wind resistant
Self-sealing square-tab strip shingle 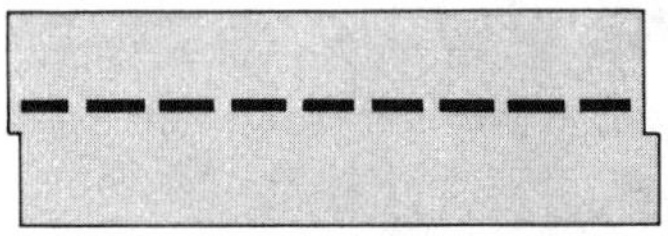 No-cutout	Various edge and surface texture treatments	200# to 300#	65 to 81	3 or 4	12" to 13¼"	36" to 40"	5" to 5⅝"	A or C - All wind resistant
Individual interlocking shingle 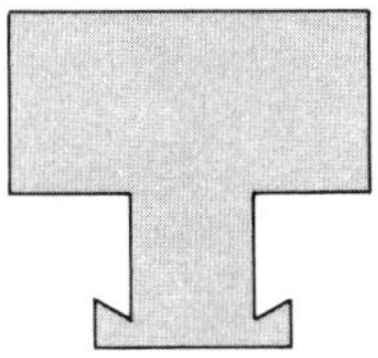 Basic design	Several design variations	180# to 250#	72 to 120	3 or 4	18" to 22¼"	20" to 22½"	—	A or C - Many wind resistant

*American Society for Testing and Materials

Courtesy of Asphalt Roofing Manufacturers Association (ARMA)

Figure 4-9 Typical asphalt shingles

exposures of 4 to 6 inches. (Exposure is the part of the shingle not covered by the next course of shingles.) The most common asphalt strip shingle is 3 feet by 1 foot laid at a 5-inch exposure.

There are also several types of individual asphalt shingles, including hex shingles, interlocking shingles and giant individual shingles. Those are described and shown at the end of this chapter in Figures 4-94, 4-95, and 4-96.

Installing Asphalt Strip Shingles

After you've laid the underlayment, drip edge and valley flashing, you're ready to install a starter course at the eaves of the roof. The starter course protects the eaves of the roof by filling in the spaces under the cutouts and joints of the first course of shingles. Without a starter course, there would only be single coverage at the eaves. Install the starter course with a ¼- to ⅜-inch overhang at the eaves and rakes.

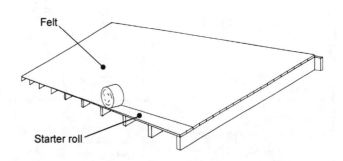

Figure 4-10 Mineral-surfaced starter roll

■ **Roll Roofing Starter Course** You can install a starter course by using a 7-inch-wide (minimum) strip of mineral-surfaced roll roofing whose color matches the shingles (Figure 4-10). Place the starter roll along the eaves with a ¼- to ⅜-inch overhang and nail the strip on 12-inch centers. Drive the nails along a line 3 to 4 inches above the eaves. If you're installing the starter roll over board sheathing, *stagger nail* to prevent splitting a board. That means don't hammer nails in a straight line along the grain of the board.

Roll roofing comes in 36-foot lengths. If you need more than one strip to cover the length of the eaves, lap the end joint at least 2 inches. Nail the underlay, then embed the overlap in roofing cement and nail it in place with three nails.

I recommend you use shingles instead of roll roofing for the starter course. That way you don't have to worry about matching colors, and the laps won't show through the overlying shingles.

Figure 4-11 Field-fabricated starter course

■ **Shingle Starter Course** Most strip shingle manufacturers recommend that you make the starter course by cutting off the shingle tabs and installing the shingles with the factory-applied adhesive along the eaves, as shown in Figure 4-11.

Asphalt Shingles

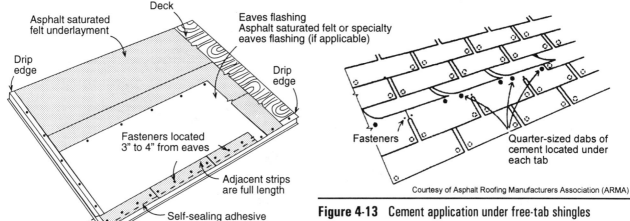

Figure 4-12 Application of starter strip

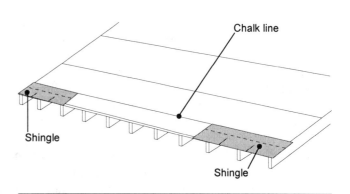

Figure 4-13 Cement application under free-tab shingles

Figure 4-14 Lining up the first course

Trim about 3 inches from the end of the first starter-course shingle to keep the joints of the first course of shingles from lining up with the joints of the starter-course shingles. That's shown in Figure 4-12. Position the starter-course shingles along the eaves with a 1/4- to 3/8-inch overhang. Drive nails into the shingles along a line 3 to 4 inches above the eaves. Position the nails so that they won't be exposed under the cutouts of the shingles in the first course. Stagger nail the starter-course shingles over board sheathing.

If you use roll roofing or shingles that don't have a factory-applied adhesive strip (free-tab shingles) for the starter course, bond the tabs of each shingle in the first course to the starter strip. Use a spot of roofing cement about the size of a quarter beneath each shingle tab. Figure 4-13 shows this. Install *all* free-tab shingles this way in high-wind areas. That includes the starter course, even when you use shingles that *do* have factory-applied adhesive strips.

■ **Start with a Straight Line** It's very important that you install the starter course and first course of asphalt shingles straight. To align asphalt shingles, nail down a shingle with the correct overhang on each end of the eaves. Snap a chalk line along the top edges of the shingles as shown in Figure 4-14. Then line up the top edges of intervening shingles along the chalk line. Repeat this alignment every third or fourth course. Measure from the eaves up to the butt position for the next course of shingles at the rakes. Install the end shingles, then snap another chalk line to align that course.

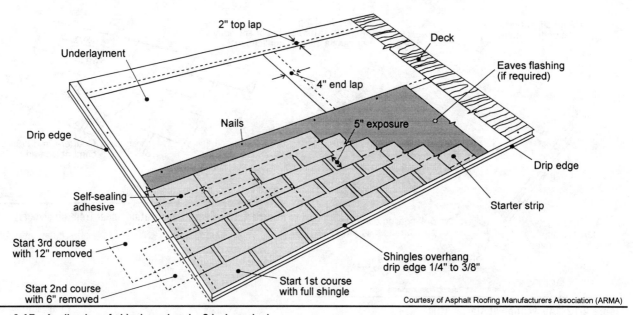

Figure 4-15 Application of shingles using the 6-inch method

You can save time by snapping all the chalk lines before you install any shingles. Snap horizontal chalk lines on 5-inch centers (assuming a 5-inch shingle exposure), allowing for the overhang at the eaves. Then snap vertical chalk lines on 6-, 12- or 36-inch centers, depending on how good you are at eyeballing a straight line. Once you've done this, you can line up shingles at the correct positions with the proper exposure without having to use the exposure gauge on your hatchet. Be sure to allow for the required overhang when you snap your chalk lines.

For example, if your exposure is 5 inches, with a ¼-inch overhang, snap your first horizontal chalk line at 11¾ inches from the edge of the eaves. Snap succeeding chalk lines 5 inches apart.

To maintain the correct exposure for square-tab strip shingles, align the butts with the top of the cutouts in the course below, since the cutouts in these shingles are 5 inches deep.

■ **Shingle Patterns** There are three basic shingling patterns used to install 3-tab asphalt strip shingles:

a) joints broken into halves, or the 6-inch pattern (half pattern)

b) the 5-inch pattern (random pattern)

c) joints broken into thirds, or the 4-inch pattern

To install the 6-inch pattern, start the first course with a full-length shingle. Remove 6 inches from the first shingle of the second course. Then remove 12 inches from the first shingle of the third course. Continue, removing an additional 6 inches from the first shingle of each course until you begin with a full shingle again on the seventh course. You can see how this works in Figure 4-15.

Asphalt Shingles

Save the full tabs you cut off and use them for hip and ridge units, filler tabs adjacent to valleys, and at the opposite ends of a gable-framed roof. The 6-inch pattern is the simplest style to install. But, because you align the cutouts every other course and the shingles vary slightly in size, you must snap chalk lines up the roof slope so you can align the edges of the shingles to keep the cutouts lined up vertically. The easiest way to install the 6-inch pattern on a gable roof is to shingle up the rake and install each course only far enough out over the deck to establish a pattern. When you get to the ridge, return to the bottom and finish out each course across the roof, working your way up the slope. This method of shingling up the rakes followed by shingling across the roof is called the *diagonal* method.

Figure 4-16 Lapping shingles over the ridge

If the top of a shingle extends beyond the centerline of the ridge, lap the shingle over the ridge and nail it on both sides of the ridge, as in Figure 4-16.

The 5-inch pattern is often called a random pattern. This pattern gives you some flexibility when you align the cutouts up the roof slope. Start the first course with a full-length shingle. Remove 5 inches from the first shingle of the second course. Then remove 10 inches from the first shingle of the third course. Continue, removing an additional 5 inches from the first shingle of each course until you begin with a full shingle again on the eighth course. (You don't start the eighth course with a 1-inch section.) Use the exposure gauge on your hatchet to measure the 5-inch increments. Figure 4-17 shows the shingle pattern this method produces. On gable roofs,

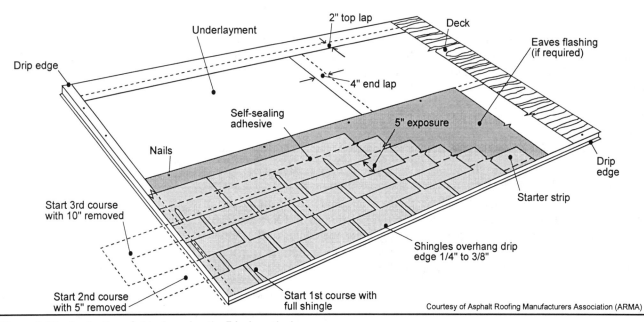

Figure 4-17 Application of shingles using the 5-inch method

Roofing Construction & Estimating

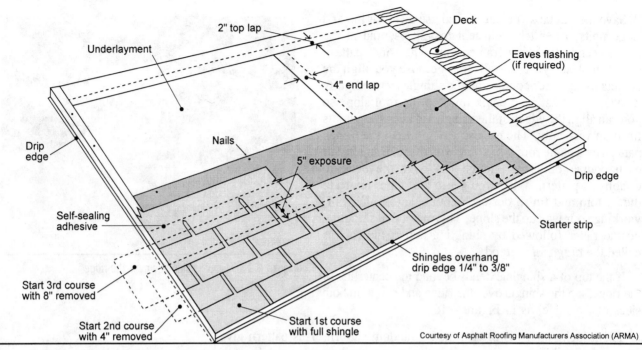

Figure 4-18 Application of shingles using the 4-inch method

shingle up the rake to the ridge, as with the 6-inch pattern, then return to the bottom and finish out each course across the roof, working your way up the slope.

You'd usually use the 4-inch pattern only on low-slope roofs ranging from 2 through 3 in 12. Trim the first shingle of each course in a multiple of 4 inches, beginning again with a full-width shingle at the 10th course. Figure 4-18 shows this pattern.

The 5-inch pattern is sometimes used on hip roofs, while the 6-inch pattern is generally used on gable roofs. Never install a shingle pattern less than 4 inches because the cutouts and joints would be so close on adjacent courses that leaks could occur.

Shingle Application

The order you follow to install shingles depends on the roof style. On gable roofs broken by dormers or valleys, start shingling at the rake and proceed toward the breaks. On simple gable roofs, start shingling at the gable end that's most visible to passers-by. On hip roofs and roofs where both gable ends are equally visible, start shingling at the center of the roof and proceed in both directions. In this case, set all your chalk lines (for the offset pattern you'll use) before you begin shingling.

On hip roofs, lap shingles over the hips from both sides, as shown in Figure 4-19. Then cut the shingle edges of the upper layer in line with the centerline of the hip. See Figures 4-20 and 4-21.

Asphalt Shingles

Figure 4-19 Lapping shingles over a hip

Figure 4-20 Trimming the hip

■ **Dormers** If there's a dormer, shingle the top of it first. Then bring the shingles of the main roof up to and alongside the dormer, all the way to the ridge of the dormer. Extend one shingle course on the main roof on one side of the dormer to a distance at least one shingle beyond the ridge of the dormer roof. Notice the top shingle on the left side of the dormer in Figure 4-22.

Snap vertical chalk lines down from the ridge starting with the edge of the extended shingle, as shown in Figure 4-22. Use those chalk lines as guides to align the shingle courses as you install them on the right side of the dormer. Slip the last shingle course under the course that's in line with the ridge of the dormer. Aligning shingles on both sides of a dormer this

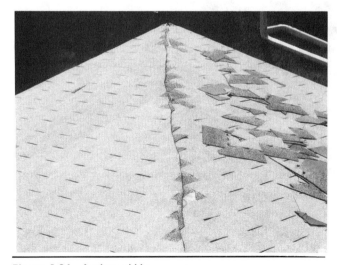

Figure 4-21 A trimmed hip

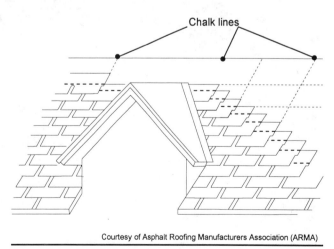

Courtesy of Asphalt Roofing Manufacturers Association (ARMA)

Figure 4-22 Tying in around a dormer

way is called "tying in." You'll have to shingle the ridge of the dormer before you finish the main roof above the dormer. We'll describe that later in this chapter, on page 92 under the heading, "Ridge and Hip Units, Cap Shingles."

No matter where you begin shingling, roofing material manufacturers recommend you apply the shingles in the diagonal pattern described earlier. Then you'll be sure you've nailed every shingle properly because you can see each one until you cover it with the next course above.

You can also use the straight-up (racking) method shown in Figure 4-23. But then you have to install some shingles under shingles you've already laid in the course above. That's shown in Figure 4-24. Since part of the underlying shingle is hidden, there's a possibility you could miss nailing that part of the shingle.

Some roofing contractors prefer the racking method because it's a more accurate way to align the shingles. You use the horizontal chalk lines and previously-laid shingle edges as guidelines. If you use this method, snap horizontal chalk lines on 5-inch centers starting at the eaves and allowing for an overhang. Then, snap two vertical chalk lines 6 inches apart. Install shingles up the roof offsetting every other course 6 inches, aligning them with the vertical chalk lines. That's shown in Figure 4-25. Then shingle the rest of the roof the same way, using the horizontal chalk lines and previously-installed shingle edges as guidelines.

Figure 4-23 The straight-up (racking) method

Figure 4-24 Installing a shingle beneath one previously installed

Figure 4-25 Vertical chalk lines used with the racking method

■ **Patterning** Offset the joints in adjacent courses of 3-tab shingles to keep water from being channeled through the joints, where it can get under the shingles. Offset the joints of laminated shingles for the same reason.

To form a random pattern, start the first course at the rake with a full-length dimensional shingle. Then remove 4 inches from the first shingle of the second course, and 11 inches from the first shingle of the third course. Start the fourth course with a full-length shingle and repeat the pattern every third course. Finish the remainder of each course with full-length shingles.

By using this method, you won't get an obvious and unattractive repeated pattern throughout the roof. If you install dimensional shingles using the 6-inch pattern, you'll get repeated diagonal trails like the ones in Figure 4-26. If you install dimensional shingles using the racking method you'll get repeated vertical trails.

Figure 4-26 Repeated diagonal pattern

■ **Shading** Asphalt shingles sold as one color won't match perfectly. Some will look lighter or darker than others. This is called "shading" and it's due to the way they were made. It can also happen if the shingles have been stored too long, or in stacks so tall that backing material of one shingle rubs off onto the face of another. If you use the racking method, (straight-up application) you'll accent the shading. Use the diagonal method of application to help blend the shingles.

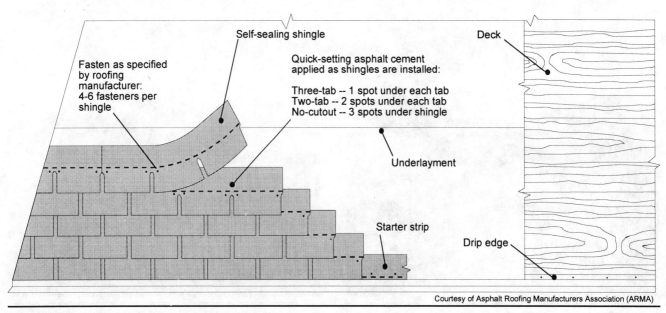

Figure 4-27 Application of shingles on steep slopes

Low or High Slopes

In general, install asphalt strip shingles only on roof slopes of 4 in 12 and steeper. You can install square-tab shingles on slopes as low as 2 in 12 (but never lower) if you follow special application procedures. The primary requirement is that you install the proper underlayment and eaves flashing (if required) to prevent damage caused by ice dams. Refer back to Chapter 3 for details on ice dams. Also, for added wind resistance, use shingles with self-sealing factory-applied adhesive strips or apply a spot of roofing cement about the size of a quarter under every shingle tab. Use cement sparingly. Too much cement can cause blisters.

Normally, you don't install asphalt strip shingles on roof slopes steeper than 21 in 12. The main problem is that the factory-applied self-sealing adhesive strip isn't very effective, especially on colder or shaded portions of the roof. However, you can install asphalt strip shingles on steeper slopes if you follow modified application procedures. Depending on the manufacturer's specifications, install each shingle with 4 to 6 fasteners.

Use roofing cement to attach shingle tabs to underlying shingles. Apply the cement in spots about the size of a quarter.

- For shingles with three or more tabs, apply a spot of cement under each tab.

- For two-tab shingles, apply two spots of cement under each tab.

- For no-cutout shingles, apply three spots of cement under the exposed portion of each shingle.

Figure 4-27 shows this.

Installing Asphalt Strip Shingles in Valleys

The three main types of valley are:

1) open (Figure 4-28)

2) closed-cut (half-lace) (look ahead to Figures 4-32 and 4-34)

3) woven (full-lace) (look ahead to Figures 4-38 and 4-39)

The valleys of aggregate-surfaced roofs are usually made of underlayment covered with aggregate embedded in bitumen. Turn back to Chapter 3 for information on valley flashing requirements. Never install a vent pipe or any other roof penetration in a valley.

■ **Open Valleys** Although Figure 4-28 shows it, I don't recommend you use open-valley construction on roofs with 3-tab asphalt shingles. Open valleys are more likely to leak than other types of valleys. The valley can get clogged by leaves, twigs, pine needles or other debris and cause a backup. Or water may be forced up under shingles adjacent to the valley during a heavy rain.

To construct an open valley, install shingles at the upper end of an open valley up to within 3 inches on each side of the centerline of the valley. Widen this distance by about 1/8-inch per foot going down the valley. You need to make this area wider because, as a stream of water flows down a valley, the stream will get wider. This widening is helpful because it lets ice free itself and slide down the valley as it melts.

Trim 1 to 2 inches from the upper corner of the last shingle in each course in the valley at a 45-degree angle. This is to direct water into the valley and not between the shingle courses. That's called "dubbing," and it's shown in Figure 4-28. In addition, you should cement the end of the shingle to the valley flashing with a 3-inch width of roofing cement. Don't allow exposed nails along the valley flashing. I also recommend dubbing-off and cementing shingle corners in closed-cut and woven valleys.

■ **Open Valleys at Dormer Roofs** Install dormer valley flashing after you've installed the shingles on the main roof deck up to a point just above the lower end of the dormer valley. Figure 4-29 shows this. Then install the valley flashing. Trim the lower part of the flashing so that it goes at least 2 inches below

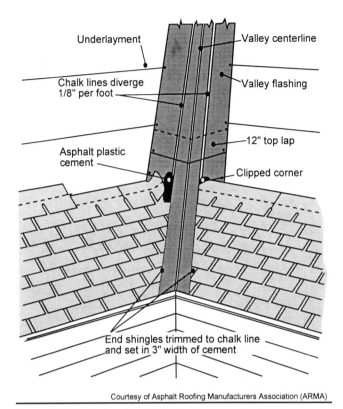

Figure 4-28 Application of shingles in an open valley

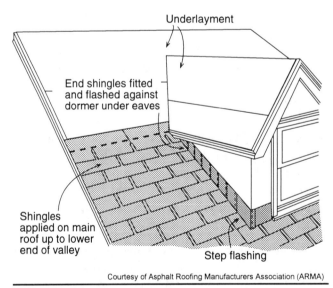

Figure 4-29 Point at which installation of open valley at dormer roof begins

Roofing Construction & Estimating

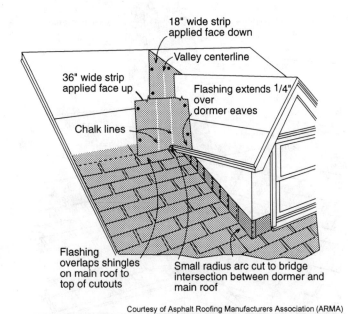

Figure 4-30 Application of roll roofing as flashing for an open valley at a dormer roof

Figure 4-31 Application of shingles in open valley at dormer roof

where the two roof decks meet. Also, trim the flashing so it overlaps the uppermost shingles (the ones you installed before) down to the top of the cutouts. In addition, cut a small arc in the flashing where the dormer and main roof decks meet, as shown in Figure 4-30. Overlap, trim, cement, and nail down the upper part of the flashing above the dormer ridge. Then install shingles over the main roof deck and dormer roof as shown in Figure 4-31.

■ **Closed-Cut Valleys** I prefer the closed-cut (half-lace) valley construction shown in Figure 4-32 over woven (full-lace) valley construction because it looks neater and more professional. And, you can usually install this type of valley faster because you can shingle each side of the valley independently.

Install each shingle course along the eaves of one side of the valley and at least 12 inches across to the other side (Figures 4-33 and 4-34). Make sure the shingle end joints are at least 10 inches from the centerline of a closed-cut or woven valley. To keep a joint from ending up in a valley, insert an individual 12-inch-wide tab within a shingle course on either side of the valley. That's shown in Figures 4-35 and 4-36. Use two fasteners to secure the end of each shingle you install across the valley.

Figure 4-32 A closed-cut valley has a neat appearance

Asphalt Shingles

Figure 4-33 Shingle at least 12 inches onto the adjoining roof plane. Note the roofing cement.

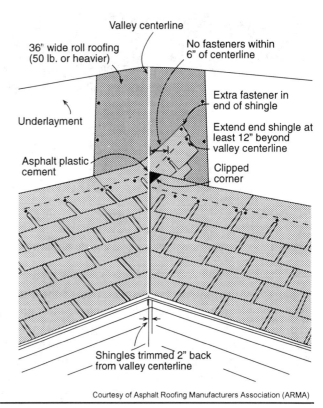

Figure 4-34 Application of shingles in a closed-cut valley

Figure 4-35 An inserted single tab prevents a joint from falling within the valley

Figure 4-36 Single 12-inch tab inserted within the shingle course

91

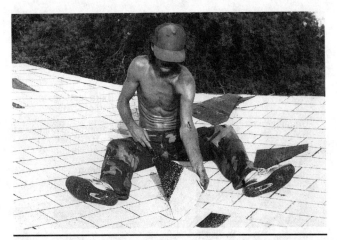

Figure 4-37 Trimming overlying shingles in a closed-cut valley

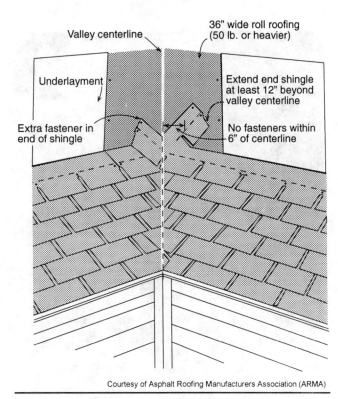

Courtesy of Asphalt Roofing Manufacturers Association (ARMA)

Figure 4-38 Application of shingles in a woven valley

Next, apply shingles to the other side of the valley, extending them beyond the valley and over the shingles you just laid. Then, trim the overlying shingles back 2 inches from the centerline of the valley, as shown in Figure 4-37. Snap a chalk line and use it for a cutting guide. Also, trim 1 inch from the upper corner of the last shingle in each course at a 45-degree angle to direct water into the valley and not between the shingle courses. In addition, cement the end of the shingle with a 3-inch width of roofing cement. Refer back to Figure 4-33.

■ **Woven Valleys** To install shingles into a woven (full-lace) valley, apply them alternately to both sides of the valley. Extend the shingles across the valley, and at least 12 inches on each side. As with a closed-cut valley, be sure the shingle end joints are at least 10 inches from the centerline of the valley. Also, secure the end of each shingle that goes across the valley with two fasteners. See Figure 4-38.

It's best to use woven valleys only when the roof slope is 3 in 12 or steeper. Even though you don't have to trim the shingles when you make a woven valley, it'll still take you longer to install it. That's because you have to work both sides of the valley at the same time. I don't like this type of valley because it doesn't look clean and professional. See Figure 4-39. What do you think?

■ **Ridge and Hip Units (Cap Shingles)** Some asphalt shingles come with a prefabricated ridge roll or prefabricated individual 12" x 12" units. The advantage of the prefab units is that they save you time — all you have to do is install them. Sometimes, when you use lami-

Figure 4-39 A woven valley

Asphalt Shingles

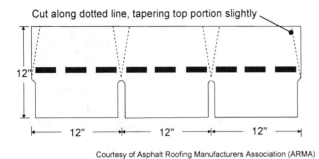

Figure 4-40 Fabrication of hip and ridge shingles from 3-tab strip shingles

Figure 4-41 Cap shingles with tapered lap portions

nated 4-tab and 6-tab shingles, you also need to use special hip and ridge shingles. You can field-fabricate the cap shingles from the same material as the rest of the roof, but as above, the factory-supplied units save time.

With 2- and 3-tab shingles, or shingles with no tabs, you can cut hip and ridge units from standard shingles. Cut a 3-tab shingle down to three 12" x 12" units as in Figure 4-40. To get a neat, professional look, taper the lap portion of each unit so it's slightly narrower than the exposed part, as shown in Figure 4-41. To make cap shingles from 2-tab or no-tab shingles, trim units to a minimum of 9" x 12". Salvage parts of shingles left over from the rakes, hips, and valleys, and make them into cap shingles.

Install the hip units before you install the ridge units. Start shingling the hips at the eaves and work up slope toward the ridge. In high-wind areas, use roofing cement to secure the first hip unit. Trim the first hip unit so its edges overhang the eaves by ¼ to ⅜ inch, depending on the overhang you allowed for the starter course. Then temporarily tack another hip unit at the top of the hip. Snap a chalk line down the hip aligned with one or both edges of the two units as a guide for intervening hip units. That's shown in Figure 4-42. Trim the top hip units where they meet at the ridge, as shown in Figure 4-43.

To cap the end of a ridge above the hips, nail down the end ridge shingle as shown in Figure 4-44 and cut about 6 inches through the center of the shingle tab. Then nail down one flap, as shown in Figure 4-45 and fold the opposite flap down into a bed of roofing cement to cover the nail and seal the hip-ridge junction. See Figure 4-46.

Figure 4-42 Snap a chalk line to line up hip units

Figure 4-43 Trim the uppermost hip units at the ridge

Figure 4-44 Install the first ridge unit above the hips

Figure 4-45 Cut the ridge unit and nail down one flap

Figure 4-46 Seal the hip-ridge juncture

Asphalt Shingles

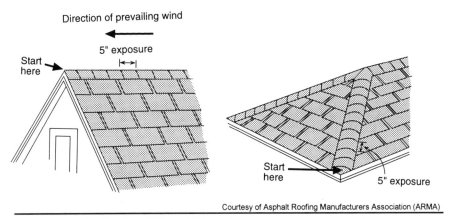

Figure 4-47 Application of hip and ridge units

On a gable or hip roof, install ridge units at opposite ends and snap a chalk line along one or both edges to align the intervening ridge units. Install ridge shingles over a gable roof beginning at the end of the roof facing into the wind, as shown in Figure 4-47. Install ridge shingles on a hip roof starting at both ends and working toward the center of the ridge. You can also follow this procedure on a gable roof. When you reach the center of the ridge, trim a shingle to use as a cap over the last ridge units. Nail the cap and cover the nails with roofing cement, as in Figure 4-48.

Install hip and ridge units at a 5-inch exposure. Secure each unit with two fasteners, one on each side. Drive the fasteners 5½ inches back from the exposed end and 1 inch up from the edge of the shingle. See Figure 4-49.

On dormers, install the ridge units starting at the front of the dormer and working toward the main roof. Extend the last unit you install at least 4 inches onto the main roof. Split the part of the shingle that extends over the main roof down the center, and nail it into place as shown in Figure

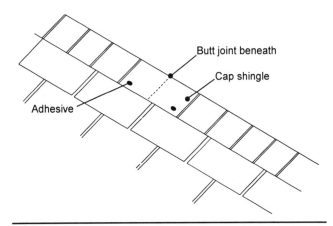

Figure 4-48 Capping the ridge units

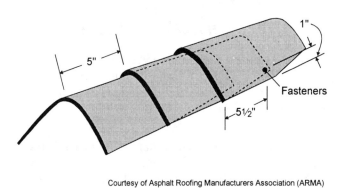

Figure 4-49 Nail location for hip and ridge shingles

Roofing Construction & Estimating

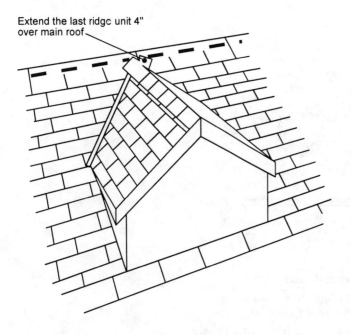

Figure 4-50 Application of ridge units over a dormer

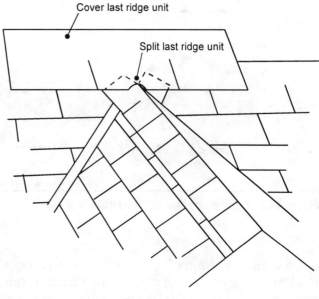

Figure 4-51 Covering the last dormer ridge unit

4-50. Then cover this last dormer shingle with shingles you apply to the main roof, as shown in Figure 4-51. If a cutout in a main roof shingle falls over the dormer ridge shingle, coat the dormer shingle with roofing cement under the main roof shingle. To provide extra waterproofing at the place where the dormer ridge and the main roof meet, install 6-inch-wide strips of water shield material between the last dormer ridge unit and the shingle beneath it.

Flashing at Chimneys and Other Vertical Structures

Any flashing turned up on a vertical surface is called a *base flashing*. Flashing built into the vertical surface and bent down over the base flashing is called *counterflashing* or *cap flashing*.

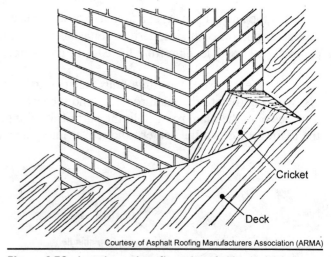

Courtesy of Asphalt Roofing Manufacturers Association (ARMA)

Figure 4-52 Location and configuration of chimney cricket

If the horizontal width of a chimney is greater than 2 feet, install a fabricated galvanized metal saddle flashing, or a wooden cricket, above the chimney, as in Figure 4-52. The cricket or saddle helps keep ice and snow from building up at the upper side of the chimney, and diverts rainwater around it. Build the saddle flashing or cricket with the same slope as the main roof.

Asphalt Shingles

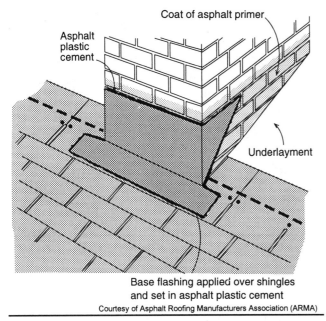

Figure 4-53 Application of base flashing at front of chimney

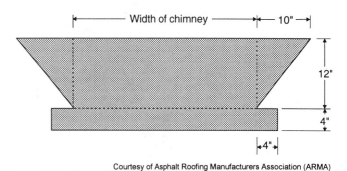

Figure 4-54 Pattern for cutting front base flashing

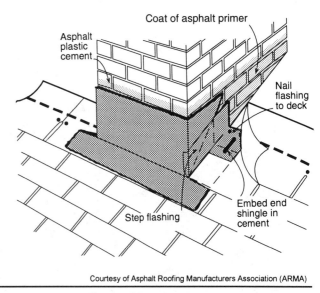

Figure 4-55 Application of base flashing at side of chimney

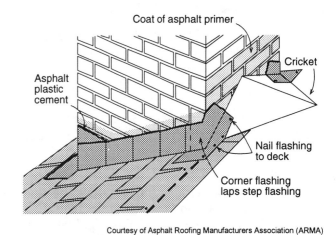

Figure 4-56 Application of corner base flashing at rear of chimney

Apply asphalt shingles up to the lower edge of a chimney before you install any metal flashing material. Then install the base flashing on the down-slope face of the chimney (Figure 4-53). Make this piece so the lower part goes at least 4 inches over the shingles and the upper section goes at least 12 inches up the chimney face. See Figure 4-54. Apply a bed of asphalt plastic cement over the shingles and masonry and set the entire flashing in it. Drive only enough nails through the flashing into mortar joints to keep the flashing in place until the cement sets. Apply a coat of asphalt primer to any masonry surface before you apply roofing cement. That seals the masonry and provides good adhesion between the cement and the masonry.

You can buy special flashing cements you can use at all temperatures and on wet or dry surfaces. This cement comes in one-gallon cans or five-gallon pails.

Install metal step flashing (baby tins) and shingles at the sides of the chimney, as shown in Figure 4-55. Step flashing installation is discussed in greater detail below.

Install the base flashing at the rear of the chimney and over the cricket, as in Figures 4-56, 4-57, and 4-58. Extend the flashing at least 6 inches onto the roof sheathing and 6 inches up the chimney.

Roofing Construction & Estimating

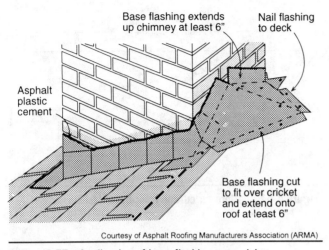

Figure 4-57 Application of base flashing over cricket

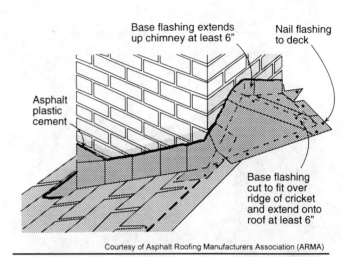

Figure 4-58 Application of base flashing over ridge of cricket

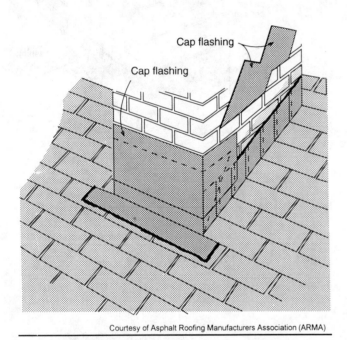

Figure 4-59 Application of cap flashing at front and side of chimney

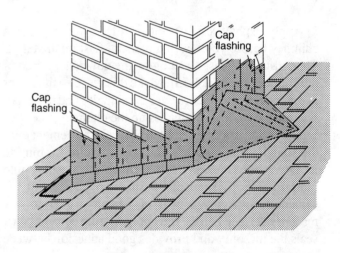

Figure 4-60 Application of cap flashing at side and rear of chimney

Install shingles over the cricket, or up to the cricket valleys. Install any shingles you apply over the cricket in a bed of asphalt plastic cement. You don't have to install shingles over a metal cricket if you can't see it from the ground or surrounding viewpoints.

Install metal cap flashing (Figures 4-59, 4-60 and 4-61) not more than 3 bricks high. Chisel and rake clean the mortar joints to a depth of $1\frac{1}{2}$ inches before you install the cap flashing. Make some mortar that's 1 part portland cement and 3 parts fine mortar sand, and refill the joints with it. Wet the joints before you apply the fresh mortar.

Asphalt Shingles

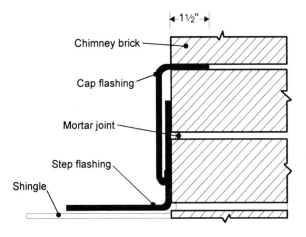

Figure 4-61 Application of cap flashing

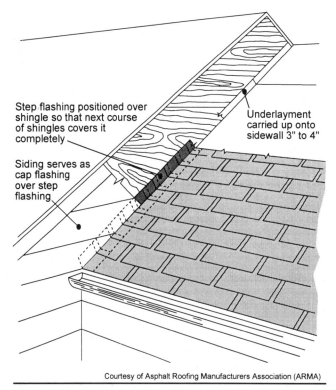

Figure 4-62 Application of step flashing against vertical side

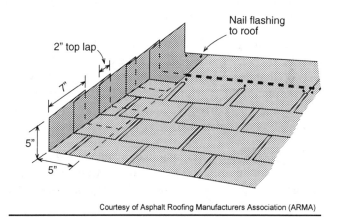

Figure 4-63 Application of step flashing

Or, you can caulk the joints after you install the flashing pieces. Install the cap flashing at the front of the chimney in one piece, as shown in Figure 4-59. Install the cap flashing at the sides and back of the chimney as individual units, beginning at the lowest point. Install each piece into mortar joints so each cap flashing unit overlaps the base flashing by at least 3 inches, as in Figure 4-60. Bend the last piece of cap flashing around the upper corners of the chimney. Use a good grade of butyl rubber sealant to seal the flashing joints at the chimney corners.

Make all exposed flashing (such as cap flashing) with its bottom edge turned under ½ inch, as in Figure 4-61. This adds stiffness against the wind and prevents snow from packing in under the flashing.

Use galvanized metal step flashing where a sloping roof meets a vertical surface such as the chimney shown in Figures 4-55 through 4-61, or the vertical side wall in Figure 4-62. The step flashing is later protected by cap flashing installed in the masonry, or by siding. The cap flashing makes a good water seal even when the roof and chimney (or wall) move independently due to expansion or settlement. You can see cap flashing in Figures 4-59 through 4-61. There is no cap flashing in Figure 4-62 because the siding takes the place of cap flashing. Overlap the step flashing joints at least 2 inches, and extend the metal under the shingles and up the chimney about 4 to 5 inches. See Figure 4-63. Lap the cap flashing down over the base flashing at least 3 inches and extend it down to within 1 inch of the finished roof. See Figure 4-61.

Step Flashing

Install step flashing as an individual piece for each course of shingles you lay by starting at the bottom and ending at the top of the chimney or side wall. Place the first piece of step flashing over the unexposed area of the shingle next to the lower edge of the chimney (refer back to Figure 4-55), or over the starter-course shingle at a side wall, as shown in Figure 4-62. Install a shingle over the step flashing so its butt is flush with the lower edge of the flashing. Install the next piece of step flashing over the shingle 5 inches above the butt. Install the next shingle so its butt is in line with the step flashing. Then, the horizontal leg of each piece of step flashing will cover the unexposed part of the underlying shingle. And, each piece of step flashing will be covered by the exposed part of the overlying shingle. See Figure 4-63. Continue this way until you've flashed and shingled the entire roof-wall intersection.

Use roofing cement to embed the end of each shingle which extends over the step flashing. Cut each piece of step flashing 10 inches wide and 2 inches longer than the shingle exposure. With 3-tab shingles, make each piece 7 inches long, providing a 2-inch lap.

Where a sloping roof and vertical wall meet, extend each piece of step flashing you install at least 5 inches up the wall and at least 5 inches under the shingles, as in Figure 4-63. Embed the horizontal leg of each piece of step flashing into roofing cement and secure it with two nails. Since the roof could eventually settle, don't nail the flashing to the wall. Cover the vertical leg of each piece of step flashing later with siding or cap flashing (in brick or stucco walls). In either case, extend the underlayment at least 3 inches up the wall, as in Figure 4-62.

These rules also apply when you install T-lock shingles (shown in Figure 4-96 near the end of the chapter), except that the length of flashing required will vary, depending on the type and size of shingle. If you install siding material correctly, it will serve as cap flashing material as in Figure 4-62.

Install continuous flashing like the one in Figure 4-64, where a sloping roof and a vertical side wall meet to form a horizontal line. One example of this is the intersection of a sloping roof and the front of a dormer. Another is a shed roof intersecting a wall, as in Figure 4-65.

To install continuous flashing, embed it into roofing cement and nail it over the last course of shingles you apply to the roof deck. Nail it above the cutouts of the shingles below it, as shown in Figure 4-65. Don't nail the flashing to the wall. This way, the roof and wall can move independently. Instead, install the flashing before the siding. If the siding's already installed, pry up the lowest siding board enough to slip the flashing beneath it.

Figure 4-64 Continuous flashing

Asphalt Shingles

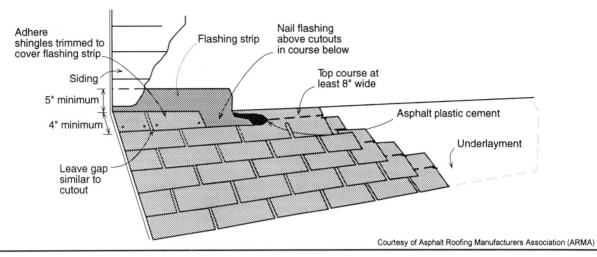

Figure 4-65 Application of flashing against vertical front wall

If the side wall is brick, use a masonry saw to remove 1½ inches of mortar from a joint at a point about 5 inches above the roof-wall intersection. Bend the top of the flashing strip so you can insert it into the joint. Close the joint with mortar or caulking compound.

If the wall finish is stucco, saw out a joint and re-pack it with mortar or caulk. After you get the flashing in, cover it with one course of shingles trimmed to fit over the flashing. Nail the shingles into place and cover each nail head with roofing cement. Use 26-gauge galvanized metal flashing and extend it at least 5 inches up the wall and 4 inches over the last shingle course, as in Figure 4-65. Where front-wall flashing turns a corner (at a dormer, for example), extend the flashing at least 7 inches around the corner. From there on, install step flashing up the slope.

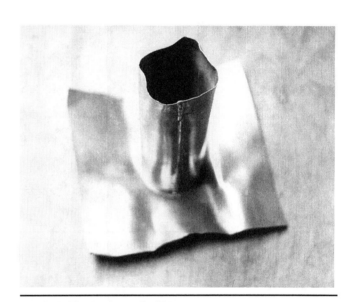

Figure 4-66 Lead vent flashing

Flashing Soil Stacks and Vents

Normally, you flash vent pipes with a one-piece lead flange and sleeve (collar) as in Figure 4-66. Turn the top of the sleeve down into the stack. Look ahead to Figure 4-73. Or you can cut a lead sleeve flush with the top of the vent pipe and counterflash it with lead extending 4 inches down the outside of the pipe and 2 inches down the inside of the pipe, as shown in Figure 4-67.

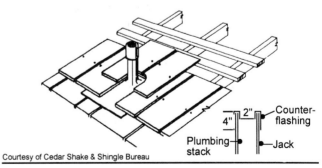

Figure 4-67 Counterflashed vent flashing

Another popular type of vent flashing is a rubber flange, which you slip down over the pipe, as in Figure 4-68. This type of flashing is often installed on metal roof decks. Vent pipes are normally 1½ to 3 inches in diameter.

When you come to a vent pipe, install shingles up to the bottom edge of the pipe, as in Figure 4-69. If the top edge of a shingle hits the pipe, notch it so it fits around the pipe, as shown in the figure. If a shingle ends up over the pipe, cut a hole in the shingle and slip it over the pipe, as in Figure 4-70.

Figure 4-68 Rubber vent pipe flange

Figure 4-69 Shingle up to the bottom of vent pipe

Slip the flange over the pipe and underlying shingle, as shown in Figure 4-71. Embed the flange and overlying shingle in roofing cement. See Figures 4-72 and Figure 4-73. Always install a full-width shingle over the pipe (Figure 4-74). Insert a single tab along the shingle course to rearrange and offset the joints so there won't be a joint above the pipe.

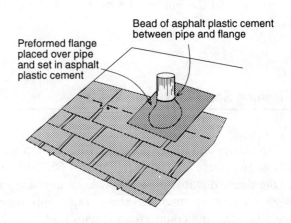

Courtesy of Asphalt Roofing Manufacturers Association (ARMA)

Figure 4-70 Application of shingle over vent pipe

Asphalt Shingles

Figure 4-71 Slip the flange over the vent pipe and underlying shingle

Figure 4-72 Embed the flange in roofing cement

Figure 4-73 Embed the overlying shingle in roofing cement

Figure 4-74 Arrange to install a full shingle over the vent pipe

Asphalt Shingles

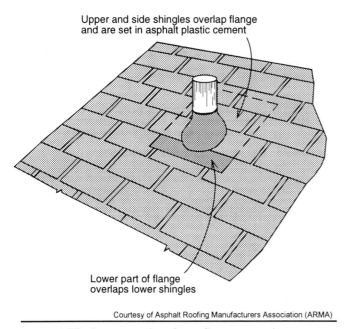

Figure 4-75 Bottom portion of vent flange exposed

Figure 4-76 Applying roofing cement over the flange of a heater vent

Continue shingling around and above the pipe, trimming successive shingle courses to fit around the pipe. Nail shingles over the vent pipe so the nails don't go into the metal flashing flange. Allow a ½-inch space between the vent sleeve and the overlying shingle so debris won't get caught between the shingle and the vent stack. Most of the debris there will be dislodged mineral granules which won't wash away easily if there's no gap. You don't need to cover the down-slope part of the flange with shingles. In fact, I recommend that you leave the bottom third of any vent flashing flange exposed, as shown in Figure 4-75. An exposed flange isn't as likely to trap debris, but it doesn't look as neat. Let your customer have the final word on this.

For added leak protection on low-sloped roofs, embed a strip of mineral-surfaced roll roofing into roofing cement under the vent pipe flange.

When you come to a larger roof penetration such as a heater vent, nail down the flange of the vent and install shingles up to the lower edge of the vent. Apply roofing cement over the flange, as in Figure 4-76. Cut shingles to fit around and above the vent as in Figure 4-77. Don't drive shingle nails through the vent flange.

Figure 4-77 Installing a course above the vent

Application	Nail length (inches)
Roll roofing on new deck	1
Strip or individual shingles on new deck	1¼
Roofing over old asphalt roofing	1½ to 2
Roofing over old wood shingles	2

Figure 4-78 Recommended nail lengths

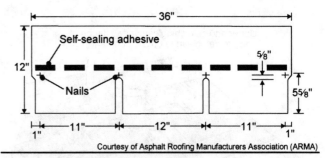

Figure 4-79 Nail locations for 3-tab strip shingle

Fasteners

Install asphalt roofing materials over solid roof sheathing. Use 11- or 12-gauge hot-dipped galvanized or aluminum roofing nails with heads that are at least ³/₈ inch in diameter and barbed or deformed shanks that are 1 to 2 inches long. Recommended lengths are shown in Figure 4-78. When re-roofing with asphalt shingles, use nails that are long enough to go at least ³/₄ inch into the sheathing. You can make sure the nails you use are long enough by checking the underside of the sheathing to see if they come through it. Allow for about 2½ pounds of nails per square when you install asphalt shingles.

When you install shingles across a roof, start nailing from the end nearest the shingle you just laid and proceed across. This will prevent buckling. Drive nails straight so the nail head doesn't damage the surface of the shingle. Don't drive nails into knotholes or cracks in the sheathing. If you have to remove a nail, seal the hole with roofing cement or remove and replace the entire shingle.

Drive nails into the shingle along a line just below the factory-applied adhesive strip. Depending on what part of the country you're building in, use at least four nails for each 3-tab shingle. When you're laying a shingle at a 5-inch exposure, drive the nails along a line 5⅝ inches above the butt edge of the shingle.

Earlier you recall we said not to nail in a straight line, in order to avoid splitting the sheathing boards. But in this case, nail placement is very important so the adhesive strip will do its job. That's not a problem with starter-course material, so stagger-nail when you can. But in this case, nail in a straight line. Drive the two outermost nails 1 inch from each end of the shingle. Center the innermost nails over each cutout, as in Figure 4-79.

Two-tab shingles also need at least four nails. When you're laying a shingle at a 5-inch exposure, drive the nails along a line 5⅝ inches above the butt edge of the shingle, and at 1 and 13 inches from each end of the shingle, as in Figure 4-80.

Asphalt Shingles

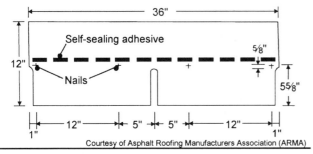

Figure 4-80 Nail locations for 2-tab strip shingle

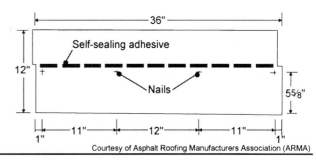

Figure 4-81 Nail locations for no-cutout strip shingle

Shingles with no cutouts also require at least four nails. When you're laying a shingle at a 5-inch exposure, drive nails $5^5/_8$ inches above the butt edge of the shingle, and 1 and 12 inches from each end, as in Figure 4-81.

Never use fewer than four nails to install each strip shingle. Some roofing contractors don't drive the fourth nail because it's hidden under the overlapping shingle above. This is called "three-nailing" and I don't recommend it.

Many building codes, especially in high wind and hurricane areas, require six nails. Check with your building inspector beforehand, and make the necessary material and labor cost adjustments to your estimate.

Stapling

I also don't recommend you use staples because they tend to come loose eventually. Roofing contractors who hand-nail shingles (and explain to their clients why they do so) have more work than they can handle, at a price that yields high profit. Many have a waiting list of customers which includes general contractors. That's because the quality and durability of their work is consistently above that of their competitors.

If you decide to use staples, use them only on new construction to fasten wind-resistant asphalt shingles with factory-applied adhesives. If the old roofing has been removed, you can use staples for re-roofing. Use galvanized staples that are at least 16 gauge with a minimum crown of $^{15}/_{16}$ inch. Make sure the staples are long enough to go at least $^3/_4$ inch into the sheathing. Locate staples the same way as roofing nails.

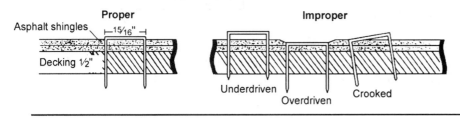

Figure 4-82 Driving staples

It's very important to hold the staple gun so the staples go in at the correct angle so the crown is practically flush with the shingle surface. And be sure to adjust the air gun so the staples go far enough into the sheathing. Figure 4-82 shows good and bad stapling.

Be aware that a pneumatic stapler has no "feel" to it. It won't be obvious when you're stapling into a joint or knothole. In warm weather, it's easy to drive a staple all the way through a soft asphalt shingle. Wind will also tear off a shingle unless you drive the crown of the staple parallel to the long shingle edge.

Number of Shingles Required per Square

Asphalt strip shingles are usually 3' x 1' and come in 3-bundle squares (for lighter-weight shingles), or 4-bundle squares (for heavier shingles). That means you need three or four bundles of asphalt strip shingles (whichever the case), laid at the recommended exposure (usually 5 inches) to cover a square of roofing surface.

Whatever the size or exposure a strip shingle is, you can figure out how many shingles you need for each square of roof surface by:

$$\text{Shingles/Square} = \frac{100 \text{ SF}}{\text{Shingle Length (in.)} \times \text{Exposure (in.)}} \times 144 \text{ sq. in./SF}$$

Equation 4-1

▼ **Example 4-1:** Assume you're using 3' x 1' asphalt strip shingles at a 5-inch exposure. Find the number of shingles you need to cover one square of roof area.

$$\text{Shingles/Square} = \frac{100 \text{ SF}}{36 \text{ in.} \times 5 \text{ in.}} \times 144 \text{ sq. in./SF}$$
$$= 80 \text{ shingles per square}$$

You shouldn't "stretch" the exposure, but you can install shingles at an exposure *less* than what's recommended. In this case, the extra shingles you need per square, in terms of a percentage-of-increase factor, are:

$$\text{Percentage-of-Increase Factor} = \frac{\text{Recommeded Exposure}}{\text{Actual Exposure}}$$

Equation 4-2

▼ **Example 4-2:** An area of 20 squares is to be covered with 3-tab strip shingles. Assume the recommended exposure is 5 inches, then find the number of shingles required to install shingles at:

a) 4½ inches, or

b) 4 inches

Asphalt Shingles

Solution:

a) Percentage-of-Increase Factor, 4½" exposure $= \dfrac{5 \text{ in.}}{4\frac{1}{2} \text{ in.}}$
$= 1.10$

Thus, you increase the shingle quantity by 10 percent:
20 squares x 1.1 = 22 squares

b) Percentage-of-Increase Factor, 4" exposure $= \dfrac{5 \text{ in.}}{4 \text{ in.}}$
$= 1.20$

Thus, you increase the shingle quantity by 20 percent:
20 squares x 1.2 = 24 squares

Number of Shingle Courses

You can determine the number of shingle courses required to cover a wall or roof by:

$$\text{Courses} = \dfrac{\text{Dimension of Structure}}{\text{Exposure}}$$

Equation 4-3

where the Dimension of Structure is the wall height, or the width of a roof section, measured from the eaves to the ridge, along the top of a common rafter.

You can also use Equation 4-3 to determine the number of hip and ridge units required (if you know the lengths of the hips and ridge).

▼ **Example 4-3:** Assume an exposure of 5 inches, then find the number of courses of 12-inch-wide asphalt shingles required to cover the roof of the building diagrammed in Figure 4-83. The roof slope is 5 in 12.

Solution: From Column 2 of Appendix A, the actual width of each side of the roof, measured along any rafter from eaves to ridge is:

Length = 13' x 1.083 = 14.1 linear feet

From Equation 4-3, the number of shingle courses required on one side of the roof is:

$$\text{Courses} = \dfrac{14.1 \text{ ft. x } 12 \text{ in. / LF}}{5 \text{ in.}}$$
$= 33.84$, rounded to 34 courses.

Double that for a total of 68 courses for both sides of the roof. Remember, this formula produces the number of courses for a *section* of roof. In this case, you multiply by 2 because both sides of the roof are the same.

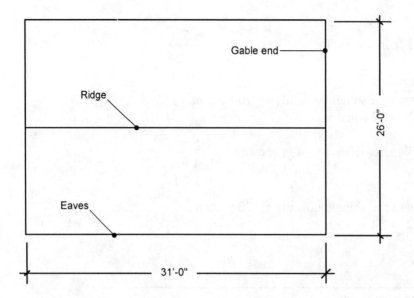

Figure 4-83 Gable roof example

Notice that the calculated answer in the previous example didn't come out to an even number of courses. Since it's impractical to install 0.84 of a course, we changed the answer to an even 34 courses, which will result in slightly less exposure throughout. We also could have used 33 courses, with slightly more exposure. Either way, use the following equation to find the exact exposure for this example.

$$\text{Exposure} = \frac{\text{Dimension of Structure}}{\text{Number of Courses}}$$

Equation 4-4

▼ **Example 4-4:** Determine the consistent exposure required to install:

a) 33 courses, or

b) 34 courses of shingles on the roof described in Example 4-3.

Solution: The consistent exposures are:

a) For 33 courses:

$$\text{Exposure} = \frac{14.1 \text{ ft.} \times 12 \text{ in./LF}}{33}$$

= 5.127 inches, or 5⅛ inches (rounded off)

Since you shouldn't "stretch" the exposure, use 34 courses in this case, with a decreased exposure.

b) For 34 courses:

$$\text{Exposure} = \frac{14.1 \text{ ft.} \times 12 \text{ in./LF}}{34}$$

= 4.98 inches, which rounds to 5 inches

Asphalt Shingles

There's another way you can install shingles on roofs or walls whose dimensions aren't evenly divisible by the shingle exposure. To do this, apply courses at the recommended exposure and decrease the exposure near the ridge so you finish with a full shingle width at the ridge. It's better to shorten than to stretch the exposure of the shingle near the ridge. This method for shortening an exposure is called "stacking" the shingles.

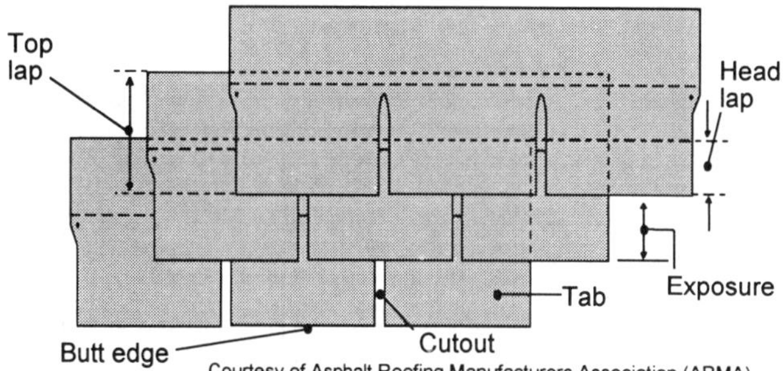

Figure 4-84 Exposure diagram

Head Lap, Top Lap and Exposure

Head lap is where shingles (or other roof coverings) are three layers thick. Top lap is where shingles (or other roof coverings) are at least two layers thick. See Figures 4-84 and 4-85. Thus, the 7-inch lap on a 12-inch-wide shingle, where the shingles are two layers thick, is the top lap. The 2-inch lap where the shingles are three layers thick is the head lap, as shown in Figure 4-84. Exposure is the part of the shingle not covered by the next course of shingles. The relationships between head lap, top lap and exposure are given in the following equations, where TL = top lap, W = width of shingle, E = exposure, and HL = head lap.

Top Lap = W - E, or   *Equation 4-5*

Top Lap = E + HL   *Equation 4-6*

Head Lap = TL - E, or   *Equation 4-7*

Head Lap = W - 2E   *Equation 4-8*

$$\text{Exposure} = \frac{W - HL}{2}$$ *Equation 4-9*

Coverage Based on Number of Shingle Plies

Shingle coverage is often designated as single, double or triple, depending on the number of plies or layers of shingles. The number of plies is defined as the number of shingle layers at the head lap. Figure 4-85 shows 3-ply construction.

If the number of plies isn't the same throughout the roof, coverage is generally considered as the number of layers installed over a majority of the roof area. Most strip shingles and T-lock shingles are designed for double coverage. To get two layers of shingles over an entire roof, overlap shingles by a bit more than half. Overlap by slightly more than two-thirds to get triple coverage.

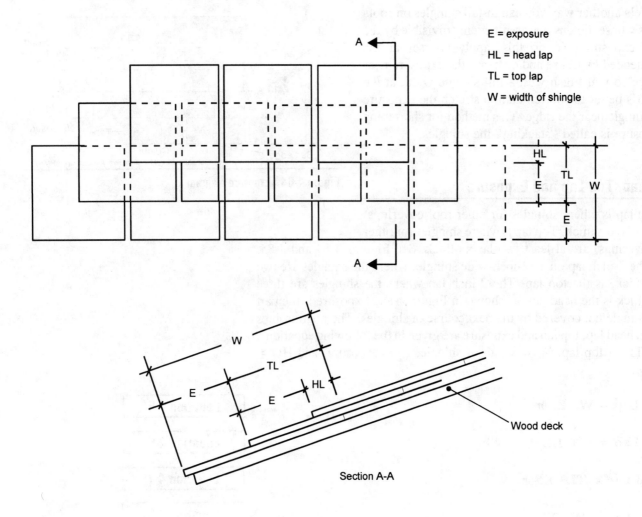

Figure 4-85 Head lap, top lap and exposure

Sometimes roof coverage specifications call for the acceptable minimum number of shingle plies.

▼ **Example 4-5:** Assume that 3-ply coverage is specified, then find the maximum exposure allowed to apply 12-inch-wide asphalt shingles with a minimum of a 2-inch head lap.

Solution: From Equation 4-9, the maximum allowable exposure is:

$$E = \frac{12 \text{ in.} - 2 \text{ in.}}{2}$$

$$= 5 \text{ inches}$$

Asphalt Shingles

Estimating Asphalt Strip Shingle Quantities

First, take off quantities for asphalt strip shingles in square feet. Then, convert these quantities into the number of bundles of shingles required. The total quantity of shingles required must include material for the starter course, hip and ridge units, cutting waste at the rakes, hips and valleys, and waste due to crew errors.

Starter Course

You can find the amount of field-fabricated starter-course material required by:

Starter Course (SF) = Eaves (LF) x Exposed Area (SF/LF) *Equation 4-10*

You can find the coverage per linear foot of shingle, by:

$$\text{Area (SF/LF)} = \frac{\text{Exposure (in.)}}{12}$$ *Equation 4-11*

▼ **Example 4-6:** Assume a 5-inch exposure, then find the coverage of starter-course material in square feet per linear foot of eaves.

Solution: The coverage is:

$$\text{Area} = \frac{5}{12} = 0.42 \text{ square feet per linear foot}$$

As a convenience, use Figure 4-86 to quickly determine the area-per-linear-foot values for shingles installed at various exposures along the eaves.

Sometimes you need to know the *number* of starter-course units. The number of eaves shingles required at any given eaves is:

$$\text{Number of Eaves Shingles} = \frac{\text{Eaves Length (ft.)}}{\text{Shingle Length (ft.)}}$$ *Equation 4-12*

Estimate eaves units carefully, section by section. The high cost of some materials, and the inconvenience and delay caused when you're short of materials, makes underestimating eaves materials very expensive.

▼ **Example 4-7:** Assume application of 10-inch-long shingles, then find the number of eaves shingles required for the roof of the building diagrammed in Figure 4-83.

Solution: The number of eaves shingles required along each eaves is:

$$\text{Number of Eaves Shingles (each eaves)} = \frac{31 \text{ feet}}{.83 \text{ feet}} = 38 \text{ (rounded up)}$$

Because there are two eaves in this example, you need to order 76 shingle units.

Exposure (inches)	Area covered (SF) per LF of shingle	Exposure (inches)	Area covered (SF) per LF of shingle
3	0.25	9	0.75
3½	0.29	9½	0.79
3¾	0.31	10	0.84*
4	0.34*	10½	0.88
4¼	0.35	11	0.92
4½	0.38	11½	0.96
4¾	0.40	12	1.00
5	0.42	12½	1.04
5⅛	0.43	13	1.08
5¼	0.44	13½	1.13
5½	0.46	14	1.17
5⅝	0.47	14½	1.21
5¾	0.48	15	1.25
6	0.50	15½	1.29
6½	0.54	16	1.34*
7	0.58	16½	1.38
7½	0.63	18	1.50
8	0.67	20	1.67
8½	0.71	22	1.84*

*For the sake of cautious estimating, I round up the answer for 4, 10, 16 and 22 inches of exposure, respectively. This results in quantities that are a bit long, but safe.

Figure 4-86 Area of coverage per linear foot of eaves

Asphalt shingle manufacturers recommend that the starter course and first course overhang the eaves and rake by ¼ to ⅜ inch. However, you don't have to increase your order to account for that since the added area is so small.

Cutting Waste at Rakes, Hips and Valleys

Don't forget to allow for shingles lost due to cutting waste at rakes, hips and valleys. Use Figure 4-87 to account for wasted material. The table doesn't include material wasted due to crew error. The table assumes 3' x 1' strip shingles laid at a 5-inch exposure.

Ridge and Hip Units

Using a 12-inch-wide shingle, you'll need 1 square foot of shingles for each linear foot of hip and ridge. Assuming a conscientious and prudent crew that uses every salvageable single-tab shingle available from material

Waste (SF per linear foot)					
Shingle type	Rake	Hip	Open valley	Closed-cut (half-lace) valley	Woven (full-lace) valley
3-tab	0.25	0.64	1.41	2.12	2.83
Other than 3-tab	negligible	0.30	0.30	1.00	1.71

Figure 4-87 Asphalt strip shingle cutting waste

Shingles salvaged (SF per linear foot)					
Shingle type	Rake	Hip	Open valley	Closed-cut (half-lace) valley	Woven (full-lace) valley
3-tab	1.00	0.50	2.00	1.00	0

Figure 4-88 Asphalt hip and ridge shingles salvaged from cutting waste

cut at the rakes, hips and valleys, you can use Figure 4-88 to determine the quantity of shingles salvaged. The table assumes 3' x 1' three-tab shingles laid at a 5-inch exposure.

The formula for allowance in hip and ridge units is the difference between the square feet required and the square feet salvaged:

Net Allowance (ridge and hip units) = SF Required - SF Salvaged   Equation 4-13

On hip roofs, you'll require more units than you've salvaged. On gable roofs, you'll salvage more units than you need. Unless you can use the excess on another roof, the salvaged units will be wasted.

A "Shortcut" Method for Determining Asphalt Strip Shingle Waste

As a "rule of thumb," some contractors add 10 percent waste on small- to average-sized gable roofs, and 15 percent on hip roofs (3-tab shingles). They add 2 percent on gable roofs and 3 percent on hip roofs for laminated shingles with prefabricated ridge and hip units. This allows for additional material required for the starter course and site-fabricated ridge and hip units (in the case of 3-tab strip shingles), cutting waste at rakes, hips and valleys, and waste due to crew error. They add a larger percentage for more complex roofs.

Cutting waste and overruns (gable roof)			
Building dimensions (LxW)	Roof slope		
	3 in 12	6 in 12	12 in 12
30 x 20	8[1] (4)[2]	8 (4)	8 (3)
40 x 30	6 (3)	6 (3)	6 (2)
45 x 30	5 (3)	6 (3)	5 (2)
50 x 30	5 (3)	5 (3)	5 (2)
60 x 30	4 (3)	4 (3)	4 (2)
70 x 30	4 (2)	4 (3)	4 (2)
80 x 30	5 (2)	5 (3)	3 (2)
50 x 40	5 (2)	5 (2)	5 (2)
60 x 40	4 (2)	4 (2)	5 (2)
70 x 40	4 (2)	4 (2)	4 (2)
80 x 40	3 (2)	3 (2)	4 (2)
90 x 40	3 (2)	3 (2)	4 (2)
60 x 50	4 (2)	5 (2)	4 (1)
70 x 50	4 (2)	4 (2)	3 (1)
80 x 50	4 (2)	4 (2)	3 (1)
90 x 50	3 (2)	3 (2)	3 (1)

(1) 3-tab shingles using site-fabricated hip and ridge units.
(2) Laminated strip shingles using a prefabricated ridge and hip roll. The roll must be taken off separately.

Figure 4-89 Cutting waste and overruns (gable roof)

I don't rely on this method of estimating waste. Remember that exclusive of crew-error waste (which varies from crew to crew and from job to job), actual waste depends on the roof type, the ratio of roof length to roof width, the roof slope, the shingle exposure, the type of shingle installed, and whether the hip and ridge units are prefabricated or site-constructed.

Waste on asphalt strip-shingle roofs varies from 1 to 8 percent on average-sized gable roofs, and from 3 to 18 percent on hip roofs. You can use Figure 4-89 and 4-90 to quickly get a "ball-park" percentage of asphalt strip shingle waste. Waste due to crew error isn't included in these figures.

Figure 4-89 lists the total percentage you add to net roof area of a *gable roof* for the starter course, cutting waste at rakes, and site-fabricated ridge units (if applicable).

This table assumes 3' x 1' strip shingles laid at a 5-inch exposure.

Asphalt Shingles

Cutting waste and overruns (hip roof)			
Building dimensions (LxW)	Roof slope		
	3 in 12	6 in 12	12 in 12
30 x 20	18[1] (10)[2]	17 (9)	14 (8)
40 x 30	14 (7)	13 (6)	12 (6)
45 x 30	13 (7)	12 (6)	10 (5)
50 x 30	12 (6)	11 (6)	9 (5)
60 x 30	11 (6)	11 (5)	9 (4)
70 x 30	11 (5)	11 (5)	9 (4)
80 x 30	10 (5)	9 (4)	8 (4)
50 x 40	11 (6)	9 (4)	9 (5)
60 x 40	9 (5)	9 (4)	9 (4)
70 x 40	9 (4)	9 (4)	8 (3)
80 x 40	9 (4)	8 (4)	6 (3)
90 x 40	8 (4)	8 (4)	6 (3)
60 x 50	9 (5)	9 (4)	7 (3)
70 x 50	9 (4)	8 (4)	7 (3)
80 x 50	9 (4)	6 (4)	6 (3)
90 x 50	7 (4)	6 (4)	6 (3)

(1) 3-tab shingles using site-fabricated hip and ridge units.
(2) Laminated strip shingles using a prefabricated ridge and hip roll. The roll must be taken off separately.

Figure 4-90 Cutting waste and overruns (hip roof)

Figure 4-90 shows the total percentage you add to net roof area of a *hip roof*, including the starter course, cutting waste at hips, and site-fabricated ridge units (if applicable). The table assumes 3' x 1' strip shingles laid at a 5-inch exposure. The table doesn't include crew-error waste.

▼ **Example 4-8:** Assume a roof slope of 5 in 12 and 3-bundle squares installed at a 5-inch exposure. How many bundles of 3' x 1' 3-tab shingles are required for the roof diagrammed in Figure 4-83? Assume also that you'll use field-fabricated 3-tab shingles for the starter course at the eaves and 12" x 12" tabs salvaged at the rakes for ridge units.

Solution: First, remember the formula for net roof area to be covered (from Chapter 1):

Actual (Net) Roof Area = Roof Plan Area x Roof-Slope Factor

Thus, Net Roof Area = 31' x 26' x 1.083
　　　　　　　　　　(from Column 2 of Appendix A)
　　　　　　　　　= 873 square feet ÷ 100
　　　　　　　　　= 8.73 squares (use this quantity to estimate labor costs)

The total eaves length is 2 x 31, or 62 linear feet.

The total area of starter-course material required is:

Area (Starter Course) = 62' x 0.42 SF/LF
　　　　　　　　　　　(Equation 4-10 or Figure 4-86)
　　　　　　　　　　= 26 square feet

The total rake length is:

LF (Rake) = 4 ea. x 13' x 1.083 (Column 2, Appendix A)
　　　　　= 56 linear feet

Cutting waste at the rakes is:

Waste (Rake) = 56' x 0.25 SF/LF (from Figure 4-87)
　　　　　　= 14 square feet

Remember, you need 1 square foot of single-tab shingles per linear foot of ridge, or 31 square feet.

From Figure 4-88, the amount of salvaged tabs at the rakes is 56 square feet. That's more single-tab units than needed, so from Equation 4-13, the net allowance for ridge units is:

Net Allowance (Ridge Units) = 56 SF - 31 SF
　　　　　　　　　　　　　= 25 square feet

Thus, the gross roof area = 873 SF + 26 SF + 14 SF + 25 SF
　　　　　　　　　　　　= 938 square feet ÷ 100
　　　　　　　　　　　　= 9.38 squares

From Chapter 1, the waste factor (excluding crew-error waste) is:

$$\text{Waste Factor} = \frac{\text{Area Covered (including waste)}}{\text{Net Roof Area}}$$

Therefore,

$$\text{Waste Factor} = \frac{9.38 \text{ Squares}}{8.73 \text{ Squares}}$$

　　　　　　　　= 1.08

As you can see, that answer agrees with the information from Figure 4-89. In that table, the nearest size to our example is a 30' x 20' roof with a 6 in 12 slope, and the table shows a waste factor of 8 percent for 3-tab shingles.

Total material required: 9.38 squares x 3 bundles/square = 29 bundles.

▼ **Example 4-9:** Work the same problem assuming use of 3' x 1' laminated strip shingles with field-fabricated shingles used for the starter course at the eaves. Also, assume using 3-bundle squares and a prefabricated ridge roll.

Solution: First, you need 31 linear feet of ridge roll to cover the length of the ridge. We already know the net roof area is 8.73 squares (from the last example). We also know there's 26 square feet of starter course material required. From Figure 4-87 we see that cutting waste at the rakes is negligible for this type roof.

Thus, the gross roof area   = 873 SF + 26 SF
                                      = 899 square feet ÷ 100, or 8.99 squares

The waste factor is 3 percent (8.99 ÷ 8.73), close enough to the 4 percent waste predicted in Figure 4-89. At 3 bundles per square, this job requires 27 bundles.

The following examples show you how to estimate roofs of any size, with any slope. If you calculate the waste factors for each example, you'll see that you can use the tables (Figures 4-89 and 4-90) with confidence.

▼ **Example 4-10:** Assume a roof slope of 5 in 12 and the use of 3-bundle squares installed at a 5-inch exposure, then find how many bundles of 3' x 1' 3-tab shingles are required for the roof of the building diagrammed in Figure 4-91. Also, assume using field-fabricated 3-tab shingles for the starter course at the eaves and 12" x 12" tabs salvaged at the hips for hip and ridge units.

Solution: The net roof area to be covered is:

Actual (Net) Roof Area = Roof Plan Area x Roof-Slope Factor

Net Roof Area = 40' x 20' x 1.083 (from Column 2 of Appendix A)
                     = 866 square feet ÷ 100
                     = 8.66 squares (use this quantity to determine labor costs)

The total eaves length is:

Perimeter = 2(L+W)

LF (Eaves)   = 2 x (40' + 20')
                = 120 linear feet

The total area of starter-course material required is:

Area (Starter Course)   = 120' x 0.42 SF/LF
                            (from Equation 4-10, or Figure 4-86)
                            = 50 square feet

From Column 3 of the Slope Factor table, each hip length is:

Length (Hip)   = 10' x 1.474
                = 14.74 linear feet

The total hip length is 4 x 14.74, or 59 linear feet.

From Figure 4-87, calculate cutting waste at the hips:

Waste (Hips) = 59' x 0.64 SF/LF
= 38 square feet

You need 1 square foot of single-tab shingles for each linear foot of ridge and hips, therefore:

Hips and Ridge = (59' + 20') x 1 SF/LF
= 79 square feet

From Figure 4-88, calculate the number of tabs salvaged at the hips for use on ridge and hips:

59' x 0.5 SF/LF = 30 square feet

You need more shingles than you salvaged, so you have to allow for additional shingles:

Net Waste (Hip and Ridge Units) = 79 SF - 30 SF
= 49 square feet

The total of the above calculations produces the gross roof area to be covered:

Gross Roof Area = 866 SF + 50 SF + 38 SF + 49 SF
= 1,003 square feet ÷ 100
= 10.03 squares

The waste factor (excluding crew-error waste) is:

$$\text{Waste Factor} = \frac{\text{Area Covered (including waste)}}{\text{Net Roof Area}}$$

$$= \frac{10.03 \text{ Squares}}{8.66 \text{ Squares}}$$

$$= 1.158$$

This job requires 31 bundles of shingles:

Bundles = 10.03 Squares x 3 Bundles/Square
= 31 bundles (10.33 squares)

▼ **Example 4-11:** Using the same dimensions as those in Example 4-10 (Figure 4-91), assume the application of 3' x 1' laminated strip shingles using field-fabricated shingles for the starter course at the eaves. Also, assume 3-bundle squares and a prefabricated hip and ridge roll. Now calculate how many bundles of 3' x 1' laminated strip shingles you need to roof the building.

Solution: The total hip and ridge length is the same as in the previous example, 79 linear feet. The net roof area is also the same, 8.66 squares, as is the total area for the starter course, 50 square feet.

Asphalt Shingles

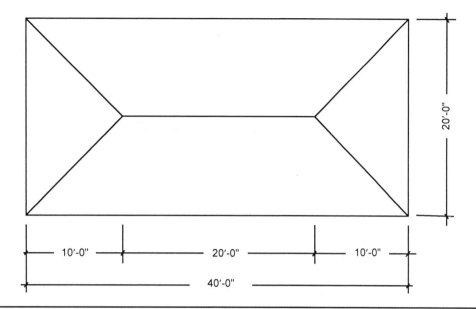

Figure 4-91 Hip roof example

From Figure 4-87 cutting waste at the hips is:

Waste (Hips) = 59' x 0.30 SF/LF
= 18 square feet

The gross roof area in this case is 866 SF + 50 SF + 18 SF

= 934 square feet ÷ 100
= 9.34 squares

This job requires 29 bundles of shingles:

Bundles = 9.34 Squares x 3 Bundles/Square
= 29 bundles (9.67 squares)

Example 4-12: Assuming a roof slope of 4 in 12 and the use of 3-bundle squares installed at a 5-inch exposure, determine the number of bundles of 3' x 1' three-tab shingles required for the roof of the building diagrammed in Figure 4-92. Also assume using field-fabricated 3-tab shingles for a starter course at the eaves and 12" x 12" tabs salvaged at the hips and valleys for ridge and hip units. Also assume the closed-cut valley method of construction.

Solution: The formula for net roof area is:

Actual (Net) Roof Area = Roof Plan Area x Roof-Slope Factor

Net Roof Area = [(50' x 22') + (22' x 11')] x 1.054
(Column 2, Appendix A)
= 1,415 square feet ÷ 100
= 14.15 squares (Use this figure to estimate labor costs.)

121

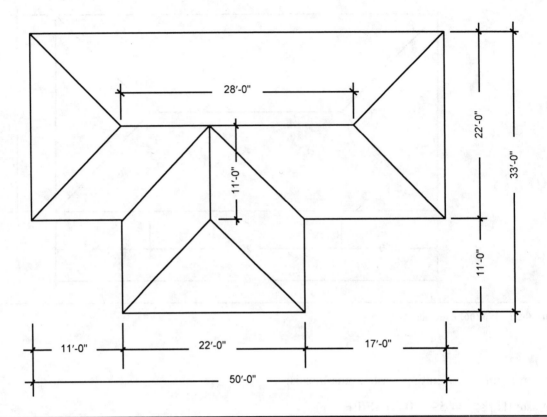

Figure 4-92 Hip and valley roof example

The total length of eaves is:

Perimeter = 2(L+W)

LF (Eaves)   = 2 x (50' + 33')
             = 166 linear feet

From Equation 4-10 or Figure 4-86, the total area of starter-course material required is:

Area (Starter Course)   = 166' x 0.42 SF/LF
                        = 69 square feet

From Column 3 of the Slope Factor table in Appendix A, the length of each hip or valley is:

Length (Hip or Valley) = 11' x 1.453
                       = 16 linear feet

There are 6 hips and 2 valleys. Use Figure 4-87 to calculate the waste allowance:

Waste (Hips)   = (6 ea. x 16' x 0.64 SF/LF)

Waste (Valleys) = (2 ea. x 16' x 2.12 SF/LF)

Add the two together:

$$= (96' \times 0.64) + (32' \times 2.12)$$
$$= 129 \text{ square feet}$$

Now, calculate the total length of ridge and hips to find the number of single-tab shingles required:

LF (Ridge and Hips) = 39' + 96'
= 135 linear feet
= 135 square feet (at 1 SF/LF)

From Figure 4-88, the quantity of tabs salvaged at the hips and valleys for hip and ridge units is:

SF salvaged at hips = (96' x 0.50 SF/LF)

SF salvaged at valleys = (32' x 1.00 SF/LF) (closed-cut valley)
= 80 square feet

Use Equation 4-13 to calculate the net waste allowance. Since we can expect to salvage 80 square feet, and need 135 square feet, we need an additional 55 square feet of single-tab shingles for the hips and ridges.

Thus, Gross Roof Area = 1,415 SF + 69 SF + 129 SF + 55 SF
= 1,668 square feet ÷ 100
= 16.68 squares

At 3 bundles per square, this job requires 50 bundles of shingles.

▼ **Example 4-13:** Work the last example, using 3-bundle, 3' x 1' laminated strip shingles with field-fabricated shingles used for the starter course at the eaves. Assume closed-cut valley construction, and a prefabricated hip and ridge roll.

Solution: The total hip and ridge length is the same as the example above, 135 linear feet. The net roof area is also the same, 1,415 square feet, as is the total area of starter-course material required, 69 square feet.

The total length of hips is 96 linear feet, so from Figure 4-87, calculate the waste allowance:

Waste (Hips) = 96' x 0.30 SF/LF
= 29 square feet

The total length of valleys is 32 linear feet, so from Figure 4-87, calculate the waste allowance:

Waste (Valleys) = 32' x 1 SF/LF
= 32 square feet

Thus, Gross Roof Area = 1,415 SF + 69 SF + 29 SF + 32 SF
= 1,545 square feet ÷ 100
= 15.45 squares

This job requires a total of 47 bundles (15.45 x 3).

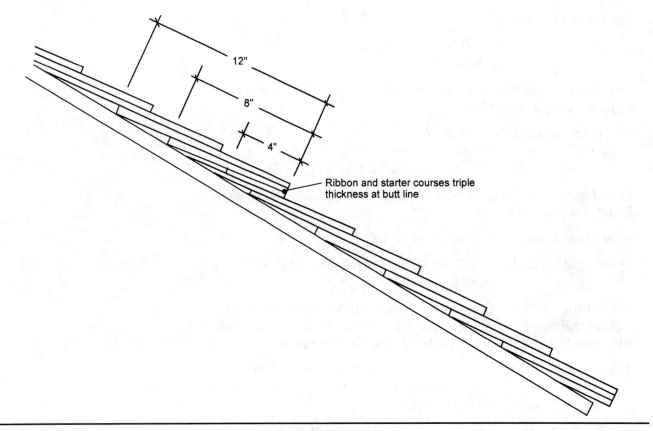

Figure 4-93 Ribbon coursing

Notice that if you calculate the waste factors for the previous two examples, you'll see that the rule of thumb for estimating waste we discussed on page 115 ("shortcut") isn't accurate. The calculated waste factors are 18 percent (16.68 ÷ 14.15) and 9 percent (15.45 ÷ 14.15). The rule of thumb method would have produced waste factors of 15 percent and 18 percent, respectively. That's probably not close enough on a large job in a competitive market.

Estimating Ribbon-Course Quantities

A popular shingle pattern, where every fifth course beginning at the eaves has a triple thickness of shingles, is called ribbon (shadow) coursing. You use a single 3-tab shingle to make the first two layers (4 and 8 inches wide, respectively). The third (top) layer is a full shingle, as shown in Figure 4-93.

Since you make the first two layers of a ribbon course from one shingle and the top layer is a full-width shingle, we must only add one layer of shingles at each ribbon course to our estimate.

Install the first ribbon course along the eaves. It's not necessary to add to the material estimate for this course however, because you can alter the shingles you use for the double starter course to serve as a ribbon course.

Use Equation 4-14 to find the total number of ribbon courses (N) required for a *gable roof*:

$$N \text{ (Gable)} = \frac{2 \times \text{Roof Dimension}}{\text{Ribbon Spacing}} - 2$$

Equation 4-14

where: Roof Dimension is the actual length of roof measured from eaves to ridge. Remember, that's the plan length times the Roof-Slope Factor (Column 1) from Appendix A.

Use Equation 4-15 to find the total number of ribbon courses (N) required for a *hip roof*:

$$N \text{ (Hip)} = \frac{\text{Roof Dimension}}{\text{Ribbon Spacing}} - 1$$

Equation 4-15

where: Roof Dimension is the actual length of roof measured from eaves to ridge.

If your result in Equation 4-14 or 4-15 isn't a whole number, round the answer up to the next whole number.

Use Equation 4-16 to find the total length of ribbon courses required for a *gable roof*:

Total Length (Gable) = N x L

Equation 4-16

where: L equals the length of the roof.

Use Equation 4-17 to find the total length of ribbon courses required for a *hip roof*:

Total Length (Hip) = $2N [(L + W) - 2(N + 1) \times \frac{\text{Ribbon Spacing}}{\text{Roof-Slope Factor}}]$

Equation 4-17

where: W is the width of the roof and the Roof-Slope Factor is from Column 1 of Appendix A.

▼ **Example 4-14:** Assume a ribbon course is installed at every fifth course, then find the total number of squares of shingles required for the gable roof described back in Example 4-8, Figure 4-83.

Solution: From Example 4-8, we know the gross roof area of that building is 938 square feet. At a 5-inch exposure, the ribbon spacing is 25 inches, or 2.083 linear feet (5 courses at 5-inch exposure).

Use Equation 4-14 to find the number of ribbon courses:

$$N = \frac{2 \times 13 \text{ ft.} \times 1.083}{2.083 \text{ ft.}} - 2$$
$$= 11.5, \text{ rounded up}$$
$$= 12 \text{ ribbon courses}$$

Use Equation 4-16 to find the total length of ribbon courses required:

LF (Ribbons) = 12 ea. x 31'
                = 372 linear feet

The formula to calculate the square feet of ribbons is similar to Equation 4-10, using Figure 4-86:

Area (SF) = Ribbons (LF) x Exposure Area (SF/LF)

Area (Ribbons) = 372' x 0.42 SF/LF
                    = 156 square feet

Thus, Gross Roof Area = 938 SF + 156 SF
                            = 1,094 square feet ÷ 100
                            = 10.94 squares

▼ **Example 4-15:** Assume a ribbon course at every fifth course, then find the total squares of shingles for the hip roof described back in Example 4-10, Figure 4-91.

Solution: From Example 4-10, we know the Gross Roof Area is 1,003 square feet. At a 5-inch exposure, the ribbon spacing is 25 inches, or 2.083 linear feet (5 courses at 5-inch exposure).

Use Equation 4-15 to calculate the number of ribbon courses:

$$N = \frac{10 \text{ ft.} \times 1.083}{2.083 \text{ ft.}} - 1$$
$$= 4.19$$
$$= 5 \text{ courses}$$

Use Equation 4-17 to find the total length of ribbon courses:

$$\text{LF (Ribbons)} = 2 \times 5 \times [(40' + 20') - 2 \times (5 + 1) \times \frac{2.083 \text{ ft.}}{1.083}]$$
$$= 10 \times [60' - 12 \times \frac{2.083 \text{ ft.}}{1.083}]$$
$$= 369 \text{ linear feet}$$

Asphalt Shingles

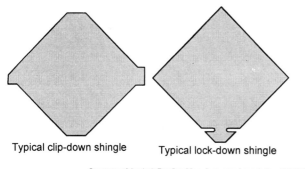

Figure 4-94 Individual "hex" shingles

Figure 4-95 Giant individual shingle

Use Figure 4-86 to find the additional square feet of shingles required for the ribbon courses:

Area (Ribbons) = 369' x 0.42 SF/LF
= 155 square feet

Thus, Gross Roof Area = 1003 SF + 155 SF
= 1,158 square feet ÷ 100
= 11.58 squares

Individual Shingles

The three most common types of individual shingles are:

1) Hexagonal, as in Figure 4-94

2) Giant, as in Figure 4-95

3) Interlocking, as in Figure 4-96

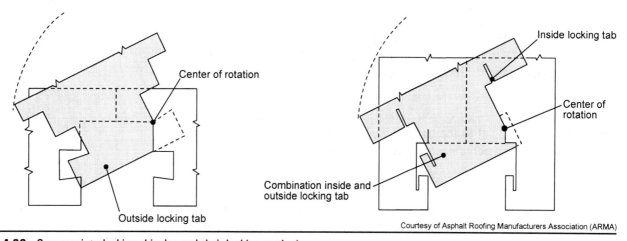

Figure 4-96 Common interlocking shingles and their locking methods

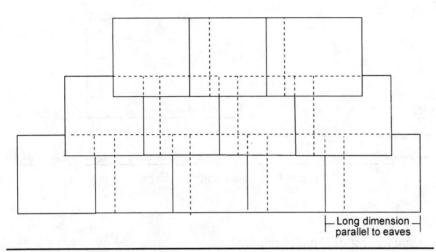

Figure 4-97 Dutch lap method

Individual clip-down and lock-down shingles are designed to withstand strong winds. You don't have to use any adhesives with these shingles, except where you've removed the locking devices, for example, at the rakes and eaves. Normally, you finish hips and ridges of individual shingle roofs with single tabs you cut from 3-tab shingles (laid at a 5-inch exposure), or a prefabricated ridge roll.

Two types of individual and hexagonal (hex) shingles are available: those locked together with a clip and those locked together via a built-in locking tab, as shown in Figure 4-94. Both types are lightweight and intended primarily for re-roofing over old roofing. They also work well on a new roof. For either application, install the shingles on roof slopes of 4 in 12 and steeper.

Use the giant individual shingles shown in Figure 4-95 for new construction or re-roofing. There are two ways to apply them: the Dutch lap method (like Figure 4-97) or the American method (Figure 4-98). For either application, install the shingles over roof slopes of 4 in 12 and steeper.

With the Dutch lap method, you install giant individual shingles with the long dimension parallel to the eaves. Apply the shingles so they overlap adjacent shingles in each course as well as the course below. See Figure 4-97.

Normally, you use the Dutch lap method to re-roof over a smooth deck that holds nails well. You can use it over new decks where single-coverage roofing will do. Install the shingles over roof slopes of 4 in 12 and steeper.

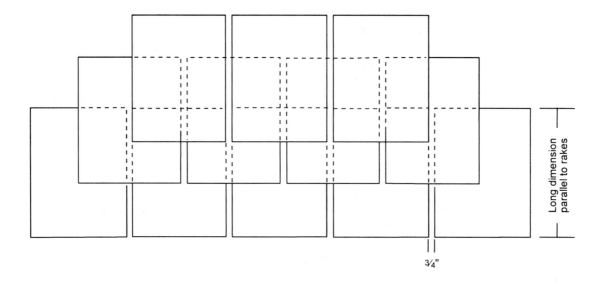

Figure 4-98   American method

With the American method, you install giant individual shingles with the long dimension parallel to the rakes. Center the shingles over the shingles of the underlying course, with ¾-inch-wide joints, as shown in Figure 4-98. You can use the American method for new construction or for re-roofing. Install the shingles on roof slopes of 4 in 12 and steeper.

Interlocking shingles (T-locks) have sets of locking tabs to make them more wind resistant. See Figure 4-96. Use them primarily for re-roofing over existing roofing. They provide single coverage and aren't recommended for new construction.

Because of their irregular shape, you may have to remove the locking tabs on shingles you install along the rakes and eaves. To prevent wind damage, cement these altered shingles to the sheathing or nail them down according to the manufacturer's recommendations.

Hexagonal Strip Shingles

Hexagonal strip shingles come most commonly in two- or three-tab units, as shown in Figure 4-99. These shingles cover the same area as 3-foot 3-tab strip shingles. Hips and ridges are usually finished with single tabs cut from 3-tab strip shingles (laid at a 5-inch exposure) or a prefabricated ridge roll, because the narrow hex tabs don't make good hip and ridge units.

The shingles described in this section are not widely used, but their varied sizes and unusual shapes make them an attractive alternative to the more traditional shingle patterns. Follow the manufacturers' guidelines for coverage requirements and installation procedures.

Estimating Asphalt Shingle Roofing Costs

Let's say you've got to cover the roof shown in Figure 4-91 under the conditions given in Example 4-10. Assume that each square of shingles costs $50.00, including sales tax. From Example 4-10, we must purchase 31 bundles of shingles, or 10.33 squares (31 ÷ 3 = 10.33). So the total material cost is 10.33 squares x $50.00/square = $517.00 (rounded off).

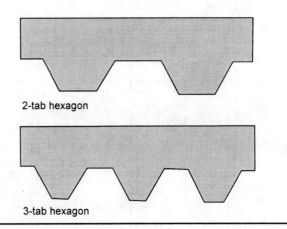

Figure 4-99 Hexagonal strip shingles

We have a roof deck area of 8.66 squares and a total of 79 linear feet of hips and ridge. According to the NCE, an R-1 crew consisting of one roofer and one laborer can install composition shingles at the rate of 1 square per 2.05 manhours, and hip and ridge units at the rate of 1 linear foot per 0.02 manhours. Assuming the roofer makes $20.00 per hour and the laborer makes $15.00 per hour, including labor burden, the average cost per manhour is: ($20 + $15) ÷ 2 = $17.50. So the labor cost will be:

Labor (installing shingles) = 8.66 squares x 2.05 manhours/square x $17.50/manhour = $311.00

Labor (installing hip and ridge units) = 79 LF x 0.02 manhours/LF x $17.50 = $28.00

The total labor cost is $311.00 + $28.00 = $339.00.

The total cost is $517.00 + $339.00 = $856.00.

Now that you know how to install asphalt shingles and estimate their quantities, let's move on to another asphalt roof covering material, mineral-surfaced roll roofing. That's the topic of the next chapter.

5 Mineral-Surfaced Roll Roofing

▶ The asphalt roofing industry has been around since roll roofing products were developed about a hundred years ago. Technically, roll roofing includes all materials such as asphalt-saturated felts, base sheets, cap sheets, and smooth-coated rolls. I've discussed asphalt-saturated felts in Chapter 3 on underlayment, and Chapter 10 covers roll roofing materials used in built-up roofing. In this chapter, I'll just talk about mineral-surfaced roll roofing — roll roofing whose exposed face is surfaced with granular material.

Roll-Roofing Materials Described

Mineral-surfaced roll roofing is inexpensive. It's also quick and easy to install. But it probably won't last more than 10 years. That's because a maximum of two layers is all that's ever installed, and you use fewer nails than for other types of roofing materials. It also blows off easily. Any weakness, even a small tear, will affect a large deck area, so it's necessary to replace damaged roll roofing to protect the deck.

Mineral-surfaced roll roofing is practical as underlayment for long-life roof coverings such as tile. As long as mineral granules protect the roll roofing base material, it'll hold up very well as underlayment.

Mineral-surfaced roll roofing is made of the same materials as asphalt shingles. Use it as a single- or double-coverage roofing material, or for valley flashing as described in Chapter 3.

Single-coverage rolls are usually 3 feet wide and 36 to 38 feet long (one factory square per roll). Rolls weigh from 40 to 90 pounds per square. The 90-pound roll is most common. For use as a single-coverage material, the exposed face is entirely covered with crushed slate embedded in the asphalt surface coating.

As a double-coverage material, 17 inches of the exposed face of the 36-inch-wide roll is surfaced with granular material. This is called *selvage roll roofing* or *split-sheet roofing*. The remaining 19 inches (selvage edge) may have any of a variety of finishes. That finish is usually granule-free and saturated, or saturated and coated with asphalt. Selvage roll roofing usually comes in 36-inch-wide rolls, and in lengths of 18 or 36 feet. Selvage rolls weigh from 55 to 70 pounds per square.

Pattern-edge roll roofing comes in 36-inch widths 42 feet long, or in 32-inch widths 48 feet long. Rolls weigh about 105 pounds per square. One side of the roll is mineral-surfaced, except for a 4-inch-wide center strip. You cut the roll in half down the center, so you get two 18-inch-wide strips with a 16-inch exposure and a 2-inch top lap. In the same way, you get two 16-inch strips with a 14-inch exposure and a 2-inch top lap from the 32-inch roll. Including normal waste, one factory square will cover 100 square feet of roof area.

Storing Mineral-Surfaced Roll Roofing

Store mineral-surfaced rolls upright in a dry, cool area. If you stack the rolls, put plywood sheets between the tiers so the roll ends don't get damaged. If you store rolls outdoors, put them on a platform raised above the ground. Cover the material to protect it from the weather. Don't store rolls in the hot sun.

Modified Bitumen Asphalt (MBA) Roofing

Asphalt mineral-surfaced roll roofing is being replaced with a roll product called MBA. It's made of a rubberized asphalt mat reinforced with fiberglass and surfaced on one side with mineral granules. Each roll weighs about 110 pounds. You install the rolls over a 45-pound base sheet and roll on cold-applied bitumen. Figure 5-1 shows this.

You can also *torch on* this material. Use a blow torch to heat the base sheet as you "walk" the MBA roll over and into the base sheet. The torch heats the asphalt in the base sheet and makes the base sheet stick to the MBA.

Mineral-Surfaced Roll Roofing

Figure 5-1 Cold-applied bitumen for roll roofing

Figure 5-2 Installing roll roofing

You can use MBA on flat roofs. Lap the sides about 3 to 4 inches and the ends about 12 inches. Adhere the laps with cold-applied bitumen. Always walk the roll down into the cement as shown in Figure 5-2.

Installing Mineral-Surfaced Roll Roofing

The biggest problem you'll have with mineral-surfaced roll roofing is its tendency to buckle. Try to install rolls when the weather is above 45° F. In cold weather, cut the rolls into shorter lengths (12 to 18 feet long) and lay them out on the ground to let them warm in the sun.

No matter what method you use to install mineral-surfaced roll roofing, these rules apply:

- Install sheets parallel or perpendicular to the eaves, allowing a ¼- to ⅜-inch overhang at the eaves and rakes.

Direction of rolls	Nailing method	Minimum roof slope allowed
Parallel to rake	Exposed	4/12
Parallel to rake	Concealed	3/12
Parallel to eaves	Exposed	2/12
Parallel to eaves	Concealed	1/12

Figure 5-3   Methods of applying single-coverage mineral-surfaced roll roofing

- Cut the sheets flush with the rake edges, or run the rolls 3 to 4 inches beyond the roof edge and cut off the excess with a hook blade.

- Install mineral-surfaced roll roofing over a solid deck — plywood is best.

- Flash valleys with a 36-inch-wide strip of mineral-surfaced roll roofing before you install the main roof.

- Don't try to bend roll roofing at a 90-degree angle. It'll crack.

- Don't cement mineral-surfaced roll roofing to the sheathing. If the roof deck shifts, the sheets may split. To prevent this when you hot-mop the sheets, nail down a base sheet before you install the mineral-surfaced roll roofing.

Valley Flashing

Starting at the low point, nail valley flashing on 6-inch centers along rows ¾ inch from each edge. If a flashing strip is too short to cover the full length of the valley, lap the upper strip over the lower by at least 6 inches. Nail the under layer down and embed the overlap into a bed of roofing cement. Apply only enough cement to securely fasten the sheets. Too much cement will cause blistering.

Mineral-Surfaced Roll Roofing (Single Coverage)

There are four methods of applying single-coverage mineral-surfaced roll roofing. They are described in Figure 5-3. The concealed-nail methods result in a more durable roof.

Exposed-Nail Method

Use this method only for temporary roofing, or on buildings such as storage sheds, because it doesn't last very long. You can install rolls parallel to the eaves, as in Figure 5-4, or parallel to the rake, like Figure 5-5.

Mineral-Surfaced Roll Roofing

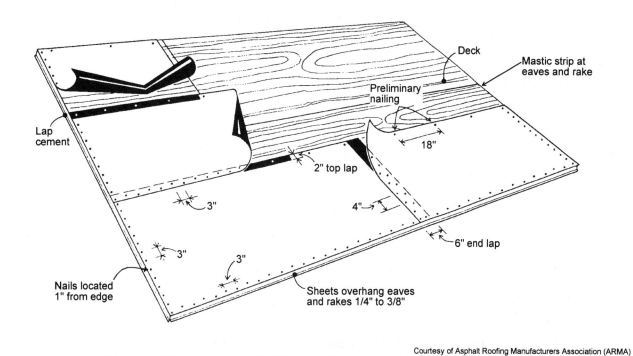

Figure 5-4 Exposed-nail method of applying roll roofing parallel to the eaves

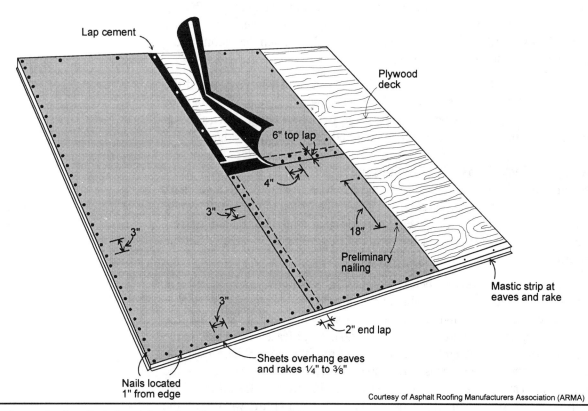

Figure 5-5 Exposed-nail method of applying roll roofing parallel to the rakes

When you install mineral-surfaced roll roofing using the exposed-nail method, nail a drip edge along the eaves and rakes. Then flash the valleys as described above.

■ **Exposed-Nail Method (Rolls Applied Parallel to Eaves)** To position the first course of roll roofing, snap a chalk line at 35¾ inches above the eaves (assuming a ¼-inch overhang beyond the eaves). Nail the top edge of the first sheet at ½ to ¾ inch from the edge on 18- to 20-inch centers. Apply a 2-inch-wide band of roofing cement along the edges of the roof, and nail the sheet on 3-inch centers along the eaves and rake.

Drive nails along the eaves and rake 1 inch from the edges of the sheet. Stagger the nails along the eaves to prevent splitting a solid deck. If you need more than one sheet to complete a course, lap the next sheet by 6 inches. Then stagger-nail the underlay on 4-inch centers in rows 1 and 5 inches from the end of the sheet. Embed the overlap in roofing cement and nail it in place. You can get special lap cement to secure laps of roll roofing. It comes in one-gallon cans or five-gallon pails. Be sure to stagger the end laps in succeeding courses.

Snap a chalk line 2 inches down from the upper edge of the first course. Position the second course on this line so it overlaps the first by 2 inches. Nail the top of the second strip on 18-inch centers as before, to temporarily secure the sheet to the sheathing. Spread a 2-inch-wide strip of roofing cement along the top edge of the underlying sheet and stagger-nail the lap on 3-inch centers at about ¾ inch from the edge of the sheet. (Staggering the nails just a little will prevent splitting the deck.) Attach the second sheet at the rakes as described previously.

Continue up the roof as shown in Figure 5-4 until you've covered the entire roof deck. Trim, butt and nail the sheets where they meet at the hips and ridge. If it's not windy, you can save time by nailing down the top edges of all the strips before you cement and nail the bottom edges.

■ **Hip and Ridge Units** To cover hips and ridges, make 12-inch-wide strips by cutting roll mineral-surfaced roofing material lengthwise. Snap a chalk line at 5½ inches on each side of the centerline of the hip or ridge. Apply a 2-inch-wide band of roofing cement inside each chalk line. Bend the strips lengthwise and lay them over the hips and ridge. In cold weather, heat the roofing with a torch before you bend it.

Install all hip coverings before you lay the ridge coverings. Starting at the low end of the hip, nail the hip cap on 3-inch centers along rows ¾ inch from each edge. If a single cap strip is too short to cover the length of the hip, overlap the lower strip with the next strip by at least 6 inches. Nail the under layer down and embed the overlap in roofing cement. Don't use too much cement as it may cause blistering. Neatly trim, nail and cement the end of the cap strip at the ridge, as shown in Figure 5-6. Install the ridge cap the same way as the hip caps. To prevent wind uplift, work with the prevailing wind at your back.

Mineral-Surfaced Roll Roofing

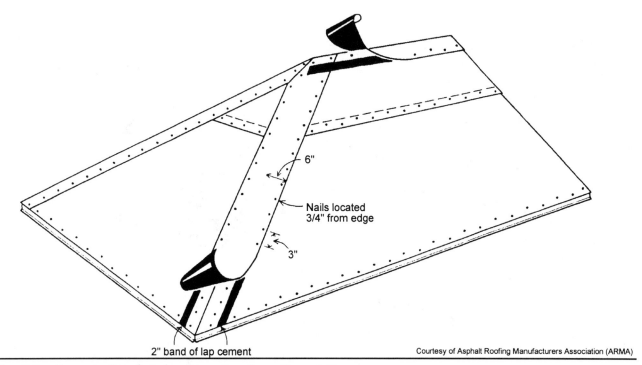

Figure 5-6 Exposed-nail method of applying roll roofing to hips and ridges

■ **Exposed Nail Method (Rolls Applied Parallel to Rake)** Apply the sheets vertically from the ridge. Fasten them with three or four nails. Then unroll toward the eaves. Now, you can use the same application methods as for horizontal installation, as far as laps, cementing, nailing and finishing of hips and ridges are concerned. See Figure 5-5.

Concealed-Nail Method

You can install mineral-surfaced roll roofing using the concealed-nail method on slopes as low as 1 in 12 as long as you install the rolls with a 3-inch top lap.

To use the concealed-nail (blind-nail) method, nail a drip edge to the deck along the eaves and rakes, then flash the valleys as described earlier in this chapter. Before you install the first full sheet, nail down a 9-inch-wide strip of mineral-surfaced roll roofing (edge strip) along the eaves and rakes on 4-inch centers at 1 inch from both edges of the strip. Allow a ¼- to ⅜-inch overhang. See Figure 5-7

■ **Concealed-Nail Method (Rolls Applied Parallel to Eaves)** Install the first roll with its lower edge and ends flush with the edge strips at the eaves and rakes. To position the first course of mineral-surfaced roll roofing, snap a chalk line 35¾ inches above the eaves (assuming a ¼-inch overhang).

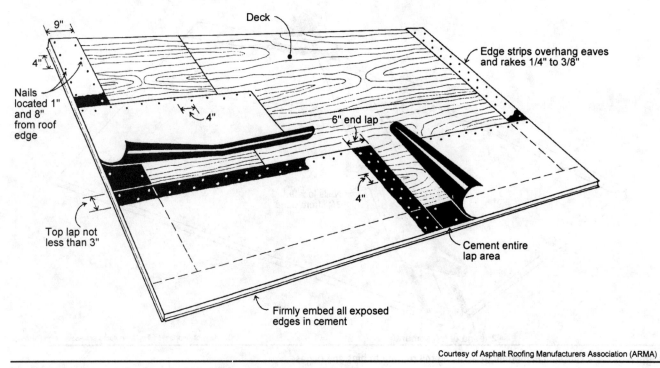

Figure 5-7   Concealed-nail method of applying roll roofing parallel to the eaves

Stagger-nail the top edge of the first sheet on 4-inch centers in rows ¾ inch and 2 inches from the edge of the sheet. Locate nails so that the next course overlaps them by at least 1 inch.

Apply roofing cement over the 9-inch strip along the edges of the roof. Embed (don't nail) the edges of the roll along the eaves and rakes into the roofing cement applied over the edge strip. If you need more than one sheet to finish a course, lap the next sheet by at least 6 inches and stagger-nail the under layer on 4-inch centers along rows 1 and 5 inches from the end of the sheet. Embed the overlap in a 6-inch-wide bed of cement.

Snap a chalk line 3 inches down from the upper edge of the first course. Position the second course at the chalk line so it overlaps the first by 3 inches. Nail the upper edge of the second course as you did the first course. Also, embed the edges along the rakes into cement and install overlaps as you did when you installed the first course. Be sure you stagger the end laps in succeeding courses.

Spread a 3-inch-wide strip of cement along the upper edge of the underlying sheet and embed the overlap into the cement (don't nail). Continue up the roof this way until you've covered the entire roof deck. Figure 5-7 shows this process. Trim, butt and nail the sheets as they meet at the hips and ridge. If it isn't too windy, you can save time by nailing down the top edges of all the strips before you cement the bottom edges.

Mineral-Surfaced Roll Roofing

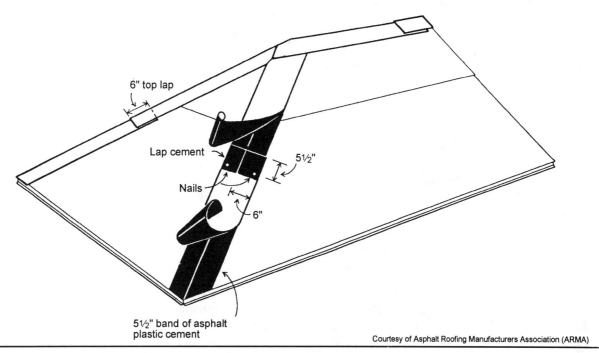

Figure 5-8 Concealed-nail method of applying roll roofing to hips and ridges

■ **Hip and Ridge Units** To make hip and ridge units, cut the roofing material across the roll into 12" x 36" strips. To cover hips and ridges, snap a chalk line 5½ inches on each side of the centerline of the hip or ridge. Apply a 5½-inch-wide band of roofing cement within the chalk lines. Bend the strips lengthwise and lay them over the hips and ridge. In cold weather, be sure to heat the roofing before you bend it.

As with the exposed nail method, install the hip units before the ridge units. Starting at the low end of the hip, embed a hip unit into the cement and secure the top with two nails 5½ inches from the top of the strip. Overlap the lower strip with the next strip by at least 6 inches. Embed the overlap into roofing cement, as shown in Figure 5-8. Use cement sparingly or it may cause blistering. Neatly trim, nail and cement the end of the cap strip at the ridge. Install the ridge caps the same way as the hip caps. To prevent wind uplift, work with the wind at your back.

■ **Concealed-Nail Method (Rolls Applied Parallel to Rake)** Apply the sheets vertically from the ridge. Fasten them with three or four nails. Then unroll toward the eaves. Now, you can use the horizontal application methods as far as edge strips, laps, cementing, nailing and finishing of hips and ridges are concerned.

Double-Coverage Mineral-Surfaced Roll Roofing

Double-coverage mineral-surfaced roll roofing is also called *19-inch selvage double-coverage roll roofing*, or *split-sheet roofing*. You can install this type of roofing on roof slopes as low as 1 in 12. You can install rolls parallel to the eaves as in Figure 5-9, or parallel to the rake, as in Figure 5-10.

To install double-coverage mineral-surfaced roll roofing, nail the drip edge along the eaves, then flash the valleys as previously described.

■ **Roofing Rolls Applied Parallel to Eaves** Before you install the first full sheet, snap a chalk line 18¾ inches above the eaves (assuming a ¼-inch overhang). Nail a 19-inch-wide selvage piece (the part without the mineral surface) of the roll along the eaves, allowing a ¼-inch overhang. Drive staggered nails on 12-inch centers along two rows 4¾ inches from the top edge, and 1 inch from the bottom edge of the strip.

To locate the first full-width course of roll roofing, snap a chalk line at 35¾ inches above the eaves. Stagger-nail the top edge on 12-inch centers in rows 4¾ and 13¼ inches from the top edge of the sheet. If a course requires more than one sheet, lap succeeding sheets at least 6 inches, and secure it the same as for single-coverage roll roofing.

Apply roofing cement at the rate of 1½ gallons per 100 square feet over the 19-inch selvage strip. Embed (don't nail) the exposure part of the first full-width roll into the cement. Don't apply more cement than necessary, because it will cause blistering.

Lay the lower edge of the second course in line with the top edge of the exposed part of the underlying sheet. Nail the top of this sheet and embed the bottom of the sheet in cement, as you did the first course. Be sure you stagger the end laps in succeeding courses. Continue up the roof this way until you've covered the entire roof deck, as shown in Figure 5-9. Trim, butt and nail sheets where they meet at the hips and ridge.

Use the mineral-surfaced part of the roll you trimmed from the starter strip at the eaves to finish at the ridge.

■ **Hip and Ridge Units** Cut the roofing material across the roll into 12" x 36" strips. Snap a chalk line at 5½ inches on each side of the centerline of the hip or ridge. In cold weather, heat the roofing before you bend it lengthwise to lay over the hips and ridge. Install roofing on the hips before you do the ridge.

Starting at the low end of a hip, nail a 12-inch-wide selvage starter strip 19 inches long along the lower end of each hip. Drive the nails on 4-inch centers 1 inch from the edges of the starter strip. Apply a 5½-inch-wide band of roofing cement over the starter strip. Embed the exposed part of the next hip strip into the cement. Secure the selvage part with nails on 4-inch centers 1 inch from the edges of the strip. Continue this up to the

Mineral-Surfaced Roll Roofing

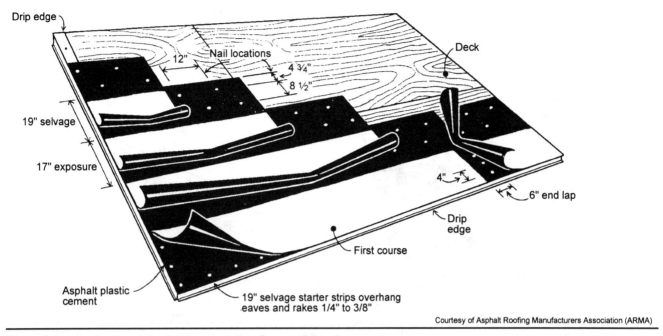

Figure 5-9 Application of double-coverage roll roofing parallel to the eaves

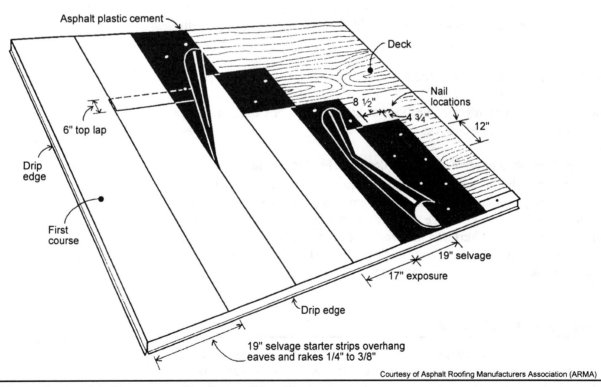

Figure 5-10 Application of double-coverage roll roofing parallel to the rake

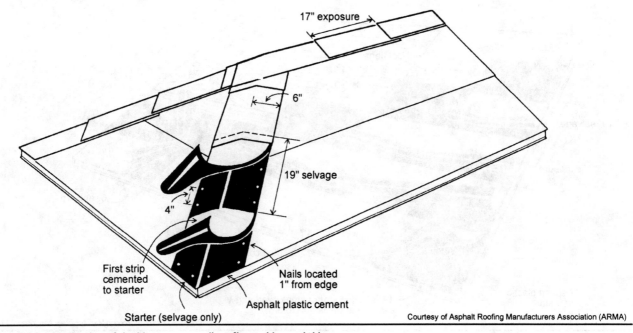

Figure 5-11 Application of double-coverage roll roofing to hips and ridges

ridge, as shown in Figure 5-11. Neatly trim the last hip unit at the ridge. Install the ridge units the same way as the hip units. To prevent wind uplift, work with the wind at your back.

■ **Roofing Rolls Applied Parallel to Rake** Use the installation method just described, except apply the sheets vertically from the ridge. Fasten them with three or four nails and unroll toward the eaves. Look back at Figure 5-10 to see vertical installation.

■ **Finishing Shed Roofs with Double-Coverage Mineral-Surfaced Roll Roofing** When a roof has no ridge, nail the selvage part of the last course along the upper edge of the roof. Then cement a granular-surfaced part of another sheet over the selvage strip. If you're not concerned about looks, use the full width of surfaced material to save trimming. Otherwise, trim the sheet to cover just the selvage of the course below. Then cement metal flashing over the upper edge of the roof.

Vent Pipes and Flashing

To work around a vent pipe with single-coverage roll roofing, cut a hole in the roll roofing and slip the roofing over the pipe. Then slip vent flashing over the pipe and embed the flashing in roofing cement. Nail the edges of the flashing on 2-inch centers. Cover the flashing with roofing cement.

Mineral-Surfaced Roll Roofing

Figure 5-12 Cut the roll to fit around the vent pipe

Figure 5-13 Apply roofing cement around the vent pipe

With double-coverage roll roofing, cut a piece out of the underlying roll to fit around the pipe, as shown in Figure 5-12. Next, cut a hole in a sheet of mineral-surfaced roll roofing to use as additional weather protection (shown in the upper left corner of Figure 5-13). Embed the roll roofing flashing in roofing cement. See Figure 5-14. Now, slip the vent flashing over the pipe and embed that in roofing cement. Then cover the vent flashing with roofing cement and install the overlying roll roofing with a hole cut to neatly fit over the vent pipe. You'll have five layers:

1) the deck

2) the bottom layer of mineral-surfaced roll roofing with a slot cut out

3) the embedded piece of coated roofing material with a hole cut to fit the vent

4) an embedded metal flashing covered with roofing cement

5) the top layer of roofing material.

If a roof intersects a vertical surface such as a wall or chimney, install continuous flashing, not step flashing. Refer back to Chapter 4 for details on flashing junctions between roofs and vertical structures.

Figure 5-14 Install roll roofing flashing for added leak protection

Cement Smudges

It's almost impossible to keep roofing cement from squeezing out from between laps and onto the roof surface. To make matters worse, someone always steps in it and tracks it across the roof. The easiest way to "hide the evidence" is to sprinkle out loose granules and rub them into the cement. You can buy tubes of granules or collect your own by rubbing two granule-surfaced shingles or pieces of surfaced roll material together.

Coverage classification	Coverage (FS/Sq.)	Coverage (SF/FS)
Single	1	100
Double	2	50

Figure 5-15 Covering capacity of roll roofing

Estimating Mineral-Surfaced Roll Roofing

First you estimate mineral-surfaced roll roofing quantities (single- or double-coverage) by the square foot. Then you change those measurements to rolls, since that's the way this material is sold. Remember, you'll lose some material due to laps. The amount depends on the installation method you use (exposed or concealed nails). Use Figure 5-15 to roughly change from square feet to rolls.

Figure 5-15 doesn't allow for additional material required for overlaps, starter strips, flashings, concealed nailing, roof layout, or waste due to crew errors.

Additional Material Waste

The type and amount of additional material waste depends on:

- single- or double-coverage installation
- applying the material parallel or perpendicular to the eaves
- exposed or concealed nailing
- the type of roof (gable or hip).

■ **Over-Cutting at the Edge of the Roof** When you apply mineral-surfaced roll roofing (single- or double-coverage; exposed- or concealed-nail method) perpendicular to the edge of a roof, you usually cut the material off beyond the edge, and trim it flush with the edge later. So, let's assume an average of 4 inches of waste beyond:

1) the gable ends of a gable roof when you apply the rolls parallel to the eaves, or

2) the eaves of a gable or hip roof when you apply the rolls perpendicular to the eaves.

This is 0.34 square feet per linear foot of roof edge. Here are some other allowances you'll have to make:

Starter-Strip Allowance: When you use the concealed-nail method on a gable or hip roof, either parallel or perpendicular to the eaves, you must also allow for additional material required for a 9-inch-wide starter strip at the edges of the roof. You'll need 0.75 square foot of material per linear foot of roof edge.

Ridge and Hip Units: No matter which application method you use, add 1 square foot of area to be covered per linear foot of ridge and hips. This assumes you install these units with the same exposure as the rest the roof.

Selvage-Strip Waste: You have to allow for waste due to the 19-inch-wide selvage strip when you install mineral-surfaced roll roofing for double-coverage.

Whether you install the 19-inch selvage strip at the eaves (rolls parallel to the eaves), or at the rake (rolls perpendicular to the eaves) of a gable roof, you can install the remaining surfaced part of the roll next to the ridge, or at the opposite rake. Therefore, selvage waste on a gable roof is minimal.

When you install rolls parallel to the eaves on a hip roof, you can only use part of the surfaced section remaining from the starter strip roll at the ridge. (You'll also have some cutting waste at the hips.)

Here's a formula for the starter-strip waste at the ridge:

Waste (Ridge) = 4 x E x (L - R - E)   Equation 5-1

where: E = exposure, L = roof length, and R = ridge length

For double coverage, with a 17-inch exposure, the equation becomes:

Waste (Ridge) = 5.67' x (L - R - 1.417')   Equation 5-2

The waste at the hips is:

Waste (Hips) = 4 x E x (Run - E) x Roof-Slope Factor   Equation 5-3

where the Roof-slope factor comes from Column 4 of Appendix A.

Substitute 17 inches for the exposure, and the equation becomes:

Waste (Hips) = 5.67' x (Run - 1.417') x Roof-Slope Factor   Equation 5-4

Find the starter-strip and hip-cutting waste for double-coverage with mineral-surfaced roll roofing *parallel* to the eaves on a hip roof by combining Equations 5-1 and 5-3 as follows:

**Waste (At Starter Strip & Hips)
   = 4 x E x [(L - R - E) + (Run - E)] x Roof-Slope Factor**   Equation 5-5

where the Roof-slope factor comes from Column 4 of Appendix A.

Again, substituting 17 inches for the exposure, the equation is:

= 5.67' x [(L - R - 1.417') + (Run - 1.417')] x Roof-Slope Factor **Equation 5-6**

When you apply roll roofing (double coverage) *perpendicular* to the eaves on a hip roof, the equation for total waste (starter-strip and hip-cutting waste) is:

= 4 x E x (Run - E) x Roof-slope factor **Equation 5-7**

where Roof-slope factor comes from Column 4 of Appendix A.

A "Shortcut" for Estimating Mineral-Surfaced Roll Roofing Waste

Use Figures 5-16 through 5-21 to estimate material overrun and waste on various sizes and slopes of roofs, depending on the coverage, roof type, direction of application and nailing method. Notice that these tables don't include waste due to roof layout (where dimensions aren't evenly divisible by roofing material exposure). You'll find the formulas for non-conforming roof layout waste following these tables.

Waste from Non-conforming Roof Layout

Regardless of roof type, coverage, direction of application, or nailing method, you may have extra material waste due to the length of:

1) The rake (material installed parallel to the eaves), or

2) The eaves (material installed perpendicular to the eaves).

For example, you're using double coverage (17-inch exposure) installed perpendicular to the eaves. If the length of the eaves is evenly divisible by the exposure, there won't be any waste. Here are some more cases:

Eaves length	Courses
28 feet, 4 inches	20
29 feet, 9 inches	21
31 feet, 2 inches	22

Mineral-Surfaced Roll Roofing

Gable roof material overrun (Roll roofing, single coverage, exposed nail method)			
Total percentage to be added to net roof area including 4-inch cutting waste at rake (roofing applied parallel to eaves), or eaves (roofing applied perpendicular to eaves). Figures include allowance for material required to make ridge units			
Building Dimensions	Roof Slope		
(L x W)	3/12	6/12	12/12
30 x 20	5 (7)	5 (7)	4 (5)
40 x 30	5 (6)	5 (6)	4 (4)
45 x 30	5 (6)	4 (6)	4 (4)
50 x 30	4 (6)	4 (6)	3 (4)
60 x 30	4 (6)	4 (6)	3 (4)
70 x 30	4 (6)	4 (6)	3 (4)
80 x 30	4 (6)	4 (6)	3 (4)
50 x 40	3 (4)	3 (4)	3 (4)
60 x 40	3 (4)	3 (4)	3 (4)
70 x 40	3 (4)	3 (4)	3 (4)
80 x 40	3 (4)	3 (4)	3 (4)
90 x 40	3 (4)	3 (4)	3 (4)
60 x 50	3 (4)	3 (4)	2 (2)
70 x 50	3 (4)	3 (4)	2 (2)
80 x 50	3 (4)	3 (4)	2 (2)
90 x 50	3 (4)	3 (4)	2 (2)
Roofing applied parallel to eaves. (Figures in parentheses are for roofing applied perpendicular to eaves.) Waste due to crew error and non-conforming roof layout not included.			

Figure 5-16 Gable roof material overrun (Roll roofing, single coverage, exposed nail method)

Hip roof material overrun (Roll roofing, single coverage, exposed nail method)			
Total percentage to be added to net roof area. Allowance is included for ridge and hip units and 4-inch waste at the eaves when roofing is applied perpendicular to eaves.			
Building Dimensions	Roof Slope		
(L x W)	3/12	6/12	12/12
30 x 20	11 (16)	11 (16)	9 (13)
40 x 30	8 (12)	8 (12)	7 (10)
45 x 30	7 (11)	7 (10)	6 (9)
50 x 30	7 (10)	6 (9)	6 (9)
60 x 30	7 (10)	7 (10)	5 (7)
70 x 30	6 (9)	6 (9)	5 (7)
80 x 30	6 (9)	5 (8)	4 (6)
50 x 40	7 (10)	6 (9)	5 (7)
60 x 40	6 (9)	6 (9)	5 (7)
70 x 40	5 (8)	5 (7)	5 (7)
80 x 40	5 (7)	4 (6)	4 (6)
90 x 40	4 (6)	4 (6)	4 (6)
60 x 50	5 (7)	5 (7)	4 (6)
70 x 50	5 (7)	5 (7)	4 (6)
80 x 50	5 (7)	4 (6)	4 (6)
90 x 50	4 (6)	4 (6)	4 (6)
Roofing applied parallel to eaves. (Figures in parentheses are for roofing applied perpendicular to eaves.) Waste due to crew error and non-conforming roof layout not included.			

Figure 5-17 Hip roof material overrun (Roll roofing, single coverage, exposed nail method)

Roofing Construction & Estimating

Gable roof material overrun (Roll roofing, single coverage, concealed nail method)			
Total percentage to be added to net roof area including 4-inch cutting waste at rake (roofing applied parallel to eaves), or eaves (roofing applied perpendicular to eaves). Also included are ridge units and a 9-inch-wide starter strip.			
Building Dimensions	Roof Slope		
(L x W)	3/12	6/12	12/12
30 x 20	18 (20)	17 (19)	14 (15)
40 x 30	14 (15)	13 (14)	11 (11)
45 x 30	13 (12)	13 (14)	11 (11)
50 x 30	12 (14)	12 (14)	10 (11)
60 x 30	11 (13)	11 (13)	9 (10)
70 x 30	11 (13)	11 (13)	9 (10)
80 x 30	11 (13)	10 (12)	8 (9)
50 x 40	10 (11)	9 (10)	9 (10)
60 x 40	9 (10)	9 (10)	8 (9)
70 x 40	9 (10)	9 (10)	8 (9)
80 x 40	9 (10)	8 (9)	8 (9)
90 x 40	8 (9)	8 (9)	7 (8)
60 x 50	8 (9)	8 (9)	7 (7)
70 x 50	8 (9)	8 (9)	6 (6)
80 x 50	8 (9)	8 (9)	6 (6)
90 x 50	8 (9)	7 (8)	6 (6)
Roofing applied parallel to eaves. (Figures in parentheses are for roofing applied perpendicular to eaves.) Waste due to crew error and non-conforming roof layout not included.			

Figure 5-18 Gable roof material overrun (Roll roofing, single coverage, concealed nail method)

Hip roof material overrun (Roll roofing, single coverage, concealed nail method)			
Total percentage to be added to net roof area. Allowance is included for ridge and hip units and 4-inch waste at the eaves when roofing is applied perpendicular to eaves.			
Building Dimensions	Roof Slope		
(L x W)	3/12	6/12	12/12
30 x 20	23 (28)	22 (27)	18 (22)
40 x 30	17 (21)	16 (20)	13 (16)
45 x 30	15 (19)	15 (18)	12 (15)
50 x 30	15 (18)	13 (16)	12 (15)
60 x 30	14 (17)	14 (17)	10 (12)
70 x 30	13 (17)	12 (15)	10 (12)
80 x 30	13 (17)	11 (14)	9 (11)
50 x 40	14 (17)	12 (15)	10 (12)
60 x 40	12 (15)	12 (15)	9 (11)
70 x 40	11 (14)	10 (12)	9 (11)
80 x 40	11 (13)	9 (11)	8 (10)
90 x 40	9 (11)	9 (11)	8 (10)
60 x 50	10 (12)	10 (12)	8 (10)
70 x 50	10 (12)	10 (12)	8 (10)
80 x 50	10 (12)	8 (10)	8 (10)
90 x 50	8 (9)	8 (10)	7 (9)
Roofing applied parallel to eaves. (Figures in parentheses are for roofing applied perpendicular to eaves.) Waste due to crew error and non-conforming roof layout not included.			

Figure 5-19 Hip foof material overrun (Roll roofing, single coverage, concealed nail method)

Mineral-Surfaced Roll Roofing

Gable roof material overrun (Roll roofing, double coverage)			
Total percentage to be added to net roof area including 4-inch cutting waste at the rake (roofing applied parallel to eaves), or eaves (roofing applied perpendicular to eaves). Also included are ridge units.			
Building Dimensions	Roof Slope		
(L x W)	3/12	6/12	12/12
30 x 20	5 (7)	5 (7)	4 (5)
40 x 30	5 (6)	5 (6)	4 (4)
45 x 30	5 (6)	4 (6)	4 (4)
50 x 30	4 (6)	4 (6)	3 (4)
60 x 30	4 (6)	4 (6)	3 (4)
70 x 30	4 (6)	4 (6)	3 (4)
80 x 30	4 (6)	4 (6)	3 (4)
50 x 40	3 (4)	3 (4)	3 (4)
60 x 40	3 (4)	3 (4)	3 (4)
70 x 40	3 (4)	3 (4)	3 (4)
80 x 40	3 (4)	3 (4)	3 (4)
90 x 40	3 (4)	3 (4)	3 (4)
60 x 50	3 (4)	3 (4)	2 (2)
70 x 50	3 (4)	3 (4)	2 (2)
80 x 50	3 (4)	3 (4)	2 (2)
90 x 50	3 (4)	3 (4)	2 (2)
Roofing applied parallel to eaves. (Figures in parentheses are for roofing applied perpendicular to eaves.) Waste due to crew error and non-conforming roof layout not included.			

Figure 5-20 Gable roof material overrun (Roll roofing, double coverage)

Hip roof material overrun (Roll roofing, double coverage)			
Total percentage to be added to net roof area including cutting waste at starter strip and hip, plus a 4-inch waste allowance at the eaves (roofing applied perpendicular to eaves) Also included are hip and ridge units.			
Building Dimensions	Roof Slope		
(L x W)	3/12	6/12	12/12
30 x 20	35 (23)	33 (23)	29 (20)
40 x 30	27 (18)	26 (18)	22 (19)
45 x 30	24 (17)	23 (15)	19 (14)
50 x 30	23 (15)	22 (15)	18 (14)
60 x 30	19 (13)	20 (13)	15 (11)
70 x 30	17 (13)	16 (13)	14 (10)
80 x 30	16 (12)	14 (11)	13 (10)
50 x 40	22 (14)	21 (14)	17 (12)
60 x 40	19 (13)	17 (12)	15 (11)
70 x 40	16 (12)	16 (11)	13 (9)
80 x 40	15 (10)	14 (10)	12 (9)
90 x 40	14 (10)	12 (9)	11 (9)
60 x 50	18 (11)	17 (11)	14 (10)
70 x 50	16 (11)	15 (10)	13 (9)
80 x 50	14 (9)	13 (9)	12 (9)
90 x 50	13 (9)	12 (9)	10 (8)
Roofing applied parallel to eaves. (Figures in parentheses are for roofing applied perpendicular to eaves.) Waste due to crew error and non-conforming roof layout not included.			

Figure 5-21 Hip roof material overrun (Roll roofing, double coverage)

But when the eaves length isn't evenly divisible by the exposure, you have to trim and throw away part of the last strip. Or you can increase the size of the top lap of each roll (and decrease the exposure) so the edge of the last roll is flush with the edge of the rake. In either case, you'll need extra material.

The waste due to non-conforming roof layout can be so small you don't have to think about it. That's true of most of the example problems that follow. However, in Example 5-3 you can see that roof layout waste can be major, even with a short ridge. So, you should always calculate waste due to non-conforming roof layout.

To find waste due to non-conforming roof layout, first find the width of the last strip installed. For rolls parallel to the eaves (gable or hip roofs), the width of the last strip is:

Last strip width (LSW) = E x the decimal part (numbers to the right of the decimal point) of the expression:

$$\frac{W/2 \times \text{Roof–slope factor}}{E}$$

Equation 5-8

where W = roof width, E = exposure length, and the roof-slope factor comes from Column 2 of Appendix A.

Here are the exposure widths for various mineral-surfaced roll roofing applications:

Exposure for double coverage: 17 inches, or 1.417 feet.

Exposure for exposed-nail, single coverage: 34 inches, or 2.83 feet.

Exposure for single coverage, concealed-nail method: 33 inches, or 2.75 feet.

For rolls perpendicular to the eaves of a hip roof, single or double coverage, the non-conforming roof layout waste is negligible. For rolls perpendicular to the eaves of a gable roof, the last strip width is:

LSW = E x the decimal part of the expression: $\frac{L}{E}$

Equation 5-9

Where L is the roof length and E is the exposure.

For single-coverage roof roofing:

1) If LSW is less than or equal to half the exposure, the waste in square feet per linear foot of ridge (rolls parallel to the eaves on gable or hip roofs) or per linear foot of a single rake (rolls perpendicular to the eaves on gable roofs), is:

Waste (SF/LF) = E - (2 x LSW)

Equation 5-10

Mineral-Surfaced Roll Roofing

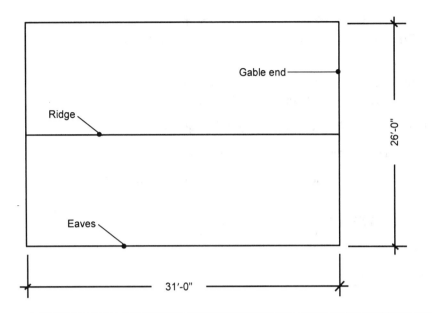

Figure 5-22 Gable roof example

2) If LSW is greater than half the exposure, the waste is:

Waste (SF/LF) = 2 x (E - LSW)   **Equation 5-11**

This equation is also true for double-coverage roofing. Waste is in square feet per linear foot of ridge (rolls applied parallel to eaves), or per linear foot of a single rake (rolls applied perpendicular to eaves).

Examples

▼ **Example 5-1:** Assume a 5-in-12 roof slope and single-coverage roll roofing (exposed-nail method) applied parallel to the eaves. Find the number of rolls required to cover the roof of the building in Figure 5-22.

First, remember from Chapter 1 that **Net Roof Area = Roof Plan Area x Roof-Slope Factor.** The roof-slope factor is from Column 2 of Appendix A.

Net Roof Area = 31' x 26' x 1.083
               = 873 square feet ÷ 100
               = 8.73 squares (Use this quantity to determine labor costs)

The total length of rake is 56 linear feet (4 x 13' x 1.083).

Cutting waste at the rakes is 19 square feet (that's 56 feet x 0.34 SF/LF where 0.34 SF/LF is the 4-inch waste for over-cutting at the eaves or rake).

The total length of ridge is 31 feet. Since you must allow 1 square foot per linear foot of ridge, allow 31 square feet for the ridge.

In Chapter 1, we defined the waste factor with the equation:

Waste Factor = Total Area Covered ÷ Net Roof Area

Therefore, waste so far, excluding waste due to non-conforming roof layout, is:

$$\text{Waste Factor} = \frac{873 \text{ SF} + 19 \text{ SF} + 31 \text{ SF}}{873 \text{ SF}}$$

$$= \frac{923 \text{ SF}}{873 \text{ SF}}$$

$$= 1.06$$

This is consistent with Figure 5-16 which predicts about 5 percent waste for a building of this approximate size and roof slope.

From Equation 5-8 (above) we know that

Last strip width (LSW) = E x the decimal part of the expression:

$$\frac{W/2 \times \text{Roof–slope factor}}{E}$$

where E is the exposure and W is the width of the roof

The expression $\frac{W/2 \times \text{Roof–slope factor}}{E}$ evaluates to $\frac{(26'/2) \times 1.083}{2.83'}$

$$= (13' \times 1.083) \div 2.83'$$
$$= 4.97$$

Now, multiply only the decimal part of 4.97 by the exposure:

$$= 0.97 \times 2.83'$$
$$= 2.77 \text{ feet}$$

Now, using Equation 5-11, you can calculate waste per linear foot of ridge:

Waste (SF/LF) = 2 x (E - LSW)

$$= 2 \times (2.83' - 2.77')$$
$$= 0.12 \text{ square feet per linear foot}$$

Waste due to non-conforming roof layout is:

Waste (Roof Layout) = 31' x 0.12 SF/LF
= 4 square feet (rounded up from 3.72)

So, total roof area to cover is:

Gross Roof Area = 873 SF + 19 SF + 31 SF + 4 SF
= 927 square feet ÷ 100
= 9.27 squares

Since this is a single-coverage job, Figure 5-15 shows we need one roll of material per square:

Rolls = 9.27 squares x 1 roll/square
= 10 rolls

▼ **Example 5-2:** Here's how to calculate this for roofing applied perpendicular to the eaves:

The net roof area from Example 5-1 is 8.73 squares. (Use this quantity to figure labor cost.)

In Figure 5-22 you see the total eaves length is 62 linear feet (2 x 31'). Cutting waste at the eaves is 21 square feet (62 feet x 0.34 SF/LF). Allowance for the ridge is 31 square feet (allowing 1 SF/LF).

When you calculate the waste factor (Waste factor = Total area covered ÷ net roof area), you get 1.06. This is consistent with Figure 5-16 which predicts a waste factor of 6 to 7 percent.

Now, from Equation 5-9, calculate the last strip width:

LSW = E x the decimal part of the expression L ÷ E

First, calculate L ÷ E:

31 ÷ 2.83 = 10.95

Now multiply only the decimal part of that number by the exposure:

= 0.95 x 2.83'
= 2.69 feet

Since LSW is more than half the exposure, waste (from Equation 5-11) per linear foot of a single rake is:

Waste (SF/LF) = 2 x (2.83' - 2.69')
= 0.14 square feet per linear foot

The length of a single rake is:

LF (One Rake) = 13' x 1.083 (length times roof-slope factor
from Appendix A)
= 14.1 linear feet

Waste due to non-conforming roof layout is:

Waste (Roof Layout) = 14.1' x 0.14 SF/LF
= 2 square feet

So, the total material required is:

Gross Roof Area = 873 SF + 21 SF + 31 SF + 2 SF
 = 927 square feet ÷ 100
 = 9.27 squares

You should order 10 squares of rolled roofing (1 roll per square, single coverage).

▼ **Example 5-3:** Now let's assume we're using the concealed-nail method with rolls parallel to the eaves for the same building as the last two examples (Figure 5-22).

You already know the net roof area is 8.73 squares (that's the amount you use to figure labor costs). Cutting waste at the rakes is 19 square feet and the material allowance at the ridge is 31 square feet.

The total length of 9-inch-wide starter strip required is:

LF (Starter Strip) = (2 x 31') + (2 x 26' x 1.083)
 = 118.3 linear feet
 = 89 square feet (118.3 x 0.75 SF/LF)

To check the waste factor:

Waste factor = $\dfrac{873 \text{ SF} + 19 \text{ SF} + 31 \text{ SF} + 89 \text{ SF}}{873 \text{ SF}}$
 = 1012 SF ÷ 873 SF
 = 1.16

This agrees with Figure 5-18, which predicts a range from 13 to 17 percent

Now, find the last strip width:

First, $\dfrac{26'}{2}$ x 1.083 = 14.08

Then, 14.08 ÷ 2.75' = 5.12

Multiply the decimal part of that answer by the exposure:

0.12 x 2.75' = 0.33

Now, because the LSW is less than half the exposure, use Equation 5-10 to calculate the waste due to non-conforming roof layout this way:

Waste (SF/LF) = E - (2 x LSW)
Waste (SF/LF) = 2.75' - (2 x 0.33')
 = 2.09 square feet per linear foot of ridge

Mineral-Surfaced Roll Roofing

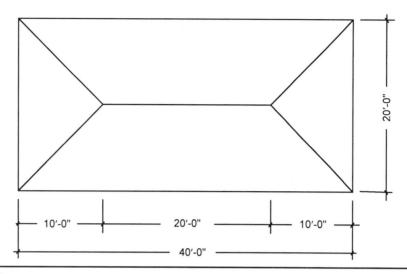

Figure 5-23 Hip roof example

Waste due to non-conforming roof layout is:

Waste (Roof Layout) = 31' x 2.09 SF/LF
 = 65 square feet

You should order a total of 1012 SF + 65 SF, or 1077 SF. For single coverage, you need 1 roll per square, so this job requires 11 rolls of material.

▼ **Example 5-4:** Now, let's try a couple of examples for a hip roof. Assume a roof slope of 6 in 12 and the installation of single-coverage roll roofing (exposed-nail method) applied parallel to the eaves, and find the number of rolls required to cover the roof of the building in Figure 5-23.

You calculate the net roof area from the equation:

Net Roof Area = Roof Plan Area x Roof-Slope Factor

Net Roof Area = 40' x 20' x 1.118 (from Column 2, Appendix A)
 = 894 square feet ÷ 100
 = 8.94 squares (use this quantity to estimate labor costs)

The total ridge length is **Length - Width**, or 20 linear feet.

From Column 3 of Appendix A, you calculate the hip length as

Run x Roof-Slope Factor

From the drawing in Figure 5-23 you see the hip is 10' x 1.5, or 15 linear feet.

There are four hips, so the total hip length is 60 linear feet. Total length of hips and ridge is 80 linear feet.

Fabrication allowance at hips and ridge is 1 SF/LF, or 80 square feet.

So far, the waste excluding allowance for non-conforming roof layout is:

$$\text{Waste Factor} = \frac{894 \text{ SF} + 80 \text{ SF}}{894 \text{ SF}}$$
$$= \frac{974 \text{ SF}}{894 \text{ SF}}$$
$$= 1.09, \text{ or } 9 \text{ percent}$$

Figure 5-17 predicts a range of 8 to 11 percent waste for a building about this size and roof slope.

The last strip width is 2.69 feet (from Equation 5-8).

From Equation 5-11, the waste per linear foot of ridge is 0.28 square feet per linear foot.

Now you can calculate waste due to non-conforming roof layout as 0.28 SF/LF x 20', or 6 square feet (rounded up from 5.6)

Thus, the total roof area is 894 SF + 80 SF + 6 SF = 980 square feet, or 9.8 squares. Since this is a single-coverage job, you need 1 roll per square (Figure 5-15), or 10 rolls of material.

▼ **Example 5-5:** This time, figure on double coverage applied parallel to the eaves.

From the above example, we already know the net roof area (8.94 squares) and the waste for ridge and hip units (80 square feet).

From Equation 5-6, cutting waste at the 19-inch-wide selvage strip and along the eaves is:

Waste (Starter Strip and Hips)
 = 5.67' x [(40' - 20' - 1.417') + (10' - 1.417')] x 1.061
 = 163 square feet

Waste excluding that for non-conforming roof layout calculates to 27 percent (1137 ÷ 894). This agrees with the range of 26 to 33 percent in Figure 5-21.

From Equation 5-8 the last strip width is 1.27 feet.

The ridge length is 20 feet, the waste per linear foot of ridge (from Equation 5-11) is 0.29 square feet per linear foot. So the waste due to non-conforming roof layout is 6 square feet (rounded up from 5.8).

The gross roof area = 894 SF + 80 SF + 163 SF + 6 SF
 = 1,143 square feet ÷ 100
 = 11.43 squares

From Figure 5-15, the material requirement is 2 rolls per square, or 23 rolls.

Estimating Mineral-Surfaced Roll Roofing Costs

Let's say you've got to cover the roof shown in Figure 5-22 with single-coverage mineral-surfaced roll roofing using the concealed-nail method with rolls parallel to the eaves. Assume that each roll of mineral-surfaced roll roofing costs $15.00, including sales tax.

From Example 5-3, you must purchase 11 rolls, at a cost of:

11 rolls x $15/roll = $165.00

The roof deck area is 8.73 squares. According to the *National Construction Estimator*, an R1 crew consisting of one roofer and one laborer can install mineral-surfaced roll roofing at the rate of 1 square per 1.03 manhours. Assuming your roofer makes $20.00 and the laborer $15.00 per hour, including labor burden, the average cost per manhour is:

($20 + $15) ÷ 2 = $17.50/manhour

So the labor cost will be:

8.73 squares x 1.03 manhours/square x $17.50/manhour = $157.36

That rounds to $157.00. So the total cost is:

$165.00 + $157.00 = $322.00

If all this is clear to you, you're ready to move on to the next chapter — on the application of wood shingles on sloping roofs.

6 Wood Shingles and Shakes

▶ You can expect a wood shingle or shake roof to last from 25 to 30 years. Some last up to 50 years. They're more expensive than asphalt shingles, but they're popular because they have a charming, rustic appearance that becomes even more attractive after a few years of weathering.

Unless you use pressure-impregnated fire-retardant (Certi-guard) shingles or shakes, a house with a wood roof costs more to insure against fire. Fire-retardant shakes or wood shingles installed over a solid deck of ½-inch-thick (minimum) plywood or equivalent material earn a Class "B" fire rating. In some parts of the country, fire-treated material is required.

Figure 6-1 shows humid locations in the United States. Don't install untreated wood shingles and shakes where heat and humidity are severe because they'll be very susceptible to decay from moss, mildew and fungus. Use treated (Certi-last) shingles on low-slope roofs, too. That's because a low-slope roof doesn't shed water as well as roofs with a higher pitch. And on roofs shadowed by overhanging trees use treated shingles because the trees will keep the roof wet longer.

Courtesy of Cedar Shake & Shingle Bureau

Figure 6-1   Locations of severe humidiity

You can get wood shingles and shakes that carry a 30-year warranty if they're pressure treated with a wood preservative at the factory. If the shingles haven't been factory-treated, you can apply a fungicidal wood preservative such as copper or zinc napthenate after the roof has weathered for a year. This waiting time lets the natural oils in the cedar leach out, so the wood is porous and soaks up more of the preservative.

In low-humidity climates, you can apply a commercial oil-based preservative to help prevent excessive dryness, which leads to curling or cracking.

Site-applied surface treatments are only temporary. Following initial application, you should re-apply a preservative about every five years. You can also extend the life of any roof by periodically removing leaves and other debris which hold moisture. This reduces the potential for growth of moss and fungi.

Wood shingles and shakes are rigid, so they're extremely resistant to wind uplift (up to 130 mph). Cedar shingles have twice the insulating value of standard asphalt shingles, four times the value of cement shingles, and five times that of slate. And they increase the overall strength of the building. Depressions from hailstones disappear after a short time, because the wood fibers recover their original shape.

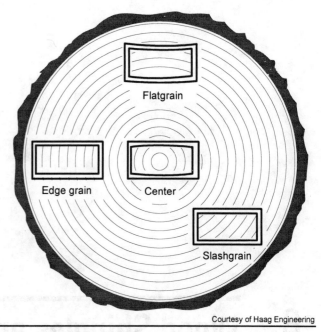

Figure 6-2 Wood shingle grain

How Wood Shingles and Shakes Are Made

Shingles are sawn, while shakes are split, or split and sawn. Shingles have a relatively smooth surface; shakes have at least one textured, natural-grain face. Wood shingles are usually pine, redwood or cedar.

Logs for wood shingles and shakes are cut into 16-, 18- and 24-inch lengths and shaped into blocks. Manufacturers try to produce blocks with an edge-grain face. (See Figure 6-2.)

Red Cedar Shingles

Red cedar trees produce fine, even-grained wood with a uniform texture and very few knots. Shingles are sawn from cedar blocks and have a straight, flat and uniform face. The five basic grades of red cedar shingles are shown in Figure 6-3. Western red cedar shingles are the most popular shingles. Red cedar shingles have several advantages, including:

- They weigh less than shakes, since they are thinner.
- They have a high crushing strength.

Grade	Length	Thickness (at butt)	# courses per bdl.	# bdls. or cartons per square
No. 1 Blue Label® - the premium grade of shingles for roofs and sidewalls. These top-grade shingles are 100% heartwood, 100% clear and 100% edge-grain. 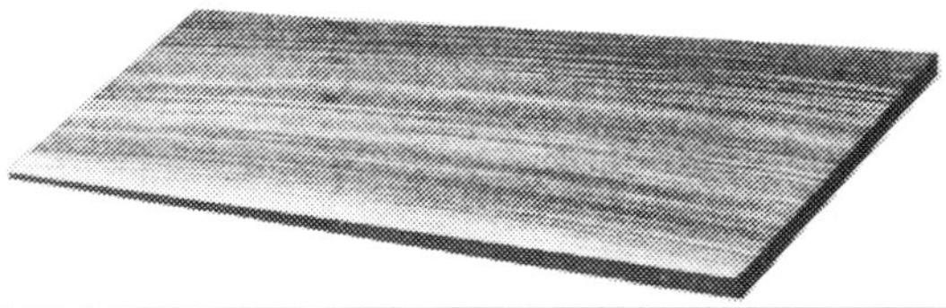	16" (Fivex) 18" (Perfections) 24" (Royals)	.40" .45" .50"	20/20 18/18 13/14	4 bdls. 4 bdls. 4 bdls.
No. 2 Red Label - a good grade for many applications. Not less than 10" clear on 16" shingles, 11" clear on 18" shingles and 16" clear on 24" shingles. Flat grain and limited sapwood are permitted in this grade. 	16" (Fivex) 18" (Perfections) 24" (Royals)	.40" .45" .50"	20/20 18/18 13/14	4 bdls. 4 bdls. 4 bdls.
No. 3 Black Label - a utility grade for economy applications and secondary buildings. Not less than 6" clear on 16" and 18" shingles, 10" clear on 24" shingles. 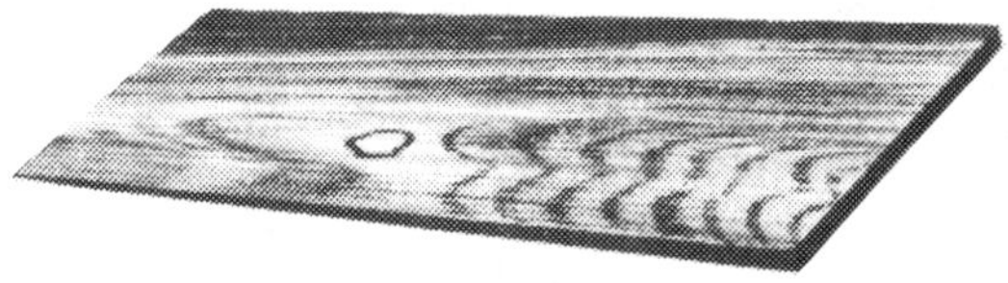	16" (Fivex) 18" (Perfections) 24" (Royals)	.40" .45" .50"	20/20 18/18 13/14	4 bdls. 4 bdls. 4 bdls.
No. 4 Undercoursing - a utility grade for starter course undercoursing. 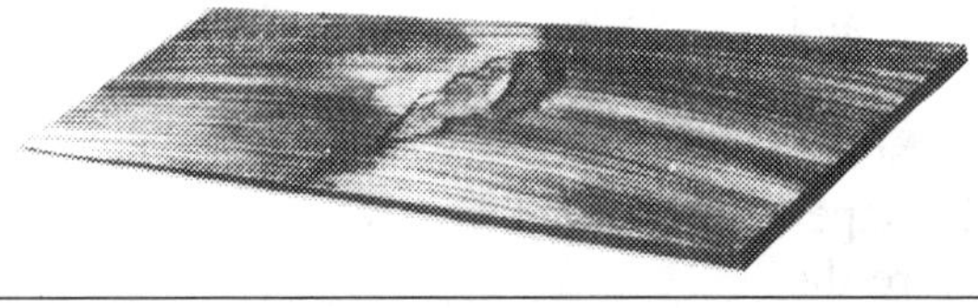	16" (Fivex) 18" (Perfections)	.40" .50"	14/14 or 20/20 14/14 or 18/18	2 bdls. 2 bdls.
No. 1 or No. 2 Rebutted and Rejointed - same specifications as above for No. 1 and No. 2 grades but machine trimmed for parallel edges with butts sawn at right angles. For sidewall applications where tightly fitting joints are desired. Also available with smooth sanded face. 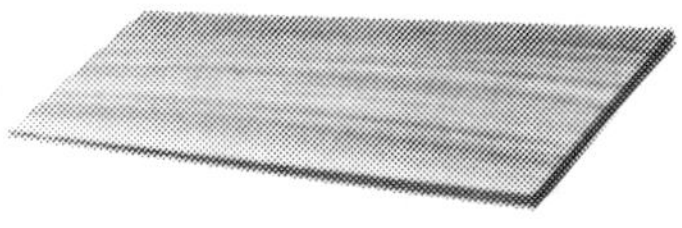	16" (Fivex) 18" (Perfections) 24" (Royals)	.40" .45" .50"	33/33 28/28 13/14	1 carton 1 carton 4 bdls.

Courtesy of Cedar Shake & Shingle Bureau

Figure 6-3 Certigrade® red cedar shingles

- They don't expand or contract significantly when the humidity changes.
- They don't absorb a lot of water.
- They have good thermal and acoustical properties.

I recommend Number 1 Grade (Blue Label) cedar shingles because they're naturally decay-resistant. Also, you'll waste less material during installation because these shingles don't break as easily as the lower grades.

Number 2 Grade (Red Label) cedar shingles are cut from wood that's free of knots and checks for the lower two-thirds of the wood from the butt. Normally, these are flatgrain and slashgrain shingles, cut from sections of the log, as shown in Figure 6-2.

Grade 3 (Black Label; utility or economy grade) shingles are clear in the lower third of the wood from the butt.

Ideally, shakes are 100 percent edge grain. But with today's grading standards, one bundle of Grade 1 (Blue Label) cedar shingles is about 73 percent clear vertical-grain (edge grain) heartwood, 20 percent flatgrain and 7 percent Number 2 Grade. Number 2 and 3 grades are OK for siding; but you'll break a lot of them because they're weak around the knots.

Don't install wood shingles on slopes less than 4 in 12 unless you use the special low-slope application described later in this chapter on page 170. You can install wood shingles on slopes between 3 in 12 and 4 in 12, provided you reduce the recommended exposure by 1½ inches.

Cedar Shakes

Figure 6-4 shows the five basic types of cedar shakes. Handsplit and resawn shakes are made with rough split faces and sawn backs. Tapersplit shakes are split along both faces so they have a rough face on both sides. Both the handsplit/resawn and tapersplit shingles have thick butts that taper to a thin head, as you can see in Figure 6-4. Straightsplit shakes are the same thickness throughout. Use them only over solidly-framed roofs because they're very heavy, especially when they're wet. One square of 16-inch shingles weighs about 144 pounds. Each square of 18-inch shingles weighs 158 pounds. The 24-inch shingles weigh 192 pounds per square. Shakes weigh about 200 to 450 pounds per square, depending on the shake length, butt thickness, and the exposure. (The shorter the exposure, the more shakes it takes to cover a square.)

Shakes have a rougher face than wood shingles so they don't shed water as easily. You don't need underlayment or interlayment with wood shingles, but you do with shakes.

Shakes have a higher insulating value, they're larger, and can be laid with a longer exposure than wood shingles. So you can cover an area faster with shakes than shingles.

Grade	Length and thickness	18" pack	
		# course per bdl.	# bdls. per sq.
No. 1 Handsplit & Resawn - these shakes have split faces and sawn backs. Cedar logs are first cut into desired lengths. Blanks or boards of proper thickness are split and then run diagonally through a band saw to produce two tapered shakes from each blank. 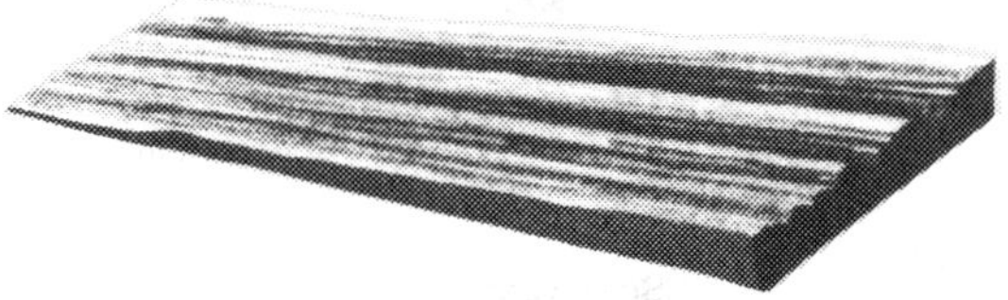	15" Starter-finish 18" x ½" Mediums 18" x ¾" Heavies 24" x ⅜" 24" x ½" Mediums 24" x ¾" Heavies	9/9 9/9 9/9 9/9 9/9 9/9	5 5 5 5 5 5
No. 1 Certi-sawn® (Tapersawn) - these shakes are sawn both sides. No. 2 and 3 are also available. For details contact the Bureau. 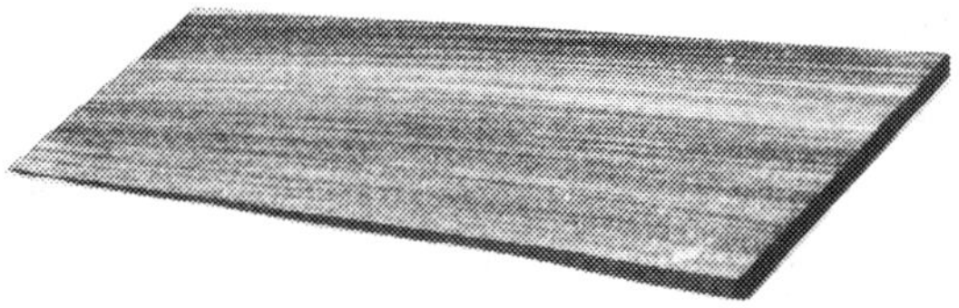	24" x ⅝" 18" x ⅝"	9/9 9/9	5 5
No. 1 Tapersplit - Produced largely by hand, using a sharp-bladed steel froe and a mallet. The natural shingle-like taper is achieved by reversing the block, end-for-end, with each split. 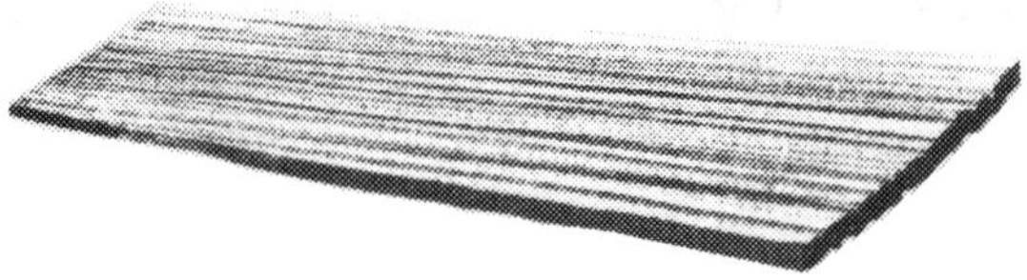	24" x ½"	9/9	5
		20" pack	
No. 1 Straightsplit - produced by machine or in the same manner as tapersplit shakes except that by splitting from the same end of the block, the shakes acquire the same thickness throughout. 	18" x ⅜" True-edge 18" x ⅜" 24" x ⅜"	14 Straight 19 Straight 16 Straight	4 5 5
No. 1 Machine Grooved - machine-grooved shakes are manufactured from shingles and have striated faces and parallel edges. Used double-coursed on exterior sidewalls. 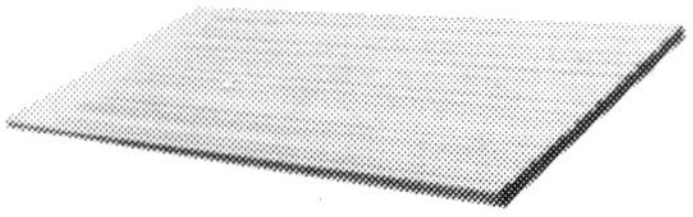	16" x .40" 18" x .45" 24" x .50"	16/17 14/14 12/12	2 ctns. 2 ctns. 2 ctns.

Courtesy of Cedar Shake & Shingle Bureau

Figure 6-4 Certi-split® red cedar shakes

Roofing Construction & Estimating

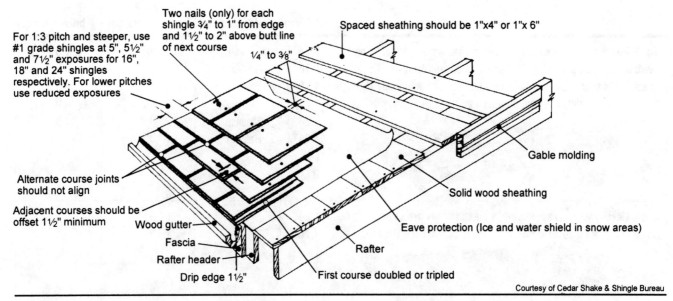

Figure 6-5 Shingles applied over spaced sheathing

In dry climates you can install hand-split shakes on a roof as flat as 4 in 12, provided you use the low-slope application method. In wet climates, I recommend a minimum slope of 6 in 12 so water will run off quickly. For more information on wood shingles or shakes, contact the Cedar Shake and Shingle Bureau, 515-116th Ave. NE, Suite 275, Bellevue, Washington, 98004-5294. Their telephone number is (206) 453-1323.

Installing Wood Shingles and Shakes

You can install wood shingles and shakes over solid or spaced sheathing, as discussed in Chapter 2. Figures 6-5 and 6-6 show installation details. In wet climates, it's better to use spaced sheathing because the increased air circulation reduces dampness. But always check the local codes, which may say you have to install solid sheathing for structural purposes, or in areas subject to wind-driven rain or snow.

After you've got the underlayment (or interlayment), drip edge and valley flashing material in place, install a starter course at the eaves of the roof. Allow a 1½-inch overhang beyond the eaves fascia and a ½- to ¾-inch overhang beyond the rake fascia. (Without a starter course, you'd only have single coverage at the eaves.)

Install the first course of wood shingles or shakes on top of the starter course. In heavy-snow regions, overlay the starter course with two layers of wood shingles or shakes. On shake roofs, use 15-inch wood shingles for

Wood Shingles and Shakes

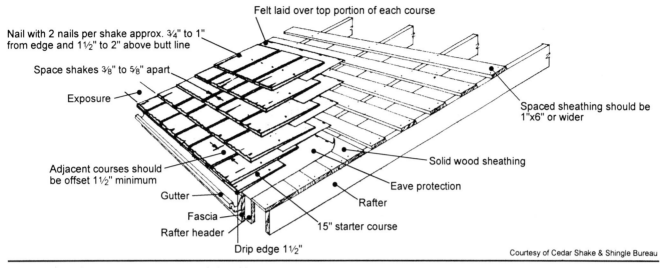

Figure 6-6 Shakes applied over spaced sheathing

the starter course and overlay them with a course of shakes. If you can't find 15-inch shakes to use as a starter course, you can cut the tops off longer shakes to make them fit.

To align the starter course and first course, nail down a shingle with the correct overhang on each end of the eaves. Then drive a nail into the butt of each shingle and stretch and tie a string between the nails. Align the butts of the intervening shingles along the string, as in Figure 6-7. Every third or fourth course, measure from the eaves up to the butts of end shingles and snap a chalk line to align the butts of the next course of shingles.

To allow for expansion when wet, space shakes $3/8$ to $5/8$ of an inch apart. Allow $1/4$ to $3/8$ inch for wood shingles. Offset adjacent shake or shingle courses by at least $1\frac{1}{2}$ inches. Don't let two joints line up directly in any three courses. In heavy-snow regions, increase the offset of adjacent courses to 2 inches. Treat knots and similar defects of a wood shingle as the edge of the shingle and align the joint in the course above at least $1\frac{1}{2}$ inches from the edge of the defect. That's shown in Figure 6-8.

Use only wood shingles that are from 3 to 8 inches wide. If a wood shingle or shake is too wide, split it with a roofer's hatchet. Saw-cutting takes too long.

Shake exposures vary, as you'll see later in Figure 6-23. Common exposures for 2-ply coverage are $7\frac{1}{2}$ inches for 18-inch shakes and 10 inches for 24-inch

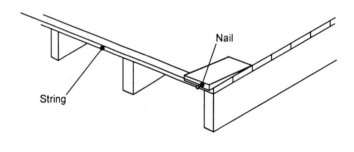

Figure 6-7 Lining up the first course of wood shingles or shakes

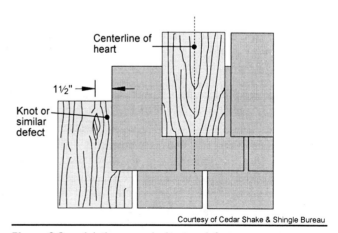

Figure 6-8 Jointing around a knot or defect

shakes. For 3-ply coverage, use exposures of 5½ inches for 18-inch shakes and 7½ inches for 24-inch shakes. Install 3-ply roofing in heavy-snow regions.

To avoid having to cut wood shingles or shakes at the ridge, shorten the exposures of the last few courses installed below the ridge. Try to reduce shake exposure approaching the ridge so you can use a 15-inch starter-finish shake immediately below the ridge. You'll have to cut these to fit if you can't get stock units.

Valleys

I prefer to shingle away from both sides of valleys and hips. It saves time because this way, you can cut all shingles adjacent to valleys and hips using the same pattern. Then you split shingles to fit where each course meets at mid-roof. Save wider shakes for finishing the valley edges.

However, if you're right-handed, you may be more comfortable working from left to right all the time. Shingle into the left side of a valley, then cut the last shingle or shake individually at the end of each course. Place a 1 x 4 in the valley as a cutting guide. Place the end shingle over the board, eyeball, and score a line, then cut the shingle with a portable power saw.

Now, shingle out from the right side of the valley. Since this is the starting point for all shingle courses, position the 1 x 4 and cut one shingle or shake and use it for a pattern to cut the rest of the starting shingles or shakes. Just be sure your starting shingles are different widths, or the joints will fall in a line. Reverse the working direction if you're left-handed.

Use the same basic principles when you shingle into a hip. When you begin shingling at a hip, use one shingle as a pattern for cutting the others. When you end a shingle course at a hip, you must cut each end shingle individually. Use the triangular pieces cut from shingles used in valleys to finish the hip edges.

Figure 6-9 shows yet another way to shingle valleys. Shingle into the valley from both sides, stopping the course before you get into the valley. Then install a pre-cut shingle at the end of the course. Finish the course by selecting (or cutting) a shingle or shake to fit between the pre-cut piece and the last full shingle or shake.

Install wood shingles or shakes to within 2 to 4 inches on each side of the centerline of a valley. Never allow joints between wood shingles or shakes to break into a valley; that is, be sure all joints have a solid shingle beneath and on top of them. And never lay shingles or shakes with the grain parallel with the centerline of a valley. Install wood shingles or shakes so they lap at least 7 inches over each side of the valley flashing. Drive nails at least 2 inches from the edge of the metal. Install wood shingles or shakes at the top of the valley at least 4 inches from each side of the centerline. Increase this distance about ⅛ inch per foot down the valley. Snap chalk lines down both sides of the valley for guidelines.

Wood Shingles and Shakes

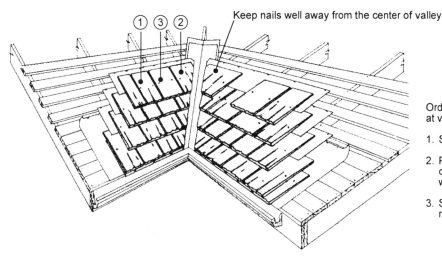

Figure 6-9 Installing wood shingles or shakes in a valley

Nailing Shingles and Shakes

Regardless of the width of a wood shingle or shake, drive two nails about ¾ to 1 inch from each side of the unit, and 1½ to 2 inches above the butt of the succeeding course. If the shake or shingle splits, drive two nails into each piece. Use stainless steel, hot-dipped galvanized nails. Bright, blue, or steel wire nails don't mix with wood preservatives. They'll quickly dissolve. And aluminum nail heads break off. Figure 6-10 shows the nail types and minimum lengths you should use to install wood shingles and shakes.

Use 5d or 6d nails for wood shingle re-roofs. Use 7d or 8d nails on shake re-roofs. When in doubt, use nails long enough to go ½ inch into the deck. But check your local code. Some building codes require ¾ inch. Use ring-shank nails when plywood sheathing is less than a half-inch thick.

Nails are better than staples because staples make twice as many punctures. Also, if you don't hold the staple gun correctly or if it's out of adjustment, you can damage the shingles. If you use staples, make sure they're 16-gauge aluminum or stainless steel with a 7/16-inch (minimum) crown. Don't use blue-steeled fasteners. Make sure the staple is long enough to go at least ½ inch into the sheathing, and that its crown is flush with the surface of the wood shingle or shake.

Shingling Around Vents

When you come to a vent pipe, use a saber saw or keyhole saw to notch the wood shingles or shakes that meet the pipe so they'll fit closely around it. Or, you can install shingles individually to fit around the pipe. Slip

Type of shingle or shake	Nail type and minimum length	
Shingles - New Roof	Type	(in.)
16" and 18" shingles	3d box	1¼
24" shingles	4d box	1½
Shakes - New Roof	Type	(in.)
18" Straight-split	5d box	1¾
18" and 24" handsplit-and-resawn	6d box	2
24" tapersplit	5d box	1¾
18" and 24" taper-sawn	6d box	2

Courtesy of Cedar Shake & Shingle Bureau

Figure 6-10 Nails used to install wood shingles and shakes

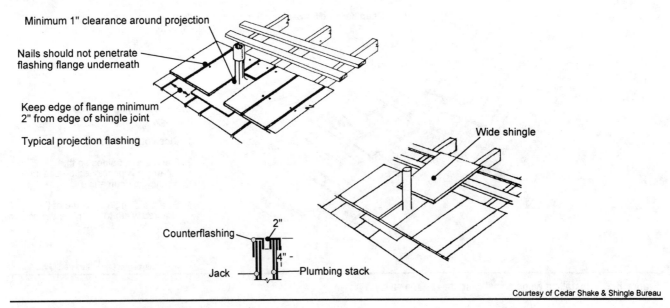

Figure 6-11 Installing wood shingles at a vent pipe

the vent flashing over the pipe and the adjacent shingles. In either case, leave the flashing area open below the pipe so that debris doesn't collect. That's shown in Figure 6-11. As added protection, install a felt strip overlaid with a wide shingle immediately above the pipe. Drop it below the butt line of other shingles in the same course if aligning it with the rest of the course would leave a gap above the pipe.

Installing Shingles or Shakes at Chimneys

Figure 6-12 illustrates flashings around chimneys. Saddle flashing goes upslope of the chimney, apron flashing goes on the downslope side. You can use a cricket flashing instead of saddle flashing. Look back at Figure 4-52 in Chapter 4 for a picture of a cricket.

Extend the apron flashing at least 3 inches up the vertical surface. It should also go at least 1½ times the shingle or shake exposure (6 inches minimum) over the roof slope. Carry cricket flashing at least 10 inches under the shingles or shakes. Extend step flashing over the roof at least 3 inches and up the chimney. It should be covered by at least 4 inches of the counterflashing. Lap each step flashing over the next piece by at least 3 inches. Install counterflashing so it extends down within 1 inch of the finished roof surface.

Dormers

At dormers, extend the apron flashing up the walls at least 3 inches under sheathing paper and at least 3 inches over the roof slope. Install step flashing as prescribed above, using shingles or siding for counterflashing. You can see that in Figure 6-13.

Wood Shingles and Shakes

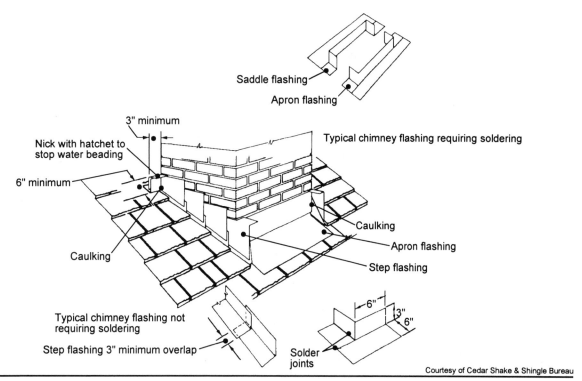

Figure 6-12 Typical chimney flashing

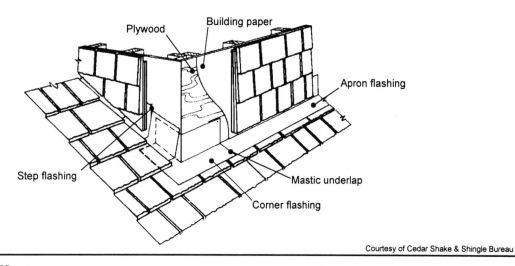

Figure 6-13 Dormer flashing

Hips and Ridge Units

You can field-fabricate hip and ridge units, but it's faster to use prefabricated units which come with alternating mitered joints and concealed nailing (staples, usually two, which hold the units together). Install the hip units before the ridge units. Install double starter units over the first course of shingles or shakes at the low end of each hip, as in Figure 6-14. Then temporarily install a hip unit at the top of the hip. Snap chalk lines on each side of the hip along the edges of the two hip units. The chalk lines serve as guides for positioning the rest of the hip units. Install hip units starting at the low end of each hip, alternating the direction of the miter joints. Trim the top hip units so they meet at the ridge with their edges flush. Use the same exposure you used for the shingles to install hip and ridge units.

Install ridge units starting at both ends of the ridge, working toward the center of the roof. Create a saddle to cover where both courses meet.

Use two 8d or 10d galvanized or aluminum nails to install each hip and ridge unit. When in doubt, use nails that are long enough to go at least ½ inch into the deck.

I recommend you apply a strip of 30-pound felt or kraft paper under hip and ridge units. It's also a good idea to install concealed metal hip and ridge flashing, as in Figure 6-15. A wood roof lasts a long time, but it's only as good as its underlayment. I consider these steps a good precaution to prevent early failure of a wood roof, especially a shake roof.

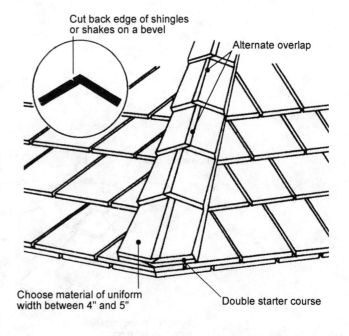

Figure 6-14 Hip and ridge application

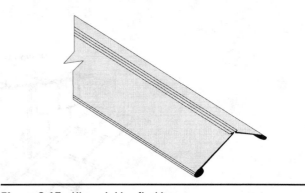

Figure 6-15 Hip and ridge flashing

Low-Slope Applications

The minimum recommended roof slope for shakes is 4 in 12, and 3 in 12 for wood shingles. However, you can apply shakes and shingles over solid sheathing to lower slopes if you first cover the sheathing using any one of the following methods:

1) Nail down a 43-pound organic sheet or a 28-pound fiberglass sheet followed by a cold-applied surface coat of asphalt or ice shield material.

2) Nail down two plies of 30-pound felt followed by cold-applied surface coat of asphalt or ice shield material.

3) Hot-mop two plies of 15-pound felt followed by a cold-applied surface coat of asphalt or ice shield material.

4) Nail down a 30-pound felt followed by hot-mopping 90-pound mineral-surfaced roll roofing.

When you use methods 1, 2 or 3, embed treated 2 x 4 spacers into the hot asphalt surface coat on 24-inch centers extending from the eaves to the ridge. If you use method 4, you can embed the 2 x 4s into cold-applied roofing cement.

Wood Shingles and Shakes

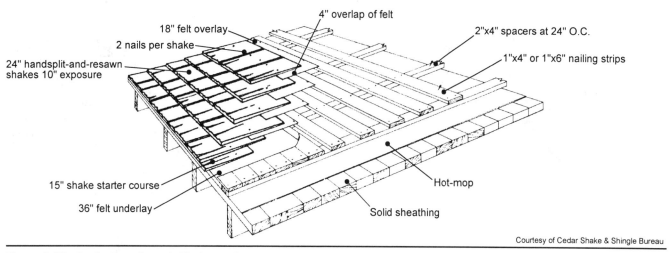

Figure 6-16 Application of wood shingles or shakes over low-slope roofs

Nail the 2 x 4s at the eaves and ridge only. This will be enough to hold the spacers in place. Also, avoid driving nails into the built-up roof because every puncture provides a path for water to leak through. If you *have* to put in additional nailing, seal the penetrations well with roofing cement.

Next, install 1 x 4 or 1 x 6 nailing strips, spaced according to the weather exposure selected (see Chapter 2, page 29 for spacing requirements). Nail the strips across the spacers to form a lattice-like nailing base. Finally, install wood shingles or shakes in the normal manner. This is shown in Figure 6-16.

Steep-Slope Applications

Wood shingles or shakes are often applied to mansard roofs over panel sheathing or spaced battens. Construction details for typical mansard roofs are shown in Figure 6-17.

Swept or Bell Eaves

You may have to soak (usually overnight) or steam the shakes or shingles you install over swept or bell eaves so you can bend them. See Figure 6-18. Install the roof in the usual manner above the sweep.

Installing Wood Shingles or Shakes over Rigid Insulation

Don't nail wood shingles or shakes through rigid insulation into the deck below. You'll create several problems:

- Since you'll have to use longer nails, they'll have thicker shanks. They'll split shingles and shakes you drive them through.

Roofing Construction & Estimating

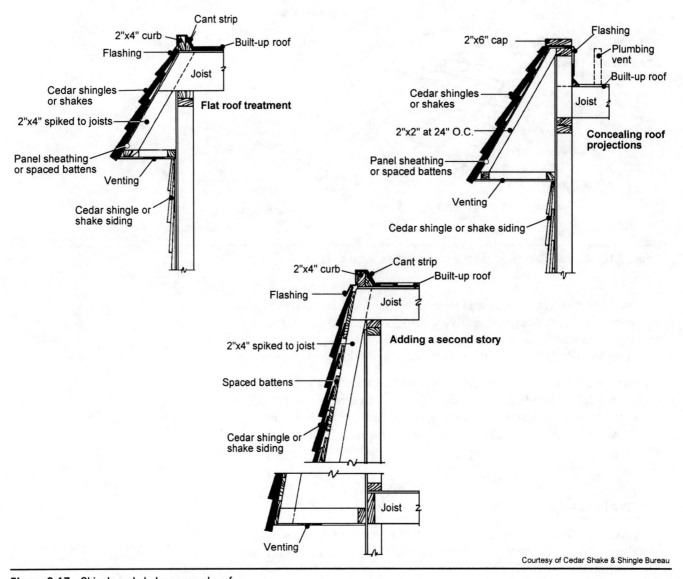

Figure 6-17 Shingle and shake mansard roofs

- The nails expand and contract, enlarging the holes in the insulation and reducing its efficiency.
- Nails allow the shingles and underlayment to move, causing leaks.

To eliminate these problems, install a plywood or *false deck* (shown in Figure 6-19) or horizontal strapping, as in Figure 6-20, immediately over the insulation. Horizontal strapping is 1 x 4s or 1 x 6s installed on centers as prescribed for spaced sheathing. Use the false deck or strapping as a nailing surface for the wood shingles or shakes, and as an air space beneath the shingles or shakes for good air circulation.

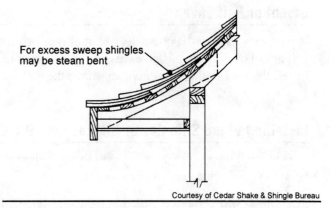

Figure 6-18 Swept or bell eaves

172

Wood Shingles and Shakes

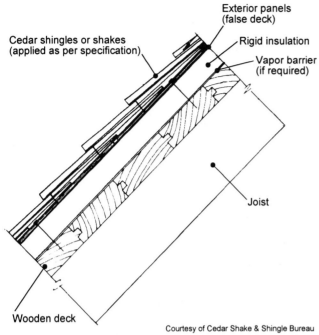

Figure 6-19 Panels over insulation

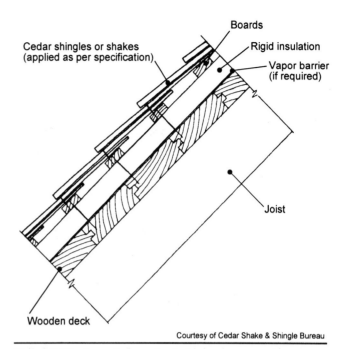

Figure 6-20 Strapping over insulation

Although you nail the false deck and strapping through the insulation into the sheathing beneath, far fewer nails penetrate the insulation. To eliminate nails penetrating the insulation or to avoid the possibility of a steep roof creeping downward under a heavy snow load, install 2 x 4s on edge from eaves to ridge between insulation panels. You can then install the false deck or strapping to the 2 x 4s, as in Figure 6-21.

Installing Wood Shingles or Shakes over a Metal Deck

Over a metal deck, fasten wood shingles or shakes to a false deck or horizontal strapping as described above. Fasten the false deck or strapping to 2-inch-wide lumber you install from eaves to ridge, as shown in Figure 6-22. You can fasten the lumber to the metal deck with screws or bolts. On decks with shallow corrugations, you might have to use clip angles (small "L" brackets) to attach the lumber to the deck. If so, nail or screw the clips to the lumber and screw or bolt the clips to the deck.

If you install a vapor barrier, put it *under* the rigid insulation. Never *choose* to install lumber between two vapor barriers. You may have to do it to comply with an architect's design. If that's the case, use preservative-treated lumber.

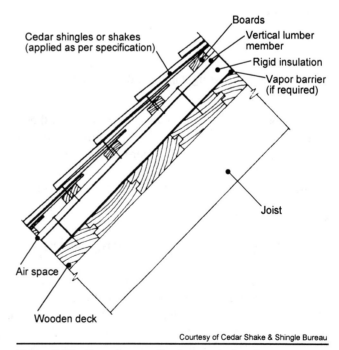

Figure 6-21 Strapping over 2 x 4s turned edgewise

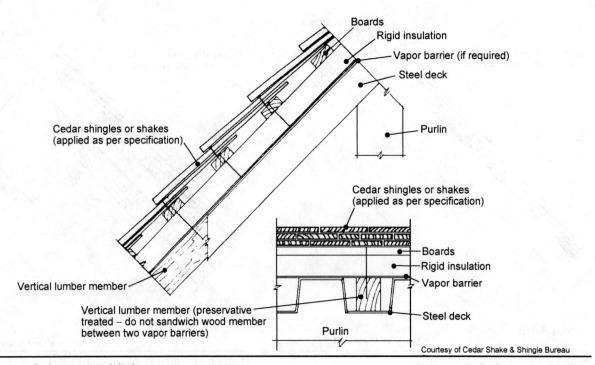

Figure 6-22 Installation over steel deck

Covering Capacity of Shakes

Shakes usually come in 5-bundle squares. Five bundles installed with a 10-inch exposure will cover 100 square feet of roof area (Figure 6-23). Exceptions are:

- True-edge straightsplit shakes, which come in 4-bundle squares (14-inch exposure)

- 18" x 3/8" straightsplit shakes, which come in 5-bundle squares (8½-inch exposure).

Shakes provide double coverage if you install them at the recommended exposures. Figure 6-23 shows that for most shakes, five bundles will cover 100 square feet (20 square feet per bundle) when laid at a 10-inch exposure. But five bundles will cover only 85 square feet (17 square feet per bundle) at an 8½-inch exposure. Use Figure 6-24 to estimate quantities.

Covering Capacity of Wood Shingles

Wood shingles generally come in 4-bundle squares. (Turn back to Figure 6-3 for exceptions.) Figure 6-25 shows the maximum recommended exposures for various wood shingle grades. Wood shingles, applied at the correct exposure, provide triple coverage. Widths will vary in a particular bundle. Lower-grade shingles are typically narrower than the higher grade ones and cost more to install.

Wood Shingles and Shakes

Grade	Length & Thickness	Approximate coverage (square feet) of 5 bundles (4 bundles in the case of true-edge straight split) of shakes when applied with ½-inch spacing, at the following exposures (in inches).									
		5½	7½	8	8½	10	11½	12	14	16	18
No. 1 Handsplit and Resawn	18" x ½"	55[a]	75[bcd]	80	85[e]	100	115	120	140[f]		
	18" x ¾"	55[a]	75bcd	80	85[e]	100	115	120	140f		
	24" x ⅜"		75[ac]	80	85	100[bd]	115[e]	120	140	160	180[f]
	24" x ½"		75[a]	80	85	100[bcd]	115[e]	120	140	160	180[f]
	24" x ¾"		75[a]	80	85	100[bcd]	115e	120	140	160	180[f]
No. 1 Tapersawn	18" x ⅝"	55[a]	75[bcd]	80	85[e]	100	115	120	140[f]		
	24" x ⅝"		75[a]	80	85	100[bcd]	115e	120	140	160	180[f]
No. 1 Tapersplit	24" x ½"		75[a]	80	85	100[bcd]	115e	120	140	160	180[f]
No. 1 Straight-split	18" x ⅜"	65[a]	88bcd	94	100[e]	118	135	141	165	188	
	24" x ⅜"		75[a]	80	85	100[bcd]	115[e]	120	140	160	180[f]
True-edge	18" x ⅜"	39	53	56	60[e]	70	81	84	98	112[f]	
15" Starter-finish	Use supplementary with shakes applied not over 10" exposure										

[a] Maximum recommended weather exposure for 3-ply roof construction.
[b] Maximum recommended weather exposure for 2-ply roof construction.
[c] Maximum recommended weather exposure for application on roof slopes between 4 in 12 and 8 in 12.
[d] Maximum recommended weather exposure for application on roof slopes of 8 in 12 and steeper.
[e] Maximum recommended weather exposure for single-coursed wall construction.
[f] Maximum recommended weather exposure for double-coursed wall construction. Use a 24-inch cedar shingle for underlay.

Figure 6-23 Covering capacity of shakes

Grade	Bundles of shakes required per square when applied with ½-inch spacing, at the following exposures (in inches).											
	5½	7½	8	8½	10	11½	12	14	16	18	20	22
No. 1 Handsplit and Resawn, No. 1 Tapersawn, No. 1 Tapersplit, and No. 1 Straight-split (24")	9.09	6.67	6.25	5.88	5.00	4.35	4.17	3.57	3.13	2.78	2.50	2.27
No. 1 Straight-split (True-edge)	10.26	7.55	7.14	6.67	5.71	4.94	4.76	4.08	3.57			
No. 1 Straight-split (18" long)	7.69	5.68	5.32	5.00	4.24	3.70	3.55	3.03	2.66			

Figure 6-24 Bundles of shakes per square

Roof Slope	Number 1 Blue Label			Number 2 Red Label			Number 3 Black Label		
	16"	18"	24"	16"	18"	24"	16"	18"	24"
3/12 - 4/12	3¾"	4¼"	5¾"	3½"	4"	5½"	3"	3½"	5"
4/12 and Steeper	5"	5½"	7½"	4"	4½"	6½"	3½"	4"	5½"

Figure 6-25 Maximum recommended exposures for wood shingles

Length and Thickness	Approximate coverage (square feet) of 4 bundles of wood shingles when applied at the following exposures (in inches)												
	3	3½	3¾	4	4¼	4½	5	5½	5¾	6	6½	7	7½
16" x 5/2"[i]	60[a]	70[bd]	75[c]	80[e]	85	90	100[f]	110	115	120	130	140	150[g]
18" x 5/2¼"		63.6[a]	68.2	72.7[bd]	77.3[c]	81.8[e]	90.9	100[f]	104.5	109.1	118.2	127.3	136.4
24" x 4/2"							66.7[a]	73.3[bd]	76.7[c]	80	86.7[e]	93.3	100[f]
Length and Thickness	8	8½	9	9½	10	10½	11	11½					
16" x 5/2"	160	170	180	190	200	210	2.20	2.30					
18" x 5/2¼"	145.5	154.5[g]	163.6	172.7	181.8	190.9	200	209.1					
24" x 4/2"	106.7	113.3	120	126.7	133.3	140	146.7	153.3[g]					
Length and Thickness	12	12½	13	13½	14	14½	15	15½	16				
16" x 5/2"	240[h]												
18" x 5/2¼"	218.2	227.3	236.4	245.5	254.5								
24" x 4/2"	160	166.7	173.3	180	186.7	193.3	200	206.7	213.3[h]				

[a] Maximum recommended weather exposure for No. 3 grades applied on roof slopes between 3 in 12 and 4 in 12.
[b] Maximum recommended weather exposure for No. 2 grades applied on roof slopes between 3 in 12 and 4 in 12.
[c] Maximum recommended weather exposure for No. 1 grades applied on roof slopes between 3 in 12 and 4 in 12.
[d] Maximum recommended weather exposure for No. 3 grades applied on roof slopes 4 in 12 and steeper.
[e] Maximum recommended weather exposure for No. 2 grades applied on roof slopes 4 in 12 and steeper.
[f] Maximum recommended weather exposure for No. 1 grades applied on roof slopes 4 in 12 and steeper.
[g] Maximum recommended weather exposure for single-coursing No. 1 grades on sidewalls. Reduce exposure for No. 2 grades.
[h] Maximum recommended weather exposure for double-coursing No. 1 grades on sidewalls.
[i] Sum of thickness; e.g., 5/2" means that 5 butts stacked have a total thickness of 2 inches.

Figure 6-26 Covering capacity of wood shingles

Referring to Figure 6-26, 16-inch wood shingles are packaged so 4 bundles will cover 100 square feet if applied at a 5-inch exposure. However, 4 bundles will cover only 90 square feet if applied at a 4½-inch exposure. The 18- and 24-inch shingles will cover 100 square feet if applied at 5½- and 7½-inch exposures, respectively. Use Figure 6-27 to estimate quantities.

Estimating Wood Shingle and Shake Quantities

In the following examples, the equations are the same as those for asphalt shingles, but the material dimensions are different.

First, you take off shingle quantities by the square foot. Then convert that figure to bundles required. The total required must include material for the starter course, hip and ridge units, cutting waste at the rakes, hips and valleys, and crew-error waste.

Length and Thickness	Bundles of wood shingles required per square when applied at the following exposures (in inches)												
	3	3½	3¾	4	4¼	4½	5	5½	5¾	6	6½	7	7½
16" x 5/2"[a]	6.67	5.71	5.33	5.00	4.71	4.44	4.00	3.64	3.48	3.33	3.08	2.86	2.67
18" x 5/2¼"		6.29	5.87	5.50	5.18	4.89	4.40	4.00	3.83	3.67	3.38	3.14	2.93
24" x 4/2"							6.00	5.46	5.22	5.00	4.61	4.29	4.00
Length and Thickness	8	8½	9	9½	10	10½	11	11½					
16" x 5/2"	2.50	2.35	2.22	2.11	2.00	1.91	1.82	1.74					
18" x 5/2¼"	2.75	2.59	2.45	2.32	2.20	2.10	2.00	1.91					
24" x 4/2"	3.75	3.53	3.33	3.16	3.00	2.86	2.73	2.61					
Length and Thickness	12	12½	13	13½	14	14½	15	15½	16				
16" x 5/2"	1.67												
18" x 5/2¼"	1.83	1.76	1.69	1.63	1.57								
24" x 4/2"	2.50	2.40	2.31	2.22	2.14	2.07	2.00	1.94	1.88				

[a]Sum of thicknesses; e.g., 5/2" means that 5 butts stacked have a total thickness of 2 inches.

Figure 6-27 Bundles of wood shingles per square

Starter Course

Look back at Figures 6-5 and 6-6 and you see that shakes and shingles are both installed with a double starter course. From Chapter 4, on asphalt shingles, we know from Equation 4-10 that:

Starter Course (SF) = Eaves (LF) x Exposed Area (SF/LF)

The starter course and first course overhang the eaves and rake to help shed water from the roof. Add a half inch (0.04 square feet per linear foot) to the actual roof size along each rake for the additional material you need for the overhang. Then add 1½ inches (0.125 square feet per linear foot) to the roof size along each eave.

Cutting Waste at Rakes, Hips and Valleys

Add 1 square foot of area to be covered per linear foot of hip or valley and 0.34 square feet of area per linear foot of rake for cutting waste and breakage. You may need to allow more than that if it's a complex roof or if you're using lower-grade shingles (lower grades break more easily). And if you're using a less experienced roofing crew, add a little more.

Ridge and Hip Units

Use the following equation to determine the number of prefabricated hip and ridge units required:

Hip and Ridge Units $= \dfrac{\text{Total LF Ridge and Hips}}{\text{Exposure (ft.)}}$ 　　**Equation 6-1**

You must also add one unit at each hip for the starter course:

Total Units $= \dfrac{\text{Total LF Ridge and Hips}}{\text{Exposure (ft.)}} + \text{Number of Hips}$ 　　**Equation 6-2**

Prefabricated wood shingle and shake ridge and hip units are sold by the bundle. Each bundle contains 40 units. If the units are field-fabricated, add 1 square foot for each linear foot of ridge and hips.

A Shortcut Method for Determining Wood Shingle and Shake Waste

As a rule of thumb, some contractors add:

- 10 percent waste to the net shingle or shake quantities on small to average-size gable roofs
- 15 percent on hip roofs with field-fabricated hip and ridge units
- 5 percent on gable roofs with prefabricated hip and ridge units
- 10 percent on hip roofs with prefabricated hip and ridge units

This includes material required for the starter course and site-fabricated ridge and hip units (if used), cutting waste at rakes, hips and valleys, and waste due to crew errors. Sometimes you'll have to increase the allowance for more complex roofs.

I prefer a more accurate, realistic way to estimate waste which depends on:

- roof type
- ratio of roof length to roof width
- roof slope
- shingle exposure
- type of shingle installed (lower grades break more easily)
- whether hip and ridge units are prefabricated or site-constructed

In Figures 6-28 and 6-29 you see that the waste on shake or wood shingle roofs, exclusive of crew errors, is from 3 to 17 percent on average-sized gable roofs, and from 6 to 33 percent on hip roofs. Use Figures 6-28 and 6-29 to get a "ball-park" percentage for waste.

Wood Shingles and Shakes

Total percentage to be added to net roof area, including double starter course, cutting waste at rakes and additional material required for overhang.											
Building dimension	3 in 12 Roof slope			6 in 12 Roof slope			12 in 12 Roof slope				
	Exposure			Exposure			Exposure					
(L x W)	5"	7.5"	10"	5"	7.5"	10"	5"		7.5"		10"	
30 x 20	8* (13**)	10* (15**)	12* (17**)	8* (13**)	9* (15**)	11* (16**)	7*	(11**)	8*	(12**)	10*	(14**)
40 x 30	6 (9)	8 (11)	9 (12)	6 (9)	8 (11)	9 (12)	5	(7)	6	(8)	7	(9)
45 x 30	6 (9)	8 (11)	9 (12)	6 (9)	8 (11)	9 (12)	5	(7)	6	(8)	7	(9)
50 x 30	5 (8)	7 (10)	8 (11)	5 (8)	7 (10)	8 (11)	4	(6)	5	(7)	6	(8)
60 x 30	5 (8)	7 (10)	8 (11)	5 (8)	7 (10)	8 (11)	4	(6)	5	(7)	6	(8)
70 x 30	5 (8)	7 (10)	8 (11)	5 (8)	7 (10)	8 (11)	4	(6)	5	(7)	6	(8)
80 x 30	5 (8)	7 (10)	8 (11)	5 (8)	7 (10)	8 (11)	4	(6)	5	(7)	6	(8)
50 x 40	4 (6)	5 (7)	6 (8)	4 (6)	5 (7)	6 (8)	4	(6)	5	(7)	6	(8)
60 x 40	4 (6)	5 (7)	6 (8)	4 (6)	5 (7)	6 (8)	4	(6)	5	(7)	6	(8)
70 x 40	4 (6)	5 (7)	6 (8)	4 (6)	5 (7)	6 (8)	4	(6)	5	(7)	6	(8)
80 x 40	4 (6)	5 (7)	6 (8)	4 (6)	5 (7)	6 (8)	4	(6)	5	(7)	6	(8)
90 x 40	4 (6)	5 (7)	6 (8)	4 (6)	5 (7)	6 (8)	4	(6)	5	(7)	6	(8)
60 x 50	4 (6)	5 (6)	6 (8)	4 (6)	5 (7)	6 (8)	3	(4)	4	(5)	4	(5)
70 x 50	4 (6)	5 (7)	6 (8)	4 (6)	5 (7)	6 (8)	3	(4)	4	(5)	4	(5)
80 x 50	4 (6)	5 (7)	6 (8)	4 (6)	5 (7)	6 (8)	3	(4)	4	(5)	4	(5)
90 x 50	4 (6)	5 (7)	6 (8)	4 (6)	5 (7)	6 (8)	3	(4)	4	(5)	4	(5)

* Indicates the use of prefabricated ridge units. These units must be taken off separately.
** Parentheses indicate the use of site-fabricated ridge units. These unit are included in the data given in this table.

Figure 6-28 Cutting waste and overruns (gable roof)

Building dimension (L x W)	3 in 12 Roof slope						6 in 12 Roof slope						12 in 12 Roof slope					
	Exposure						Exposure						Exposure					
	5"		7.5"		10"		5"		7.5"		10"		5"		7.5"		10"	
30 x 20	18*	(28**)	21*	(31**)	23*	(33**)	17*	(27**)	20*	(30**)	23*	(33**)	15*	(23**)	17*	(26**)	20*	(29**)
40 x 30	14	(22)	16	(24)	19	(27)	13	(21)	15	(23)	17	(25)	12	(19)	14	(21)	16	(23)
45 x 30	12	(19)	14	(21)	17	(24)	11	(18)	13	(20)	15	(22)	9	(15)	11	(17)	12	(18)
50 x 30	11	(18)	13	(20)	15	(22)	10	(17)	12	(19)	14	(21)	9	(15)	11	(17)	12	(18)
60 x 30	10	(16)	12	(18)	14	(20)	10	(16)	12	(18)	14	(20)	8	(13)	10	(15)	11	(16)
70 x 30	9	(15)	11	(17)	13	(19)	9	(15)	11	(17)	13	(19)	8	(13)	10	(15)	11	(16)
80 x 30	9	(15)	11	(17)	13	(19)	7	(12)	9	(14)	10	(15)	7	(12)	9	(14)	10	(15)
50 x 40	11	(17)	13	(19)	15	(21)	9	(15)	11	(17)	12	(18)	9	(14)	11	(16)	12	(17)
60 x 40	9	(15)	11	(17)	12	(18)	9	(14)	11	(16)	12	(17)	8	(13)	10	(15)	11	(16)
70 x 40	8	(13)	10	(15)	11	(16)	8	(13)	10	(15)	11	(16)	7	(11)	8	(12)	9	(13)
80 x 40	8	(13)	10	(15)	11	(16)	7	(12)	9	(14)	10	(15)	6	(10)	7	(11)	8	(12)
90 x 40	7	(12)	9	(14)	10	(15)	7	(11)	9	(13)	10	(14)	6	(10)	7	(11)	8	(12)
60 x 50	9	(14)	11	(16)	12	(17)	9	(14)	11	(16)	12	(17)	8	(11)	8	(12)	9	(13)
70 x 50	8	(13)	10	(15)	11	(16)	8	(12)	10	(14)	11	(15)	7	(11)	8	(12)	9	(13)
80 x 50	8	(12)	10	(14)	11	(15)	6	(10)	7	(11)	8	(12)	6	(10)	7	(11)	8	(12)
90 x 50	7	(11)	9	(13)	10	(14)	6	(10)	7	(11)	8	(12)	6	(9)	7	(10)	8	(11)

Total percentage to be added to net roof area, including double starter course, cutting waste at hips and additional material required for overhang.

* Indicates the use of prefabricated ridge units. These units must be taken off separately.
** Parentheses indicate the use of site-fabricated hip and ridge units. These unit are included in the data given in this table.

Figure 6-29 Cutting waste and overruns (hip roof)

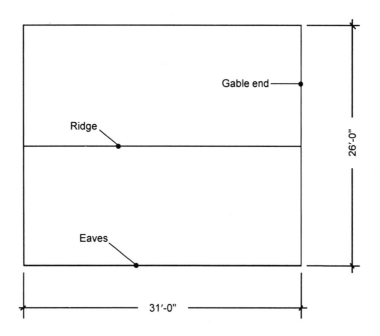

Figure 6-30 Gable roof example

▼ **Example 6-1:** Assume a roof slope of 5 in 12, and find the number of bundles of No. 1 grade (16" Fivex) wood shingles required to cover the roof of the gable roof in Figure 6-30. Also assume a double starter course at the eaves and the use of prefabricated ridge units:

From Figure 6-25 you see that the maximum recommended exposure for this 5 in 12 roof is 5 inches. Therefore, use the following equation (from Chapter 1) to find the net roof area:

Actual (Net) Roof Area = Roof Plan Area x Roof-Slope Factor

Net Roof Area
= (31' x 26' x 1.083) (from Column 2 of Appendix A)
= 873 square feet ÷ 100
= 8.73 squares (use this figure to estimate labor costs)

The total eaves length is 2 x 31, or 62 linear feet.

The total area of starter-course material required is 62' x 0.42 SF/LF (5-inch exposure divided by 12), or 26 square feet from the equation:

Starter Course (SF) = Eaves (LF) x Exposure Area (SF/LF)

The job requires 4 x 13' x 1.083, or 56 linear feet of material for the rakes (4 rakes at 13 feet each, times the roof-slope factor from Appendix A).

Figure waste for breakage at the rakes at 19 square feet (56 LF times 0.34 SF/LF).

Waste allowance for the overhang at the rakes is 56' x 0.04 = 2.24 square feet (½ inch per actual linear foot of rake).

Waste allowance for the overhang at the eaves is 62' x 0.125 = 7.75 square feet (1½ inches per linear foot of eaves).

Therefore, total waste for rakes and eaves is 10 square feet.

Add these to get the gross roof area:

= 873 SF + 26 SF + 19 SF + 10 SF
= 928 square feet ÷ 100
= 9.28 squares

Divide the net roof area by that answer to confirm the waste allowance:

$$\text{Waste Factor} = \frac{9.28 \text{ Squares}}{8.73 \text{ Squares}} = 1.06$$

Figure 6-28 predicts a range from 6 to 8 percent waste, so this is correct.

Figure 6-27 says this material comes in 4 bundles per square, so the job requires 38 bundles of shingles (9.28 x 4).

The ridge length is 31 linear feet, so from Equation 6-1, here's how to figure how many prefabricated ridge units to order:

$$\text{Ridge Units} = \frac{31'}{0.42' \text{ (5–inch exposure divided by 12)}} = 74 \text{ each}$$

Since these come 40 to a bundle, the job requires 2 bundles of ridge units.

If you had used field-fabricated ridge units, you'd need an additional 31 square feet (1 SF/LF of ridge), or 959 square feet. That's 9.59 squares, or 39 bundles (at 4 bundles per square).

Now, here's an example showing how this works for a hip roof:

▼ **Example 6-2:** Assuming a roof slope of 6 in 12, find the number of bundles of No. 2 grade (18" Perfections) wood shingles required for the hip roof in Figure 6-31. Assume a double starter course at the eaves and lower ends of the hips, and the use of prefabricated ridge and hip units.

Figure 6-25 says that the maximum recommended exposure for this type of shingle on a 6 in 12 roof is 4½ inches.

The net roof area is 40' x 20' x 1.118 (Column 2 of Appendix A), or 894 square feet. Base your labor cost on 8.94 squares of material.

The total eaves length is 2 x (40' + 20'), or 120 linear feet.

The total area of starter-course material required is 120' x 0.38 (4½-inch exposure divided by 12), or 46 square feet.

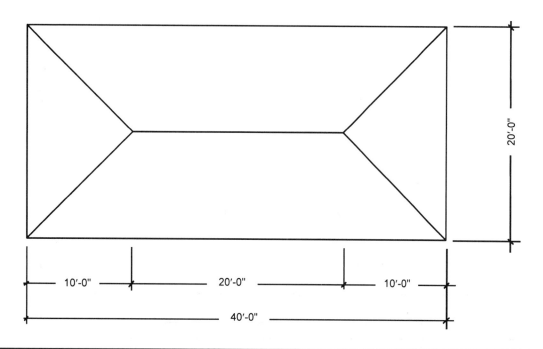

Figure 6-31 Hip roof example

The length of each hip is 10 x 1.5 (from Column 3 of Appendix A) or 15 linear feet. So there's a total of 60 linear feet of hip.

The cutting waste at the hips is 1 square foot per linear foot, or 60 square feet.

The additional material required for the 1½-inch overhang at the eaves is 15 square feet (120' x 0.125 square feet per linear foot).

Therefore, the gross roof area is 894 SF + 46 SF + 60 SF + 15 SF, which equals 1,015 square feet, or 10.15 squares of material.

Figure 6-29 predicts a range of waste from 13 to 17 percent. This example calculates to 14 percent (1015 ÷ 894).

From Figure 6-27, you see that these shingles come 4.89 bundles per square, so you have to order 50 bundles (10.15 squares x 4.89).

The total length of ridge and hips is 80 linear feet, so from Equation 6-2 the job requires 215 units (80 divided by 0.38, plus 4 units).

There are 40 units per bundle, so the job requires 6 bundles (215 ÷ 40).

For field-fabricated hip and ridge units, add another 80 square feet based on 1 square foot per linear foot of ridge and hip.

Roofing Construction & Estimating

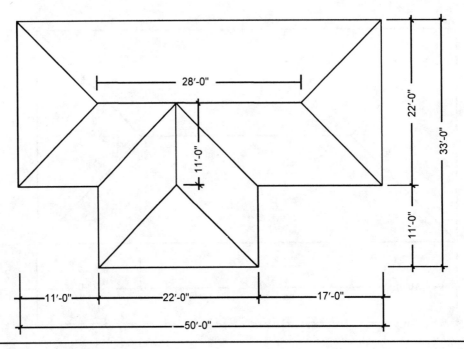

Figure 6-32 Hip and valley roof example

Here's an example for a hip and valley roof:

▼ **Example 6-3:** Assume a roof slope of 6 in 12, and find the number of bundles of No. 3 grade (24" Royals) wood shingles required for the roof in Figure 6-32. Figure on a double starter course at the eaves and lower ends of the hips, and the use of prefabricated ridge and hip units.

Figure 6-25 calls for a 5½-inch exposure for this job.

The net roof area is [(50' x 22') + (22' x 11')] x 1.118 (the roof-slope factor), or 1,415 square feet for 14.15 squares of shingles (use this figure to estimate labor costs).

Eaves length is 166 linear feet (2 x (50' + 33'))

The total area of starter-course material required is 76 square feet (166' x 0.46 SF/LF), where 0.46 is the 5½-inch exposure divided by 12.

The length of any hip or valley from Column 3 of Appendix A is 16.5 linear feet, so the total length of hips and valleys is 132 linear feet (8 ea. x 16.5').

Cutting waste at the hips and valleys is 1 square foot per linear foot, or 132 square feet.

Waste at the eaves overhang is 21 square feet (166' x 0.125 SF/LF).

Therefore, the gross roof area is:

Gross Roof Area = 1,415 SF + 76 SF + 132 SF + 21 SF
= 1,644 square feet ÷ 100
= 16.44 squares

The waste factor, excluding crew-error waste and breakage, is 16 percent (1644 ÷ 1415).

Figure 6-27 shows that we require 5.46 bundles per square for these shingles, so this job requires 90 bundles (16.44 x 5.46).

The total ridge length is 39 linear feet (28' + 11') and the total hip length is 99 linear feet (6 x 16.5') for a total of 138 linear feet.

The number of prefabricated ridge and hip units required is 306 units (138 ÷ 0.46, plus 6 units).

At 40 units per bundle, the job requires 8 bundles (306 ÷ 40).

Staggered Patterns

Staggered (serrated) wood shingle and shake patterns are popular. You can calculate material quantities required for these patterns the same way as for conventional patterns in the previous examples. This is true because the only difference is that, beginning with the second row of shingles, every other shingle is gauged up or down with an offset of around 1 inch, producing two different exposures in each row.

Dutch Weave Patterns

Another popular wood shingle application is the Dutch weave pattern (also referred to as a *thatch*, or *shake-look* roof). Increase the quantity (field shingles only) for a conventional wood shingle roof by 25 percent for this pattern. This is because you need an additional shingle under every fourth shingle in each course. Offset the double-thick set of shingles from other shingles by no more than 1 inch.

Sidewall Shakes and Wood Shingles

Shake and wood shingle siding is durable, attractive and requires no paint. Wood sidewall shingles are manufactured from red cedar or white cedar and sold in Number 1, 2 and 3 grades. A wide variety of cut patterns are available. They come in lengths of 16, 18 and 24 inches.

All wood shingles and shakes you use in roofing can also be used on sidewalls. But there are some shingles and shakes which are specifically manufactured for sidewalls. They include No. 4 undercoursing and rebutted

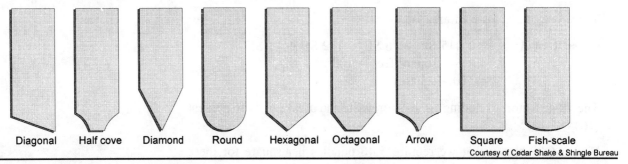

Figure 6-33 Fancy-butt red cedar shingles

and rejointed shingles which are shown in Figure 6-3. For double-coursing on sidewalls, use No. 1 machine-grooved shakes which are described in Figure 6-4.

Fancy-butt red cedar shingles are 5 inches wide and 16 or 18 inches long. A 96-piece carton will cover 25 square feet when laid at a 7½-inch exposure. Fancy-butt shingles can also be custom-made. Figure 6-33 shows some examples of fancy-butt shingles.

Installing Sidewall Shingles and Shakes (New Construction)

Before you install wood shingles or shakes over new construction, apply 15-pound felt to the walls. Start at the base of the wall and use a 2-inch side lap and a 6-inch end lap. Wrap the felt 4 inches each way beyond both inside and outside corners. Install metal flashing above window and door openings and caulk around the openings. I recommend that you also install metal flashing over all inside corners. Be sure all door and window casings are in place and caulked before you start the shingle or shake application. That's shown in Figure 6-34.

Find the number of courses required by measuring the height of the wall from a point 1 inch below the top of the foundation to the top of the wall. Divide the height into equal parts corresponding closely to the recommended exposure. Don't exceed the maximum recommended exposure. Transfer the measurement and the number of courses to a story pole (a straight 1 x 3 or 1 x 4) to lay out the courses on the walls, as in Figure 6-35.

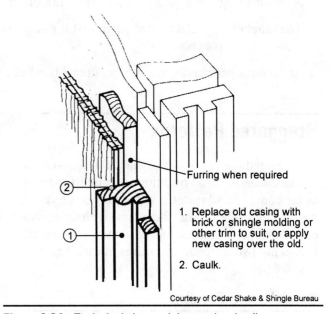

Figure 6-34 Typical window and door casing detail

Figure 6-35 Story pole

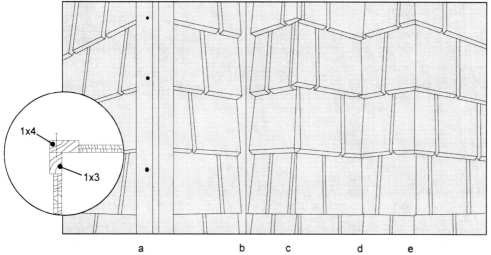

Figure 6-36 Corner details

a) Shingles butted against corner boards
b) Shingles butted against square wood strip on inside corner, flashing behind
c) Laced outside corner
d) Laced inside corner with flashing behind
e) Mitered corner

Courtesy of Cedar Shake & Shingle Bureau

Whenever possible, align the butt lines with the tops and bottoms of windows and other openings. Then you won't have to cut so many shingles to fit. Also try to line up the courses so the exposure of the final course matches those below. Use the story pole to make marks at opposite ends of each wall. Place the marks on corner boards, or on the felt. Figure 6-36 shows alternative methods for installing shingle siding at corners.

To align the first course of shingles, shakes, or undercoursing, nail down a shingle at the correct location at each end of the wall. Tack a nail into the butt of each shingle and stretch and tie a string line between the two shingles. Align the butts of the other shingles along the string. You can align succeeding courses the same way, or snap a chalk line over the preceding course to line up the butts of the next course.

When you install shingles at corners, miter, weave, or butt them against the trim boards, as shown in Figure 6-36. Mitered corners provide the least protection against wind-driven rain, and it'll take a long time to install them. Woven corners provide more protection and are quicker to install. The fastest method and most weather-resistant corner treatment is to butt shingles up to trim boards. Use 2 x 2s at inside corners and a prefabricated outside corner made up of a 1 x 3 and a 1 x 4. For added weather protection, install a bead of caulk between the shingles and corner boards.

Shingle away from corners, especially if you install mitered or laced corners, and trim shingles or shakes to fit around door and window openings. Allow a 1/8- to 1/4-inch joint space between shingles or shakes.

You can install units with tight-fitting joints if you prime and stain them first. Offset the joints of adjacent courses at least 1½ inches, as shown in Figure 6-37.

Sidewall Shingling Methods

The method you use to install shingles on sidewalls depends on the coursing style you're using, and the material under the shingles.

To install single coursing over nailable sheathing, start with a double starter course. For economy, you can use undercourse-grade shingles overlaid with a higher-grade outer course. Drive nails about 1 inch above the butt line of the succeeding course and ¾ inch from each edge, as in Figure 6-38. If a shingle or shake is wider than 8 inches, drive two additional nails about 1 inch apart near the center. This rule applies no matter which coursing style you use. Use galvanized ring-shank nails to install sidewall shingles or shakes.

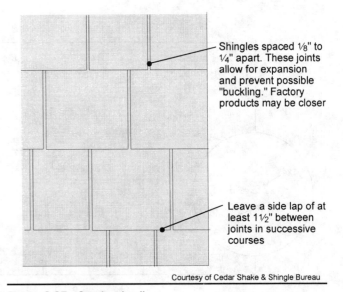

Figure 6-37 Spacing detail

The double-coursing method provides a wide exposure and deep shadow lines. It's an economical application because the wide exposure requires fewer shingles. You'll need undercoursing, but you can use lesser-grade inexpensive shingles.

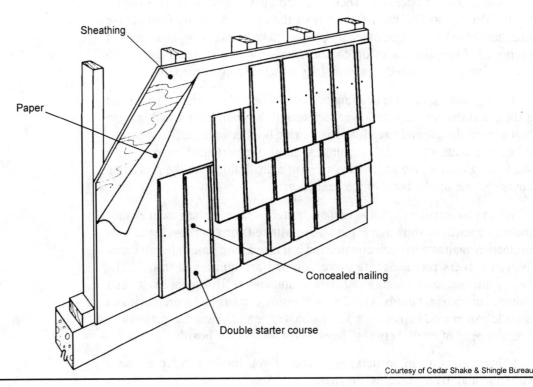

Figure 6-38 Single coursing

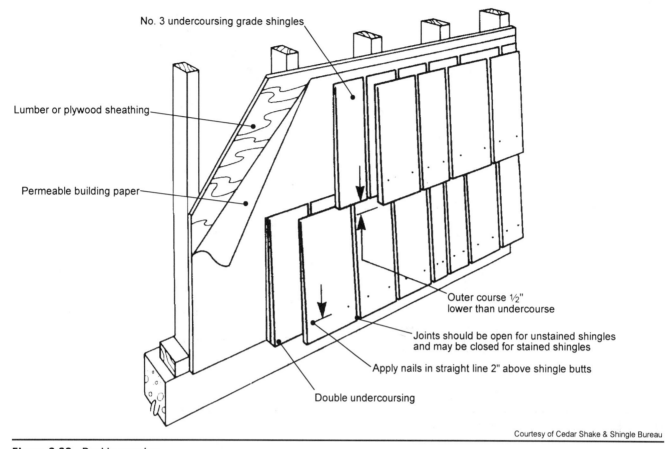

Figure 6-39 Double coursing

To install double coursing over nailable sheathing, install a triple starter course using two undercourses and one outer course. Apply all outer courses ½ inch lower than the undercourse. Install each undercourse unit with one nail or staple driven into the center of the unit. Face-nail each exposed shingle or shake to the undercourse units with two casing nails driven about 2 inches above the butt line and ¾ inches from each edge, as shown in Figure 6-39.

To install double coursing over non-nailable sheathing, first fasten lath strips to the studs with 7d or 8d nails. Space the lath strips vertically so the distance between their centers is the same as the shingle exposure. Then rest the butts of the undercoursing on the lath strips you nailed to the studs.

For the first course, use two undercourses resting on double lath strips overlaid with an outer course. Nail or staple the undercourse units to the sheathing using two fasteners per unit. Apply all outer courses ½ inch below each lath and face-nail each outer course to the laths with two small-headed 5d nails (Figure 6-40).

To install staggered single coursing over nailable sheathing, set the butts of the shingles or shakes below a horizontal line you set with a story pole. Be sure you don't go above the horizontal line, or the exposure of the

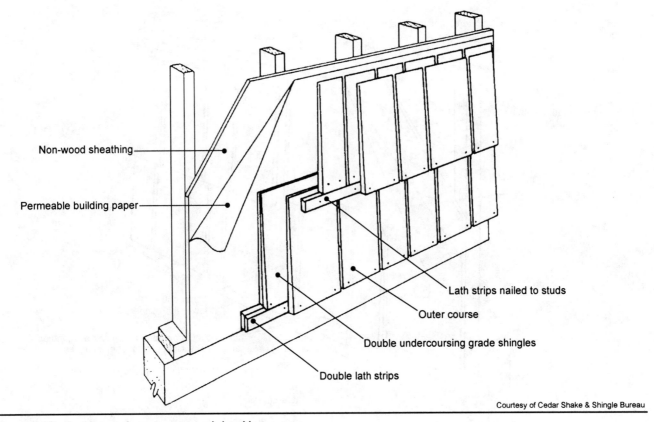

Figure 6-40 Double coursing over non-wood sheathing

underlying shingle will be too great. Don't vary the distance from the butts to the horizontal by more than 1 inch for 16- and 18-inch shingles, or more than 1½ inches for 24-inch shingles. That's shown in Figure 6-41.

Ribbon coursing gives a double shadow-line effect. You can install ribbon coursing over nailable sheathing by raising the outer course units about 1 inch above the undercoursing. Use Number 1 grade units for the undercoursing. You can see this installation in Figure 6-42.

Installing Sidewall Shingles and Shakes over Existing Walls

If you remove exterior sheathing and felt from an existing wall, be sure to take out all nails and anything else sticking out before you install new shakes or shingles. Apply new 15-pound felt over the old sheathing. This is called *re-walling*.

Usually, you can install new shakes or shingles over an existing exterior finish (*over-walling*). This saves time because you don't have to get rid of the old wall covering. But you still have to remove old door and window casings and install new ones. Figure 6-34 in the section on new construction shows this.

Wood Shingles and Shakes

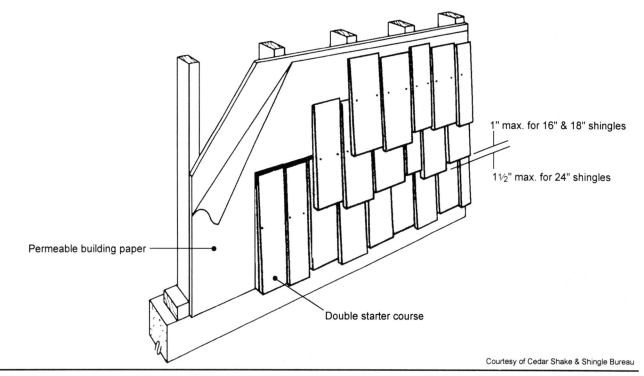

Figure 6-41 Staggered coursing

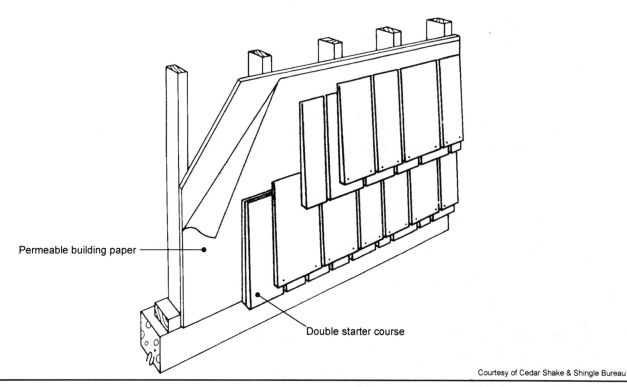

Figure 6-42 Ribbon coursing

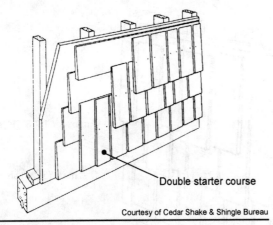

Figure 6-43 Over beveled siding detail

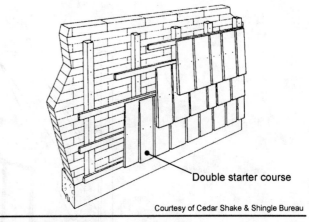

Figure 6-44 Over masonry detail

■ **Over Siding** You can install new shakes or shingles directly over old smooth wood siding the same way you would over any nailable sheathing. In this case, you don't have to apply roofing felt first, assuming there's underlayment beneath the existing siding.

To install shakes or shingles over beveled siding, fill in the low points of the existing wall with lumber or plywood strips (called *horse feathers*). This gives you more area to use as a nailing surface. Then install the shakes or shingles the same as you would over sheathing.

You can also nail the shingles or shakes to the high points of the bevels of each course of the old wall, as in Figure 6-43. Or you can nail the units to alternate siding courses, as long as you don't go over the recommended shingle exposure.

■ **Over Masonry** To install shingles or shakes over existing masonry walls, first use masonry nails at the masonry joints to install 2 x 4 furring strips vertically. Then install 1 x 3 or 1 x 4 nailing strips horizontally across the furring strips spaced at centers the same as the exposure distance of the shingles or shakes, as shown in Figure 6-44.

■ **Over Stucco** Over stucco, install horizontal nailing strips on centers equal to the exposure of the shingles or shakes. Use nails long enough to go into the underlying sheathing. That's shown in Figure 6-45.

Roof Junctures

All roof junctures must be weathertight. For flashing, use at least 26-gauge galvanized steel painted on both sides with metal or bituminous paint. If you have to bend the flashing strips at a sharp angle (greater than 15 degrees), paint the flashing *after* you make the bends. (Sheet metal shops have special machines to bend flashing materials.)

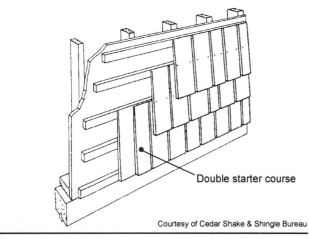

Figure 6-45 Over stucco detail

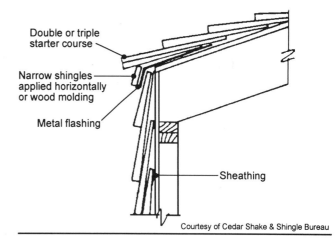

Figure 6-46 Convex roof juncture

■ **Convex Juncture** For a convex juncture, cover the top 4 inches of the final course of the wall and 8 inches of the roof with flashing, as shown in Figure 6-46. Install a narrow strip of shingles or shakes, or a wood molding strip over the downward leg of the flashing. Then install a double or triple starter course on the roof with a 1½-inch overhang at the eaves. Complete the roof in the normal way.

■ **Concave Juncture** For a concave juncture, cover the top of the roof slope and the bottom 4 inches of the wall with flashing, as shown in Figure 6-47. The flashing can be either under or over the last course(s). After you've finished the roof part of the juncture, install a double starter course at the bottom of the wall. Then finish the remaining wall as described above.

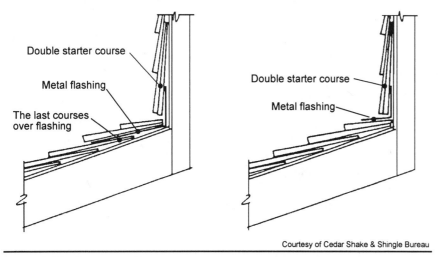

Figure 6-47 Concave roof juncture

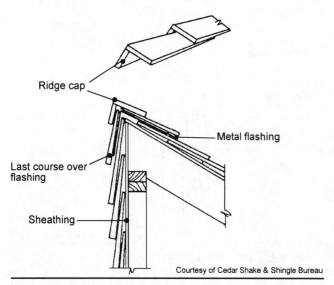

Figure 6-48 Apex roof juncture

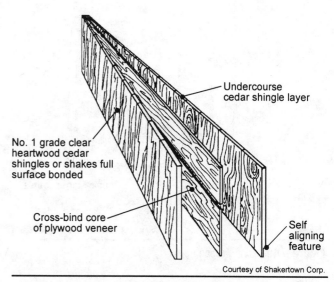

Figure 6-49 Sidewall and mansard panels

■ **Apex Juncture** Cover the top 8 inches of the roof and the top 4 inches of the wall with flashing, as in Figure 6-48. Finish the wall before you begin to shingle the roof. Then cover the juncture with prefabricated ridge units. The staples joining the prefab ridge units are flexible, so you can bend them to fit any angle.

■ **Shingle Sidewall and Mansard Panels** To save time, install prefabricated panels faced with cedar shingles bonded to various backing materials as shown in Figure 6-49. These panels are self-aligning. They're usually 8 feet long and 18 inches wide (for 14-inch exposure) or 9 inches wide (for 7-inch exposure). Panel faces come in a variety of styles and textures, and in a straight- or staggered-butt design.

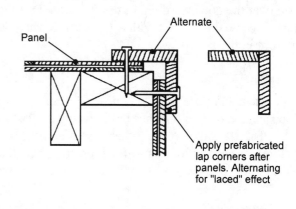

Figure 6-50 Lap corners

You can also get prefabricated outside lap corners (Figure 6-50) and flush outside corners (Figure 6-51). The 18-inch and 9-inch panels come in 4-, and 8-panel bundles, respectively. Either panel size covers 37 square feet per bundle. This equals 2.7 bundles per square.

Depending on local building code requirements, you can install the panels over sheathing, or nail them directly to studs, as shown in Figure 6-52.

Figure 6-53 shows how to apply panels over a mansard. Use 30-pound roofing felt under all panel installations.

You can get additional information from the Shakertown Corporation. They're at 1200 Kerron Street, Winlock, WA, telephone (206) 785-3501.

Figure 6-51 Flush corners

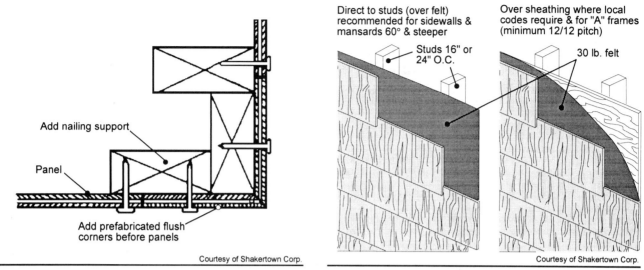

Figure 6-52 Siding panel applications

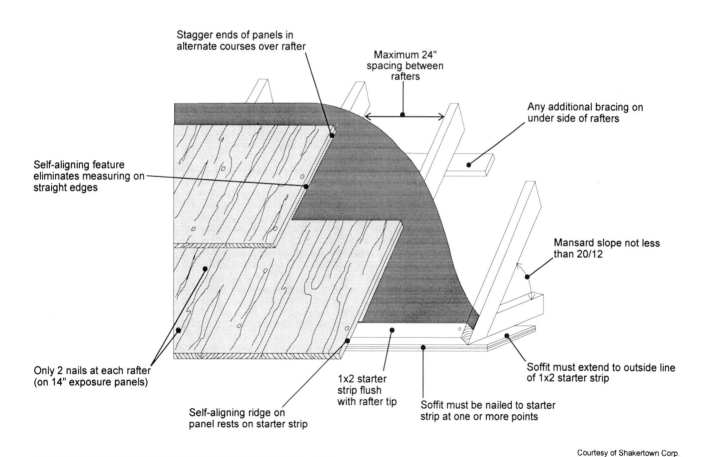

Figure 6-53 Mansard application

Estimating Wood Shingle Roofing Costs

Let's say you've got to cover the roof shown in Figure 6-30 under the conditions given in Example 6-1. Assume that each square of wood shingles costs $120.00, including sales tax. From Example 6-1, you need 9.3 squares of shingles. So the total material cost is:

9.3 squares x $120.00/square = $1,116.00

You must also buy two bundles of ridge units at $30.00 per bundle, for a total cost of $60.00. This brings the total material cost to $1,176.00.

There's a roof deck area of 8.73 squares and 31 feet of ridge units. According to the *National Construction Estimator*, an R1 crew consisting of one roofer and one laborer can install wood shingles at the rate of one square per 2.77 manhours, and ridge units at the rate of one linear foot per 0.03 manhours. If you pay your roofer $20.00 and the laborer $15.00 per hour including labor burden, the average cost per manhour is:

($20.00 + $15.00) ÷ 2 = $17.50

So the labor cost will be:

Labor (installing shingles) = 8.73 squares x 2.77 manhours/square x $17.50/manhour = $423.00

Labor (installing ridge units) = 31 LF x 0.03 manhours/LF x $17.50/manhour = $16.00

The total labor cost is:

$423.00 + $16.00 = $439.00

The total cost is:

$1,176.00 + $439.00 = $1,616.00

7 Tile Roofing

A well-constructed clay or concrete tile roof should last more than 50 years. It costs more to install than most other types of roofing because the materials are more expensive and require a strong frame to support the heavy load of the tiles. But it needs very little maintenance and no preservatives. Also, tile is fireproof, so fire insurance costs less for the entire life of the building.

Clay and concrete tiles have similar physical properties and installation methods. Concrete tile is cheaper to make than clay tile and it can be made anywhere. Clay tile must be made near a clay quarry, so freight cost to the building site is often higher than for concrete tile.

Clay and concrete tiles come in a variety of shapes, sizes and colors. Figure 7-1 shows some of them. Concrete tiles are usually 16½ inches long and 13 inches wide. Clay tiles come in many different sizes, as shown in Figure 7-2.

Concrete tiles weigh about 900 pounds per square when installed with a 3-inch top lap. It takes about 90 concrete field tiles laid with a 3-inch top lap to cover a square of roof area.

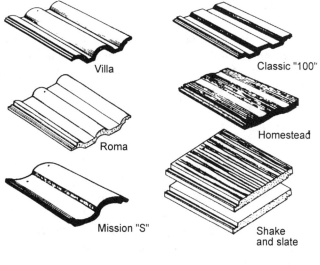

Courtesy of Monier Roof Tile

Figure 7-1 Concrete roof tile types

	Dimensions (inches)	
Tile type	Width	Length
Flat	5, 7 or 8	12, 15 or 24
English	8	13¼
French	9	16
Spanish	9¼	13¼
Roman	6 to 8	13
American	9	14
Lanai	8	14
Mission	8	14½ or 18
Norman	7	15

Figure 7-2 Common sizes of clay roofing tiles

Tile type	Approximate weight (lb./sq.)
Flat	800-1600
French	900-1000
Spanish	850-900
Mission	1250-1350
Roman	1100-1200
Greek	1250
English	800
American	800
Lanai	800
Norman	1600

Figure 7-3 Clay tile weights (Add 1000 lb./sq. for mortar.)

Clay tiles can take from 85 to 330 field tiles per square. Clay tiles weigh about 800 to 1,600 pounds per square, depending on the style, thickness and exposure. Weights are shown in Figure 7-3.

A concrete tile has baffles (weather checks) on its underside and water locks on its sides to keep wind-driven moisture and water runoff from getting under the tile. The weather check nearest the butt of the tile is called a *nose lug*. These are shown in Figure 7-4.

Some clay tiles interlock while others are joined with mortar. Concrete tiles have several sets of interlocking lugs. The top lap and exposure depend on which of the lugs you interlock with the head lug of the underlying tile. That makes it easy to keep the courses evenly spaced. If you shorten the exposure of the last few courses, you don't have to trim the top course at the ridge.

Clay tile comes glazed or unglazed. Concrete tile is colored either on its surface with a cementitious pigmented surface coating, or throughout its body. Surface-coated tile comes in many colors and deep shades. Body-colored tile has a more limited color range and its colors are more subdued. Color-coated tile is more resistant to discoloration from moss and fungus. I don't recommend color-coated tile where temperatures often range from mild days to below-freezing nights. The coating is susceptible to spalling. Body-colored tile, which is colored throughout, withstands short freeze/thaw cycles better.

Underlayment Under Tile Roof Coverings

There are three basic underlayment systems for tile roofs:

1) One-ply nonsealed system

2) One-ply sealed system

3) Two-ply sealed system

These underlayment systems are discussed in detail in Chapter 3.

One-ply underlayment systems are normally used over slopes of 4 in 12 and steeper. You must seal a 1-ply system (1-ply sealed system) over slopes between 4 in 12 and 6 in 12. On slopes steeper than 6 in 12 you can use a 1-ply nonsealed system.

It's possible to install a 1-ply sealed underlayment system over slopes as low as 2½ in 12, if you put a counter batten system over the underlayment. I'll discuss the counter batten system in detail in the following section.

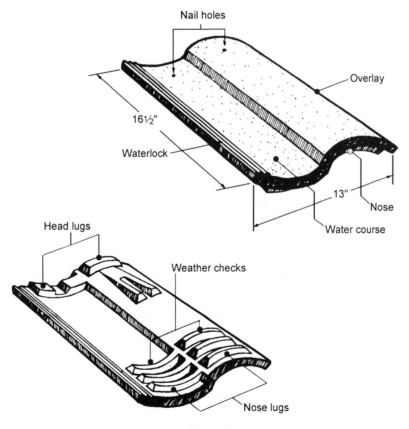

All measurements are approximate

Courtesy of Monier Roof Tile

Figure 7-4 Views of a concrete tile

You can use the 2-ply sealed underlayment system over roof slopes of 2½ in 12 and steeper. You *must* use this system when you embed the tiles in mortar.

You can direct-nail tiles over any of the underlayment systems just discussed. But you should use roofing cement to seal all nail penetrations into a sealed underlayment system. You don't have to seal nail penetrations into an unsealed underlayment system.

You can install tiles with a 2-inch top lap over any sealed underlayment system, but you have to use a 3-inch top lap over an unsealed system.

Installing Roof Tiles

Install roofing tile over solid sheathing. Some building codes specify plywood instead of solid board sheathing. After you've got the underlayment, drip edge and valley flashing material in, install a starter course at the eaves of the roof. Allow from ¾ inch to 2 inches overhang at the eaves, depending on the type of tile and whether there'll be gutters. You use a shorter overhang with gutters, because otherwise, a heavy runoff might pour right over, instead of into, the gutter.

Hold field tiles back 1 to 4 inches from the rake, depending on the type of rake tiles you'll use. We'll discuss that in detail later in the chapter.

To align the first course, nail down a tile with the correct overhang at each end of the eaves. Stretch a string between the top edges of the end tiles and snap a chalk line. Align the top edges of the rest of the tiles along the chalk line. Check alignment about every third or fourth course by measuring up from the eaves to the butts of end tiles and snapping a chalk line to align the butts of the next course of tiles.

Battens and Counter Battens

You can fasten roofing tiles directly to the sheathing or to battens you install over the underlayment. You can install battens on roof slopes of 2½ in 12 and steeper. They're required on slopes of 7 in 12 and steeper. Figure 7-5 shows tile installed over battens.

Beginning at the ridge, install battens at least 1½ inches from each side of the ridge, as in Figure 7-6. Work your way down the slope, spacing the battens according to the tile coursing requirements. For example, if concrete tiles are 16½ inches long with a 3-inch top lap, space the battens on 13½-inch centers.

When you install battens on roof slopes less than 6 in 12, use 4-foot-long battens with ½-inch drain slots between the ends of adjacent battens. Or shim the battens every 4 feet with ¼-inch moisture-resistant lath strips or a decay-resistant material such as an asphalt cap sheet or asphalt shingle, as in Figure 7-7.

Tile Roofing

Figure 7-5 Flat tiles installed over battens

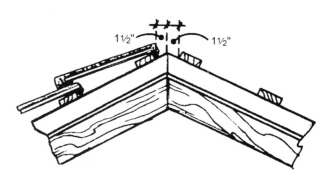

Note: Measurement should be made from the apex of the roof to the face edge of the first batten. Heads of the two tiles should not be more than 2" apart.

Courtesy of Monier Roof Tile

Figure 7-6 Battens at the ridge

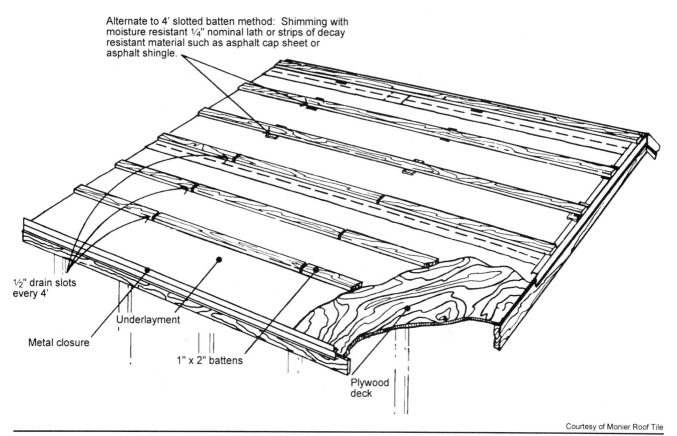

Courtesy of Monier Roof Tile

Figure 7-7 Slotted batten system

201

For battens over a sealed underlayment system, seal all nail penetrations with roofing cement. Use pressure-treated lumber and secure the battens to the sheathing using 6d corrosion-resistant nails you drive at 24-inch centers. Use nails at least ¾ inch long, or at least long enough to go into the underside of the sheathing.

A counter batten system of 1 x 2s (minimum) mounted vertically from ridge to eaves on 16-inch centers is required over single-ply underlayment systems on roof slopes less than 4 in 12, as shown in Figure 7-8. You install the 1 x 4s horizontally across the counter battens, according to the tile coursing requirements. For example, use counter battens on roofs in heavy-snow areas or over cathedral ceilings, as shown in Figures 7-9 and 7-10. A counter batten system is also recommended over roofs in damp climates where good ventilation is especially important.

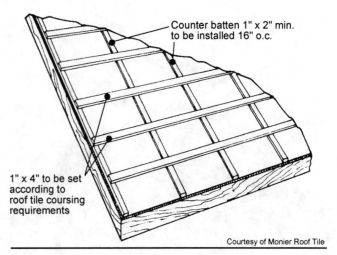

Figure 7-8   Counter batten system

The Starter Course

Raise the starter course to keep water runoff away from the eaves fascia. To do this, extend the fascia board 1½ inches above the top of the sheathing, as in Figure 7-11. Or install a 2 x 2 wood starter strip on top of the sheathing, as in Figure 7-12. In either case, you should install an 8-inch wide wood cant strip behind the raised wood. You can use a strip of antiponding metal flashing nailed to the deck along the upper edge of the flashing to provide positive drainage. That's shown in Figures 7-11 and 7-12. You can omit the drip edge along the eaves when you install antiponding flashing.

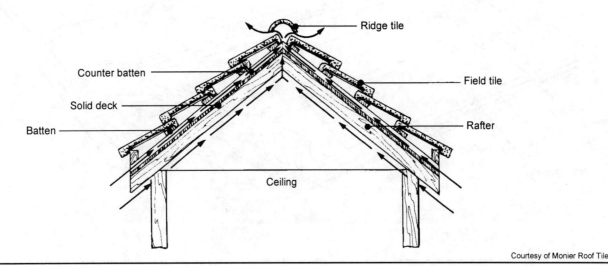

Figure 7-9   Counter battens for a well-ventilated roof

Tile Roofing

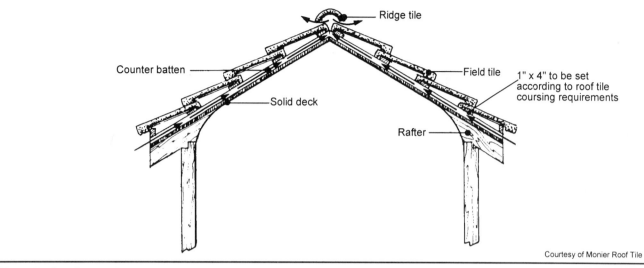

Figure 7-10 Counter battens over a cathedral ceiling

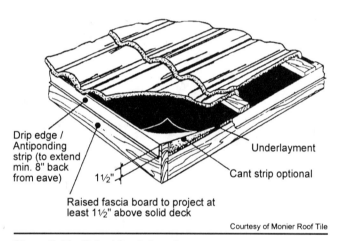

Figure 7-11 Raised fascia board

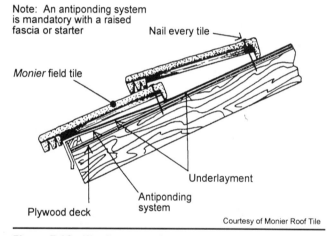

Figure 7-12 Wood starter strip

You can also install a 26-gauge metal eaves closure like the one in Figure 7-13, or a prefabricated synthetic (EPDM) rubber eaves closure strip, as shown in Figure 7-14.

Flat tiles are often elevated by under-eaves tiles installed along the eaves, as in Figure 7-15. That's to divert water away from the eaves, which helps preserve the fascia board.

You can also raise tiles by installing them in a bed of mortar as shown in Figure 7-16. Pack the spaces under the front of the eaves tiles with color-matched Type M mortar. Point (smooth) the mortar to form an

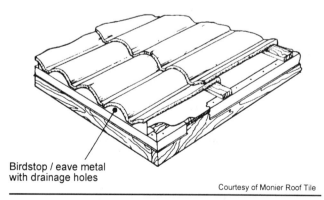

Figure 7-13 Metal eaves closure

203

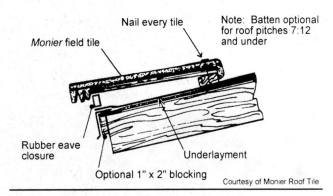

Figure 7-14 Rubber eaves closure

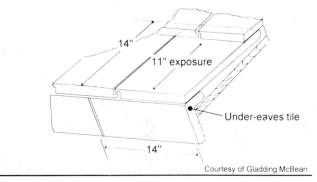

Figure 7-15 Under-eaves tiles

even, flat surface, as in Figure 7-17. Before the mortar sets up, punch holes ("bird holes") for air circulation and water drainage. That's shown in Figure 7-18.

Fastening Roofing Tiles

The way you fasten roofing tile depends on:

- mortar-set vs. direct-nail installation
- roof slope
- anticipated wind exposure and speed
- roof height above the ground
- local building code requirements

Using the Mortar-Set Method

For the mortar-set method, follow these rules:

1) On roof slopes from 4 in 12 to 6 in 12, you must nail each eaves tile with one nail in addition to using mortar.

2) On slopes from 6 in 12 to 7 in 12, install the eaves tiles as described above and nail every third tile in every fifth course in addition to using mortar.

3) On roof slopes of 7 in 12 and steeper, install battens and nail every tile in addition to using mortar.

4) On roofs 40 feet or less above the ground (measured to the eaves) in areas of wind velocities up to 80 miles per hour, follow the fastening guidelines in Figure 7-19.

5) On roofs 40 feet or less above the ground in areas of wind velocities between 80 and 110 miles per hour, follow the fastening guidelines in Figure 7-20. Use corrosion-resistant hot-dipped galvanized nails long enough to penetrate the roof deck $3/4$ inch.

Figure 7-16 Setting eaves tiles in mortar

Tile Roofing

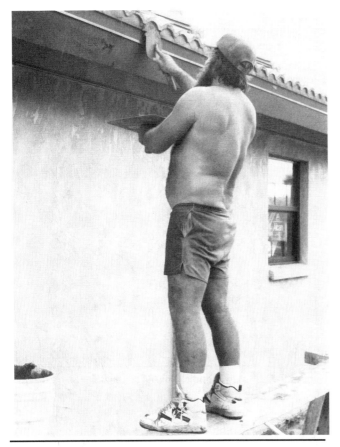

Figure 7-17 Pointing mortar at the eaves

Figure 7-18 Forming "bird holes"

Roof slope	Field tiles	Eaves course	Perimeter tiles[2]
3 in 12 through 12 in 12	Nail every tile	Nail and clip[1]	Nail every tile
Over 12 in 12	Nail and clip[1]	Nail and clip[1]	Nail and clip[1]

Courtesy of Monier Roof Tile

Use 10d nails
[1] You can use two nails per tile instead of clips.
[2] Perimeter nailing areas include the distance equal to three tiles (but not less than 36") from the edges of hips, ridges, eaves, rakes, and major roof penetrations.

Figure 7-19 Fastening requirements for tiles in wind velocities up to 80 MPH (roof height not exceeding 40 feet)

Roof slope	Field tiles	Eaves course	Perimeter tiles[2]
All slopes	Nail and clip[1]	Nail and clip[1]	Nail and clip[1]

Courtesy of Monier Roof Tile

Use one 12d nail or two 10d nails.
[1] You can use two nails per tile in lieu of clips.
[2] Perimeter nailing areas include the distance equal to three tiles (but not less than 36") from the edges of hips, ridges, eaves, rakes and major roof penetrations.

Figure 7-20 Fastening requirements for tiles in wind velocities between 80 MPH and 110 MPH (roof height not exceeding 40 feet)

Roof slope	Field tiles	Eaves course	Perimeter tiles[1]
All slopes	2 nails and clip	2 nails and clip	2 nails and clip

Use two 10d nails.

[1] Perimeter nailing areas include a distance of three tile widths (but not less than 36") from the edges of hips, ridges, eaves, rakes and major roof penetrations.

Courtesy of Monier Roof Tile

Figure 7-21 Fastening requirements for tiles in wind velocities between 80 MPH and 120 MPH (roof height exceeds 40 feet)

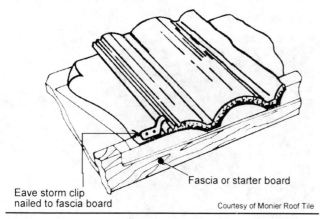

Figure 7-22 Eaves storm clip nailed to raised wooden fascia

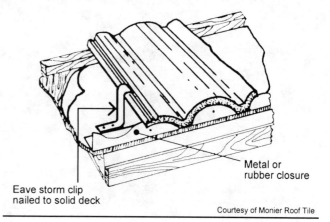

Figure 7-23 Eaves storm clip nailed to sheathing

6) On roofs more than 40 feet above the ground in areas of wind velocities between 80 and 120 miles per hour, follow the fastening guidelines in Figure 7-21.

The type of storm clip you use depends on the brand of tile you use, the location of the tile, the type of fascia installed and the fastening system.

- For a raised fascia, nail eaves-tile storm clips to the fascia board, as in Figure 7-22.

- For a metal or rubber eaves closure, nail eaves-tile storm clips to the sheathing, as in Figure 7-23.

- For battens, nail field-tile storm clips to the back of each batten, as in Figure 7-24.

- If you don't use battens, nail field-tile storm clips to the sheathing, as in Figure 7-25.

With the mortar-set method, wet each tile before you lay it, so the tile doesn't draw water out of the mortar, weakening the bond. Embed each tile in mortar applied with a filled 10-inch masonry trowel. Place the mortar (Type M) under the pan or flat part of each tile (Figure 7-26). Don't place the mortar under the lugs or heads of the tiles since this can result in tilted or crooked tile. Apply mortar from the head of the tile below; up to 2 to 4 inches from the head of the tile you're setting. Be sure the mortar meets the head of the lower course of tile and the underside of the tile you're setting.

Tile Roofing

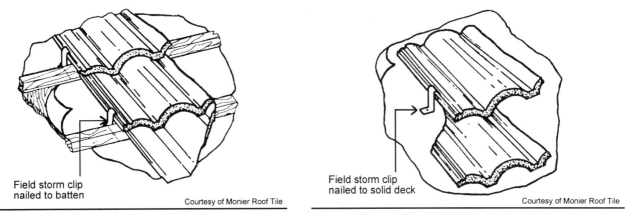

Figure 7-24 Field storm clip nailed to back of batten

Figure 7-25 Field storm clip nailed to sheathing

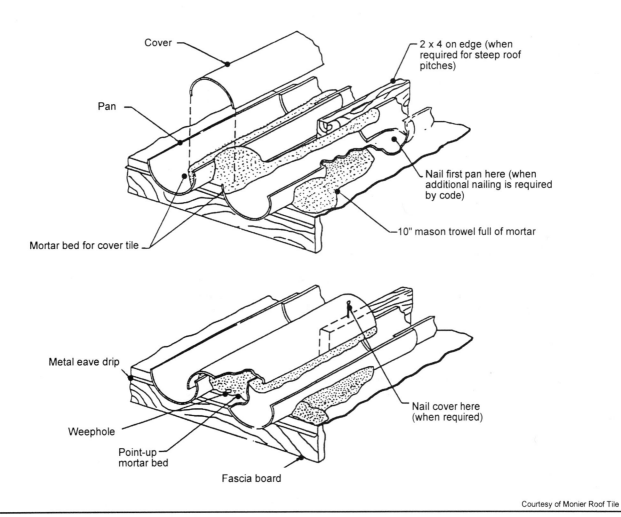

Figure 7-26 Setting field tiles in mortar

Install two-piece barrel tile like the ones in Figure 7-27 with mortar applied under the center of the pan with the narrow end of the pan facing down slope. Bond the pan you're setting to the pan in the previous course. Place mortar along each inside edge of the pans and set the covers with the wide end facing down slope.

Concrete tile manufacturers recommend that you install roofing tile using the straight-bond method, that is, with the joints of every other course aligned. The staggered- or cross-bond method where joints are offset laterally by a set distance is usually used to install flat tiles. But where heavy snow loads are likely, you should use the straight-bond method for flat tile also. That provides the maximum distance between joints from one course to the next and cuts down the likelihood of leaks where joints fall close together.

Figure 7-27 Barrel tile

To prevent tiles, which are rigid, from cracking, it's important that tiles can move independently of each other, especially if you've nailed or clipped them down. So, allow at least a 1/16-inch gap between vertical joints at the side interlock of each tile.

You can leave a 2-inch top lap over any sealed underlayment system, but you must seal all nail penetrations. Use a 3-inch top lap over an unsealed underlayment system. In this case, you don't have to seal the nail penetrations. Use a 4-inch top lap in hurricane areas where the rafters are longer than 20 feet on roof slopes ranging from 4 in 12 to 7 in 12. You can increase the top lap to provide even spacing so you don't have to trim tiles at the ridge.

Installing Rake Tiles

When you'll install standard rake (barge) tiles, adjust the field tile position (beginning at the eaves course) so the tiles are held back from the rake 1 to 2 inches to accommodate the rake tiles. Figure 7-28 shows this. When you install barrel rake tiles, hold the field tiles back 2 to 4 inches from the rake, as in Figure 7-29.

Rake tiles come with two nail holes on the vertical leg of the tile, as shown in Figure 7-30. Install each tile with two galvanized nails. Be sure the nails go at least 3/4 inch into the framing. Set the horizontal leg of the rake tiles over the field tiles into a bed of mortar or sealant. See Figures 7-28 and 7-29.

Install the rake tiles from the bottom of the rake and work up-slope, butting each rake tile against the following course of field tile, as in Figure 7-31. If there's no gutter, install the first rake tile in line with the eaves tile course.

Tile Roofing

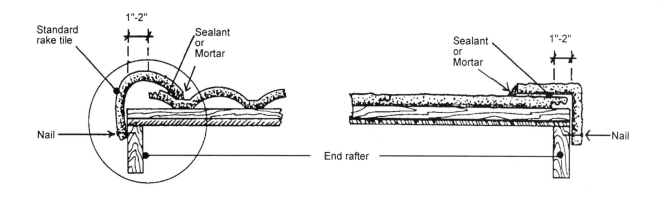

Figure 7-28 Standard rake tiles

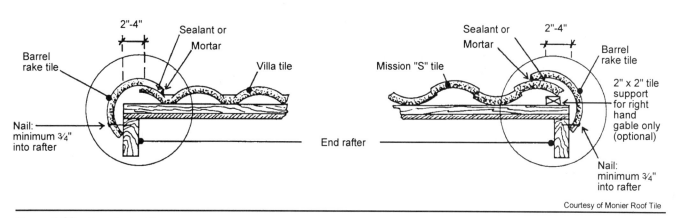

Figure 7-29 Barrel rake tiles

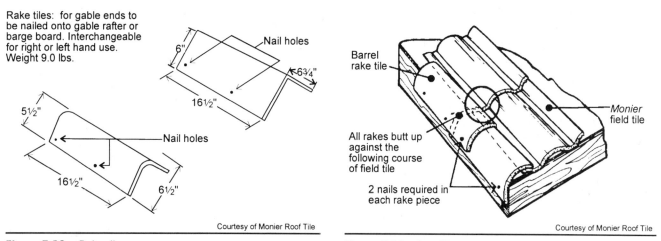

Figure 7-30 Rake tiles

Figure 7-31 Installing rake tiles

209

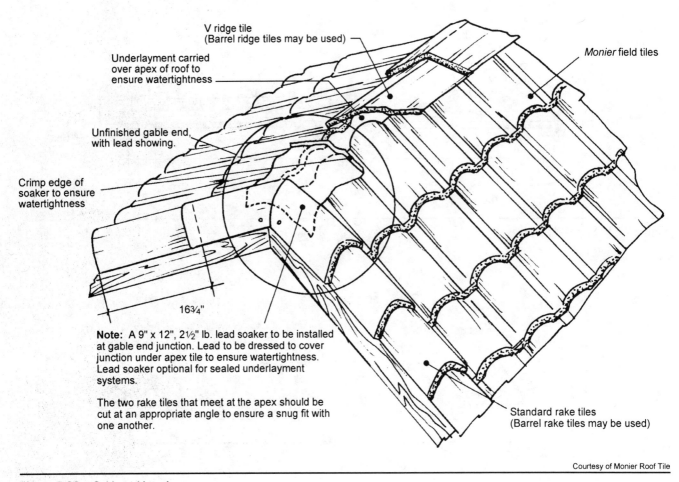

Figure 7-32 Gable end junction

Miter the rake tiles for a snug fit at the peak of the rake. Install the top rake tiles over a 9" x 12" lead soaker (a piece of flashing to help prevent leaks) when you're laying the tiles over an unsealed underlayment system. The lead soaker is optional over a sealed underlayment system. Crimp the inward edge of the soaker to ensure watertightness. These details are shown in Figure 7-32.

Instead of rake tiles, you can install prefabricated metal rake flashing with a 1-inch water guard (water return) as shown in Figure 7-33. Fasten the metal to the sheathing with clips like the ones in Figure 7-34 nailed at 24-inch centers. Never drive nails into the pan area of the flashing (the part that contacts the deck).

You can also install a mortar finish, as shown in Figure 7-35, along the rakes if you're laying flat tiles over a sealed underlayment system.

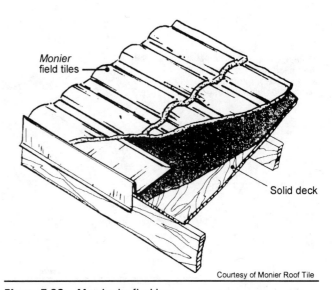

Figure 7-33 Metal rake flashing

Installing Hip and Ridge Tiles

When you come to a hip, miter the tile to form a straight edge over the hip. Cut the tile with a power saw equipped with a Carborundum blade, or trim the tile with a chipper like the one in Figure 7-36. With this tool, you press down on the handle, and the point chips the tile. Space field tiles so there's a 2-inch gap over the ridge. Do the same for field tiles at the hip, as in Figure 7-37 A.

Set hip and ridge units into color-matched Type M mortar, but don't let the mortar cover the center of the hip or ridge. Extend the field tile beyond each bed of

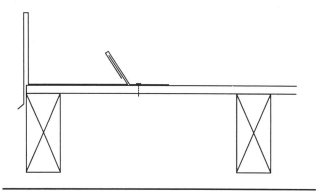

Figure 7-34 Metal fascia clips

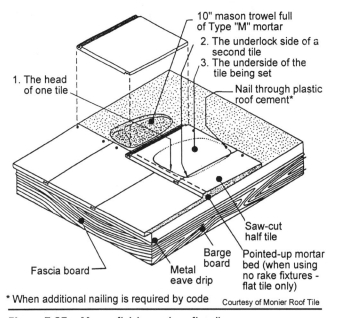

Figure 7-35 Mortar finish at rakes, flat tile

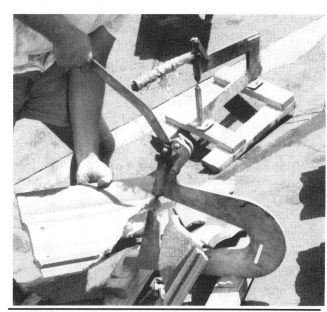

Figure 7-36 Chipper

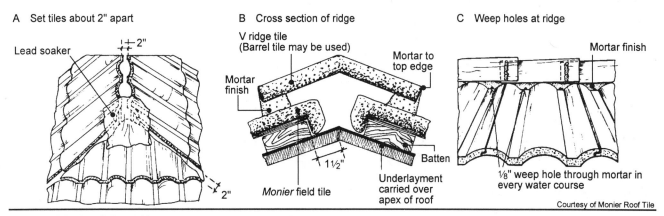

Figure 7-37 Roof tile at ridge

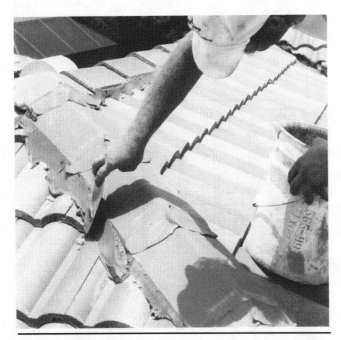

Figure 7-38 Pointing mortar at a hip

Figure 7-39 Finishing mortar with a dry sponge

mortar by about an inch, as shown in Figure 7-37 B. Before the mortar sets, point the mortar to match the contours of the tiles, as in Figure 7-38. Then rub the mortar with a dry sponge to remove the rough edges. That's shown in Figure 7-39.

Over an unsealed underlayment system, punch 1/8-inch weep holes in the mortar beneath ridge units at every water course. See Figure 7-37 C.

You can use V-ridge tiles or barrel tiles like those in Figure 7-40 to cover hips and ridges. Install hip tiles before you install ridge tiles.

You can install a prefabricated hip starter like the one in Figure 7-39 at the low end of a hip, or you can use an unaltered standard hip-ridge tile. You can also install a field-mitered hip-ridge tile and cover the rough edges with mortar.

Install a prefabricated apex trim tile (wye) like the one in Figure 7-41 at the hip-ridge juncture. Over an unsealed underlayment system, install a 9" x 12" lead soaker at the juncture (Figure 7-37 A).

Lap hip and ridge tiles at least 2 inches. Over an unsealed underlayment system, apply a continuous bead of mastic to the lapped part of the ridge tiles.

■ **Dry Hip and Ridge System** Instead of using mortar, you can install a "dry" hip and ridge system over sealed or unsealed underlayment. To do this, begin by install-

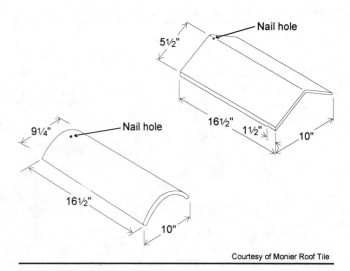

Courtesy of Monier Roof Tile

Figure 7-40 Ridge and hip tiles

Figure 7-41 Apex trim tile (wye)

Figure 7-42 A "dry" hip

ing a 2 x 3 (for standard hip-ridge tiles) or a 2 x 4 or 2 x 6 (for barrel hip-ridge tiles) along the centerlines of the hips and ridge, as shown in Figure 7-42. Toenail the boards into the deck or hip rafter.

Then cover the boards with a pressure-sensitive hip-ridge sealer to act as a wind block, as in Figure 7-43. The sealer has an adhesive back that's covered with release paper. It comes in 50-foot rolls, 9 or 12 inches wide. Available colors include terra cotta, dark brown, black and aluminum. Lap the ends of the wind block at least 2 inches.

When you've got the sealer in, nail the hip-ridge tiles to the boards through the wind block material. Use one nail per tile.

Valleys

The type of valley flashing in an open valley on a tile roof depends on the underlayment system. Over an unsealed underlayment system, install preformed 26-gauge galvanized iron flashing or 16-ounce copper with a $2^{1}/_{2}$-inch water diverter and 1-inch water guards (Figure 7-44 A). Use nails 6 inches on center or clips nailed 24-inches on center. Don't drive nails into the flashing. Treat a valley that ends at a roof plane this way.

If you need more than one piece of flashing, lap the upper piece over the lower by at least 6 inches and seal the lap with roofing cement. Use flashing at least 16 inches wide.

Install a 12" x 24" lead ridge saddle over the flashing at the valley-ridge juncture, as shown in Figure 7-44 B. When the lower end of a valley ends at a roof plane, lap the valley metal over a 24" x 16" lead soaker (skirt) at

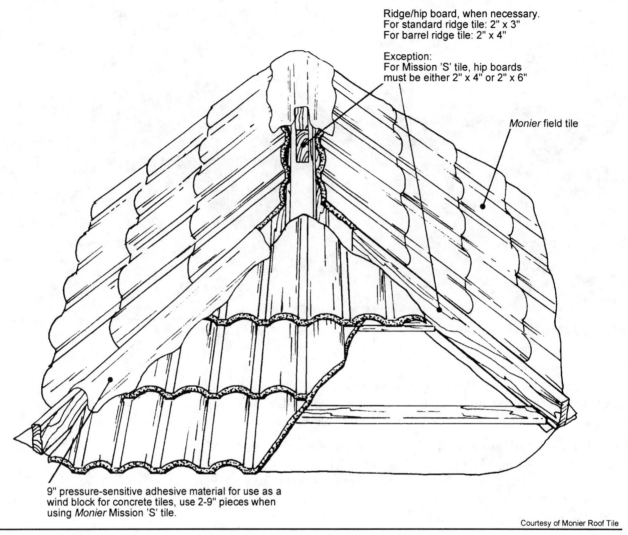

Figure 7-43 Wind block material at hips and ridge

the lower end of the valley to send water runoff back onto the field tile. That's shown in Figure 7-44 C. Or you can extend the valley metal over the top of the field tile and omit the skirt.

Over a 30/90 hot-mop sealed underlayment system, you can install standard roll-metal valley flashing sandwiched between the 30-pound felt and 90-pound mineral-surfaced roll roofing.

Over a sealed underlayment system, you can use valley flashing without water guards, as shown in Figure 7-45. Nail the flashing to the sheathing, driving nails on 6-inch centers along rows 1 inch from each edge. Seal all nail holes with a strip of 30-pound felt embedded in roofing cement. That's also shown in Figure 7-45. Again, lap pieces of flashing at least 6 inches, and seal the lap with roofing cement.

Tile Roofing

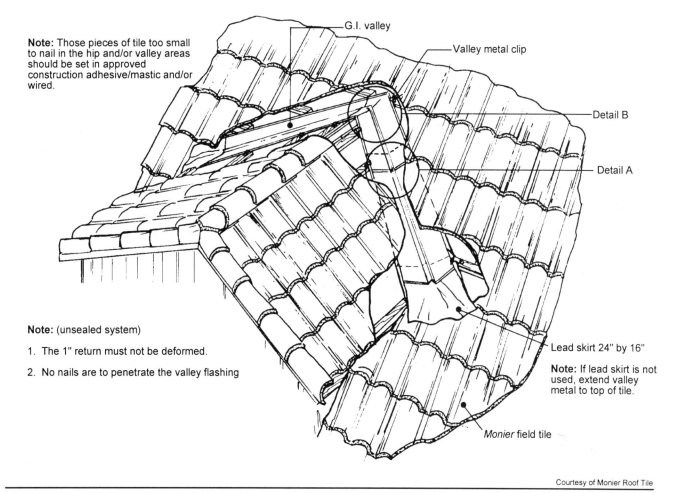

Courtesy of Monier Roof Tile

Figure 7-44 Roof tile in a valley (unsealed underlayment system)

215

Roofing Construction & Estimating

A Valley flashing for sealed underlayment system (Detail A)

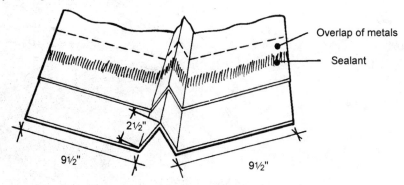

B Tile in valley over sealed underlayment

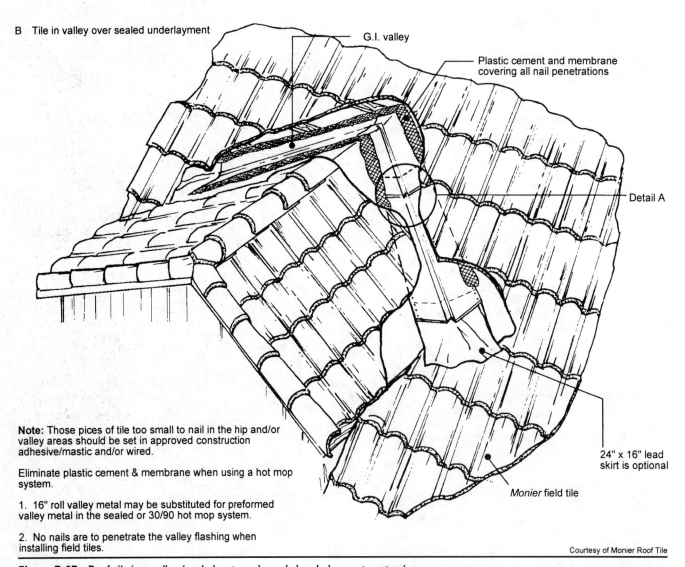

Note: Those pieces of tile too small to nail in the hip and/or valley areas should be set in approved construction adhesive/mastic and/or wired.

Eliminate plastic cement & membrane when using a hot mop system.

1. 16" roll valley metal may be substituted for preformed valley metal in the sealed or 30/90 hot mop system.

2. No nails are to penetrate the valley flashing when installing field tiles.

Courtesy of Monier Roof Tile

Figure 7-45 Roof tile in a valley (sealed or two-ply sealed underlayment system)

■ **Open Valley Installation** To secure tiles adjacent to an open valley, snap chalk lines at least 2 inches on each side of the centerline of the valley. Apply a bed of mortar along the outside edge of each chalk line and embed trimmed tiles in the mortar to form straight borders, as shown in Figure 7-46.

While the mortar under the valley tiles sets up, place two 2 x 4s on edge down the center of the valley. Apply color-matched mortar between the tiles and the 2 x 4s, pointing the mortar to match the tile contours. After the initial set (when the mortar has gotten "crunchy") slide a trowel along the outside edges of the 2 x 4s to separate them from the mortar. Remove the boards and point the mortar. Then rub the mortar with a dry sponge to remove the rough edges.

Some roofing contractors install hip-ridge tiles over a valley. I don't recommend this because a leak can trap water under the tiles in the valley.

A   Lay trimmed tiles

Flashing at Vertical Walls

The type of flashing you use along the juncture of a sloping roof and a vertical wall depends on the type of underlayment system you've used. Over an unsealed underlayment system, install continuous flashing with a 1-inch water guard on the horizontal leg. That's shown in Figure 7-47 A. Secure the flashing to the sheathing with clips nailed at 24-inch centers. Don't drive nails into the pan of the flashing.

B   Place two 2 x 4s in center of valley

Figure 7-46   Roof tile in an open valley

Begin at the low end of the wall and work up-slope to install the flashing over the underlayment. If you need more than one piece of flashing, lap the upper piece at least 4 inches over the lower one and seal the lap with roofing cement. Extend the flashing at least 6 inches over the underlayment and 5 inches up the wall, as shown in Figure 7-47 B.

When the lower end of the wall ends at a roof plane, you can:

1) Extend the flashing down to the eaves, or

2) Lap the pan flashing over a lead apron to send water runoff back onto the field tile. In this case, the underlayment can handle runoff during the most extreme rainy conditions. Both are shown in Figure 7-48.

Over a sealed underlayment system, you can install continuous flashing without a water guard. Nail the flashing to the sheathing, driving nails on 6-inch centers along a row 1 inch from the edge of the horizontal leg. Seal all nail holes with a strip of 30-pound felt embedded in roofing cement, as

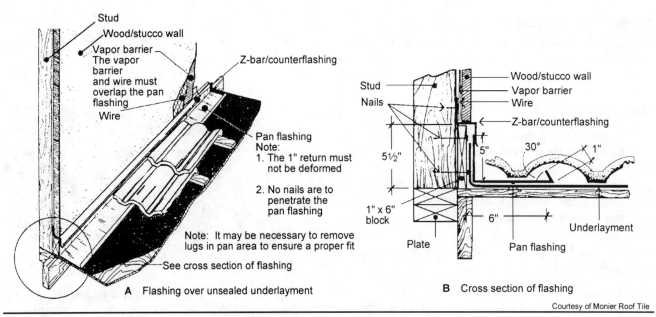

Figure 7-47 Continuous flashing over unsealed underlayment

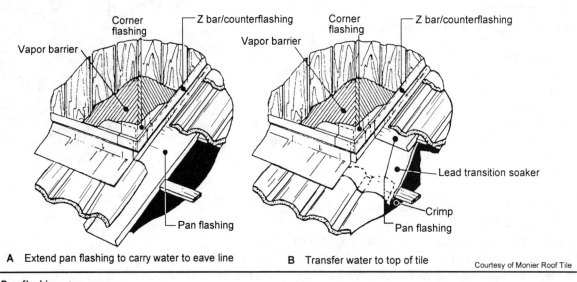

Figure 7-48 Pan flashing at corner

shown in Figure 7-49 A. Begin at the low end of the wall and work up-slope. Lap flashing at least 4 inches, and seal the laps with roofing cement. Extend the flashing at least 6 inches over the underlayment and 5 inches up the wall, as shown in Figure 7-49 B.

There's no need to send runoff water to the eaves or over the field tiles when the lower end of the wall ends at a roof plane. That's because the sealed underlayment system handles the runoff. However, cut the flashing with tin snips and bend it around the corner, as in Figure 7-50, to make sure the flashing at the bottom corners of the wall is watertight.

Tile Roofing

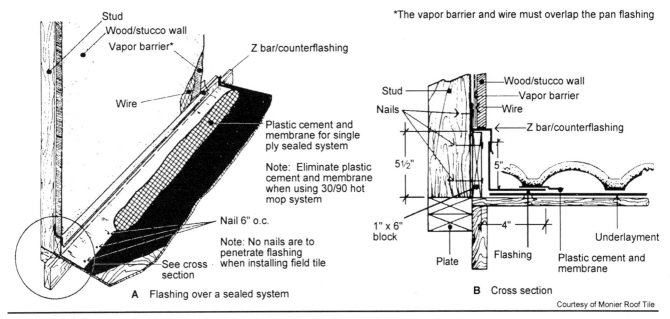

Figure 7-49 Continous flashing over sealed underlayment

There are several ways to install counterflashing on vertical walls. You can use a J bead flashing like the one in Figure 7-51. Or use the Z-bar flashings as shown in Figures 7-47 B and 7-49 B.

The flashing method for the front of a dormer also depends on the type of underlayment system. On an unsealed underlayment system, install continuous flashing that overlaps the field tiles, as in Figure 7-52. On a sealed underlayment system, install the flashing as shown in Figure 7-53.

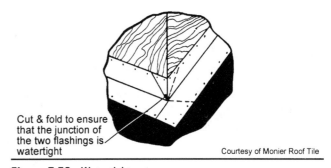

Figure 7-50 Watertight corner

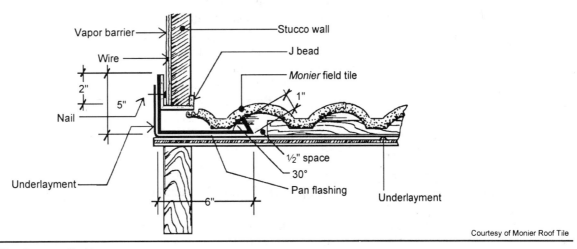

Figure 7-51 J bead counterflashing

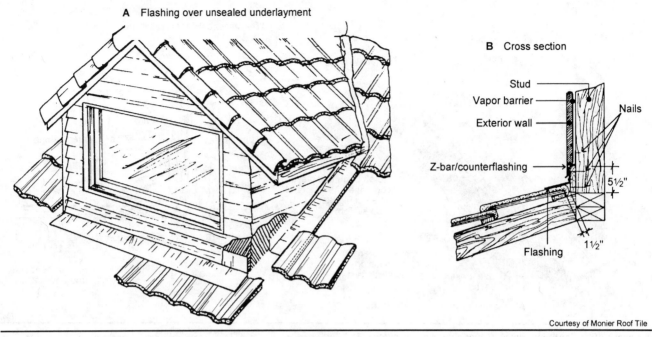

Figure 7-52   Flashing the front of a dormer over unsealed underlayment

Flash around chimneys as shown in Figure 7-54. If the horizontal width of the chimney is greater than 48 inches, install a galvanized iron cricket at the upper side of the chimney. Flashing and counterflashing is enough if the chimney is less than 48 inches wide.

Seal the flashing to the deck with plastic cement and roofing membrane. Refer back to Figure 7-48 for ways to apply the flashing at the sides of the chimney, and Figures 7-52 and 7-53 for the front. Be sure the counterflashing laps over the base flashing by at least 2 inches.

Flashing Soil Stacks and Vents

The method you use to seal a vent pipe depends on the type of underlayment system you've installed. Over an unsealed underlayment system, cut tiles to fit around the pipe, then install a lead vent jack with an 18" x 18" flange (or skirt). Form the flange to the contours of the tile courses covered. Be sure to extend the flange under the tile course immediately above the pipe. This is shown in Figure 7-55. On flat tiles, you can substitute a flat, corrosion-resistant metal flange for lead. Be sure to crimp the up-slope and side edges of the metal flange up at a 30-degree angle as shown in Figure 7-56. You can use pliers to crimp the metal, or have it done at a sheet metal shop. Secure the base flange to the tiles with roofing cement.

Over a sealed underlayment system, embed the vent jack into a bed of roofing cement over the 43-pound base sheet (single-ply sealed system) or the 30-pound felt (30/90 hot-mop sealed system). Then nail the flange along the edges. Cover the nails with felt and roofing cement (single-ply sealed system) or hot-mop 90-pound mineral-surfaced roll roofing over the flange (30/90 hot-mop sealed system). See Figure 7-57.

Tile Roofing

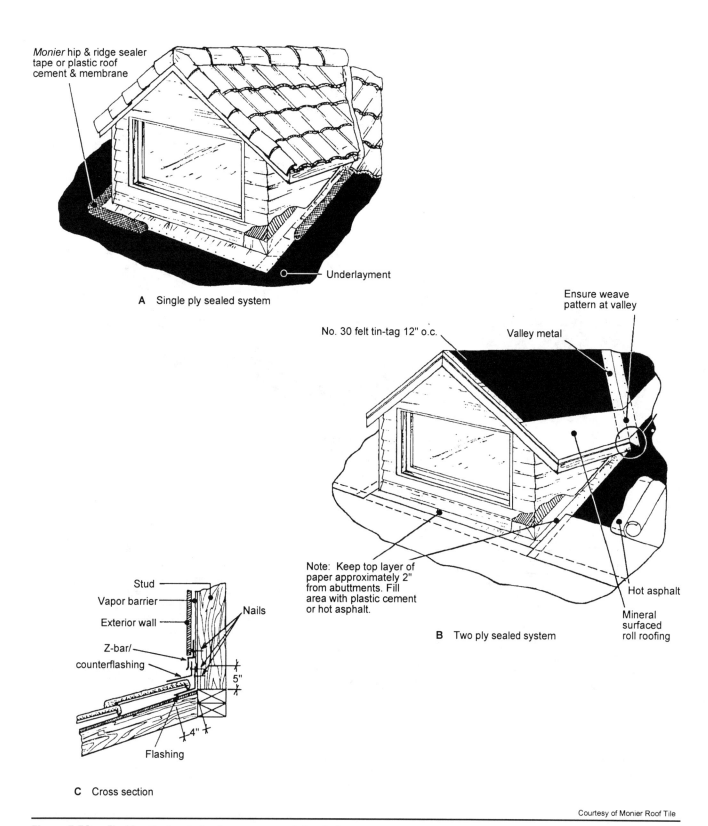

Figure 7-53 Flashing the front of a dormer over sealed underlayment

Roofing Construction & Estimating

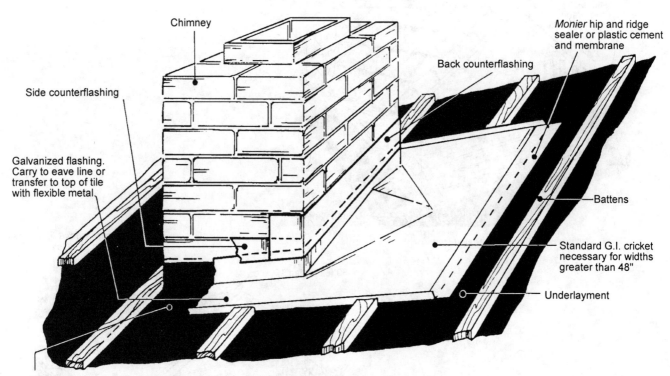

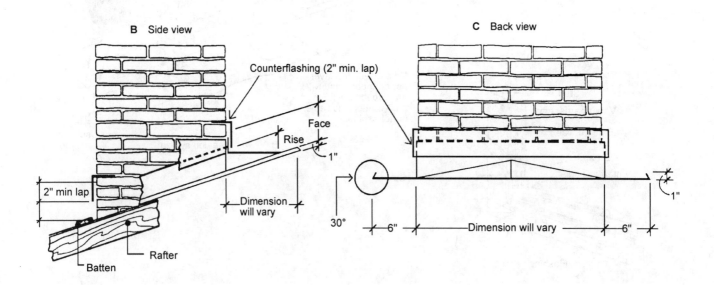

Figure 7-54 Flashing a chimney

Courtesy of Monier Roof Tile

Tile Roofing

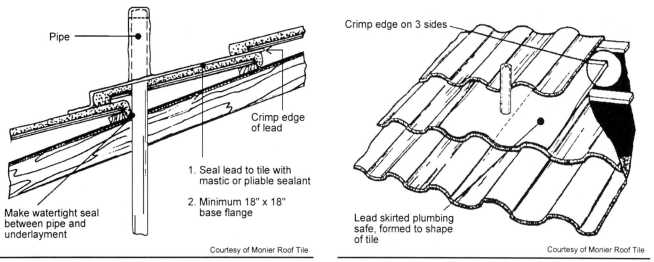

Figure 7-55 Flashing a vent pipe over an unsealed underlayment system

Figure 7-56 Flange with crimped edges

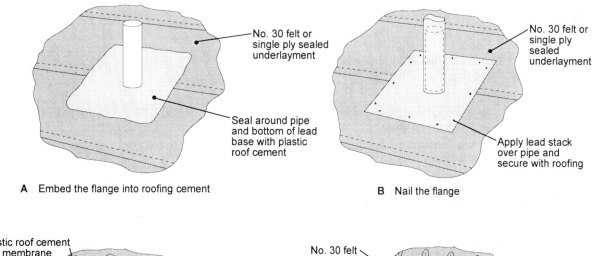

A Embed the flange into roofing cement

B Nail the flange

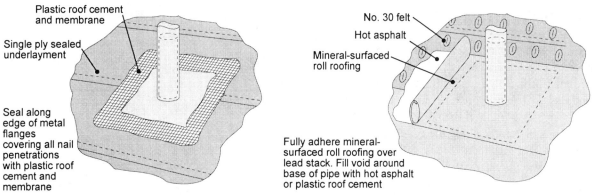

C Cover nails with felt and roofing cement

D Cover flange with mineral-surfaced roll roofing

Figure 7-57 Installing vent jacks

223

A Cut tiles to fit around penetrations

B Fill voids with mortar

Figure 7-58 Fitting tiles around penetrations

Cut tiles to fit over and around penetrations, as shown in Figure 7-58 A. Fill the voids with color-matched mortar, pointed to match tile contours as the roofer has started to do in Figure 7-58 B.

Paint all metal penetrations to match the color of the tile. Apply paint after you've finished the hot-mop work, but before you install the tile. You don't want to ruin the paint with any splashing asphalt. Also, you won't have to walk on the tiles so much after you install them.

Replacing Broken Tiles

To replace a broken or damaged field tile, break the tile with a hammer so it's easier to get out. If the tile was nailed in, drive the nails flush with the sheathing or batten with a hammer and flat bar. Then apply a minimum of 1 square inch of tile adhesive to the tile in the course below. Also apply ⅜-inch beads of cement along the edges of the new tile and the one it will lay next to. Then install the new tile over the adhesive. See Figure 7-59.

When you have to replace smaller tiles adjacent to a hip or valley, pry up the nose of the tile in the course above the damaged tile with your hand and a mortar trowel and apply a ⅜-inch bead of adhesive along the head of the replacement tile. Then set the new tile.

Figure 7-59 Replacing the tile

Estimating Tile Quantities

When you estimate tile roofs, consult the manufacturer's information for the number of field tiles per square, number of squares per pallet, and the number of tile accessories packaged per box. Keep in mind that field tile is usually sold only in full pallets, and fittings in full boxes. Also, shipping charges for tiles are by the full truckload.

It takes about 90 concrete field tiles to cover a square of roof area when you use a 3-inch top lap. Remember, over a sealed underlayment system, you only need a 2-inch top lap. In hurricane areas where the rafters are more than 20 feet long and the roof slopes from 4 in 12 to 7 in 12, you must use a 4-inch top lap. Take off tile quantities by the square foot, and then convert to the number of tiles required.

You calculate the actual number of tiles required by beginning with the number of tiles per manufacturer's square and using a percentage factor. The formula is the same as for asphalt shingles in Chapter 4, Equation 4-2:

$$\text{Percentage-of-Increase Factor} = \frac{\text{Recommended Exposure}}{\text{Actual Exposure}}$$

▼ **Example 7-1:** Ninety field tiles of the type shown in Figure 7-4 will cover a square of roof area when laid with a 3-inch top lap. To find the number of field tiles required if the tiles are installed with a top lap of 2 inches:

$$\text{Percentage-of-Increase Factor} = \frac{16.5 \text{ in.} - 3 \text{ in.}}{16.5 \text{ in.} - 2 \text{ in.}}$$

$$= \frac{13.5 \text{ in.}}{14.5 \text{ in.}}$$

$$= 0.93$$

To cover a square of roof area, you need 0.93 x 90 tiles, or 84 tiles.

To find the number of field tiles required if the tiles are installed with a top lap of 4 inches:

$$\text{Percentage-of-Increase Factor} = \frac{16.5 \text{ in.} - 3 \text{ in.}}{16.5 \text{ in.} - 4 \text{ in.}}$$

$$= 1.08$$

You need 1.08 x 90 tiles, or 98 tiles.

When you estimate tile roofs, add at least 3 percent to field tile quantities to account for waste and breakage.

Estimating Accessories and Mortar

The types of accessories you need depend on the type of field tile you've installed and the roof design. Tile accessories often include:

- rake tiles, as in Figure 7-30
- hip and ridge tiles, as in Figure 7-40
- ridge start and end, under-eaves tiles, as in Figure 7-15
- apex trim tiles, as in Figure 7-41
- hip starters like those in Figure 7-39

You'll also need portland cement mortar for tile installation and to seal eaves, valleys, hips and ridges. Fill mortar is normally made of 1 part portland cement mortar (Type M) or 1½ parts high-strength masonry mortar (Type S) and 4 parts sand. Add color to exposed mortar to match the tile color.

Another suggested mix is ½ part portland cement, 1 part masonry cement, 4 parts sand and an oxide for color. The addition of an acrylic polymer binder is optional. I recommend this mix because of the added strength provided by the binder.

You can use plastic cement or silicone sealant instead of cement mortar to install accessories such as rake tiles. That way you don't have to spend time mixing the mortar. Just snap the end off a tube and apply. The cost is about the same when you consider the time you'll save. You can also add plastic or silicone adhesives to mortar for extra strength when you install ridge tile.

You may also need various types of eaves closures, storm clips, hip-ridge wind block material and flashings.

▼ **Example 7-2:** Assume a roof slope of 5 in 12, a 2-inch eaves overhang, a 10-inch-wide flat tile laid at a 15-inch exposure and the use of roofing tiles sold at 110 pieces per square. Find the quantities for field tiles, rake tiles and under-eaves tiles required for the gable roof in Figure 7-60. Also assume a 3 percent waste factor for field tiles.

From the roof-slope factor in Appendix A, you calculate the net roof area (including a 2-inch overhang at the eaves) as:

Net Roof Area = 31' x 26'4" x 1.083 (the roof-slope factor)
                 = 884 square feet ÷100
                 = 8.84 squares

For this job, you need 8.84 squares at 110 tiles/square, times 1.03 (for the 3 percent waste factor), or 1,002 tiles.

From the roof-slope factor table, the rake length is 13 feet times 1.083, or 14 linear feet.

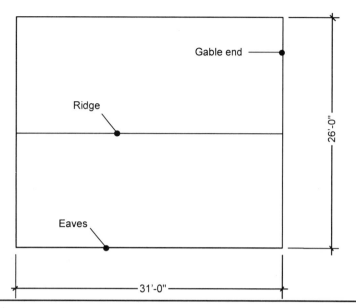

Figure 7-60 Gable roof example

The formula from Chapter 4 for shingle courses also works in this situation. Here's the formula:

Equation 4-3: Number of Courses $= \dfrac{\text{Dimension of Structure}}{\text{Exposure}}$

Just substitute rake length for dimension of structure:

Number of Rake Tiles (Single Rake) $= \dfrac{14 \text{ ft.}}{1.25 \text{ ft.}}$

$= 12$ tiles

There are four rakes, so you need a total of 48 tiles. Notice you round up for each rake, then multiply the result.

The eaves length is 31 linear feet. Use the same formula to find the number of under-eaves tiles:

Number of Under-Eaves Tiles (Each Eave) $= \dfrac{31 \text{ ft.}}{10 \text{ in. (tile width)}}$

$= 38$ tiles

(Note that 10 inches is the same as 0.83 feet.) Double the 38 tiles to get total tiles for both eaves (76 tiles).

Again, the formula also works for the ridge tiles:

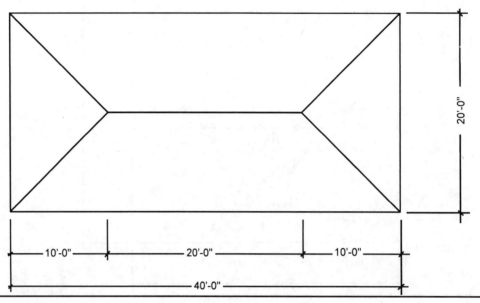

Figure 7-61 Hip roof example

$$\text{Number of Ridge Tiles} = \frac{31 \text{ ft.}}{15 \text{ in.}}$$
$$= 25 \text{ tiles}$$

▼ **Example 7-3:** Assume a roof slope of 6 in 12, and the conditions given in Example 7-2. Find how many field tiles, under-eaves tiles and hip and ridge tiles you need for the roof of the building diagrammed in Figure 7-61. Assume a 3 percent field tile waste.

Using the roof-slope factor from Appendix A, the net roof area, including the 2-inch overhang at the eaves, is:

Net Roof Area = 40'4" x 20'4" x 1.118
= 917 square feet ÷100
= 9.2 squares

You'll need 1043 tiles: 9.2 squares x 110 tiles per square x 1.03 (3 percent waste).

Also, order 49 under-eaves tiles for each of the long eaves (40'4" ÷ 10" = 49), and 25 under-eaves tiles for each of the short eaves (20'4"÷ 10" = 25), for a total of 148.

The ridge is 20 feet long, and each of the hips is 15 feet long (10 foot run times the hip factor of 1.5 from Appendix A), for a total of 80 linear feet for ridge and hips. Therefore, you need 64 ridge and hip tiles (80' ÷ 15" = 64). Of these 64 tiles, four are hip starters. And don't forget the two apex trim tiles where the ridge joins the hips.

Estimating Total Tile Roofing Costs

Let's say you've got to cover the roof shown in Figure 7-60 with "S"-shaped Spanish clay roofing tile. Assume that field tiles cost $75.00 per square, including freight and sales tax. Allow $0.80 per linear foot for ridge units and $1.40 per linear foot for rake units.

From Example 7-2, you must purchase 8.84 squares of field tile, 31 linear feet of ridge tile and (4 x 14' =) 56 linear feet of rake tile. Therefore we have a material cost of:

Field tiles = 8.84 square x $75.00/square = $663.00
Ridge tiles = 31 LF x $0.80/LF   =   25.00
Rake tiles   = 56 LF x $1.40/LF   =   78.00

The total material cost is: $766.00

According to the *National Construction Estimator*, an R1 crew consisting of one roofer and one laborer can install field tile at the rate of 1 square per 3.46 manhours. The crew can install ridge and rake tiles at the rate of 1 linear foot per 0.047 manhours. Assuming the roofer makes $20.00 per hour and the laborer makes $15.00 per hour, including labor burden, the average cost per manhour is: ($20.00 + $15.00) ÷ 2 = $17.50 per manhour. Therefore the labor cost is:

Field tiles = 8.84 square x 3.46 manhours/square x $17.50/manhour
             = $535.00

Ridge tiles = 31 LF x 0.047 manhours/LF x $17.50/manhour
             = $25.00

Rake tiles   = 56 LF x 0.047 manhours/LF x $17.50/manhour
             = $46.00

The total labor cost is: $535.00 + $25.00 + $46.00 = $606.00

So the total cost (excluding overhead and profit) is:

$766.00 + $606.00 = $1372.00

Now, let's move on to the next chapter on slate roofing.

8 Slate Roofing

▶ Like tile, slate roofing is among the more expensive types of roofing: The material itself costs more, it takes longer to install, and slate thicker than 3/8 inch requires a stronger supporting frame. But with the extra cost comes some important advantages.

- If it's well constructed, a slate roof has a normal life expectancy of more than 50 years.
- A slate roof is fireproof, so insurance premiums are lower.
- It's also waterproof, and resists damage from extremes in the weather.
- Slate requires very little maintenance and never needs any kind of preservative.
- The higher cost of a good slate roof is usually reflected in a higher value for the home.
- Slate comes in a wide range of attractive textures and colors.

These advantages make slate a very desirable roofing material, especially with some architectural styles and in some communities.

Slate Size, Color and Texture

Most slates manufactured in the United States are quarried in Virginia, Maine, Maryland, Pennsylvania and Vermont. Slates are manufactured by splitting slate blocks into standard thicknesses of 3/16, 1/4, 3/8, 1/2, 3/4, 1,

L x W	L x W	L x W
26 x 14	18 x 12	14 x 9
24 x 14	18 x 11	14 x 8
24 x 12	18 x 10	12 x 10
22 x 12	18 x 9	12 x 9
22 x 11	16 x 12	12 x 8
20 x 14	16 x 10	12 x 7
20 x 12	16 x 9	12 x 6
20 x 11	16 x 8	10 x 8
20 x 10	14 x 12	10 x 7
	14 x 10	10 x 6

Figure 8-1 Common slate sizes

1¼, 1½, 1¾ and 2 inches. The most common thickness is 3/16 inch (commercial standard slate). The maximum practical thickness for roofing slate is 2 inches, but slate that thick is rarely used in residential construction.

Natural stone can't be split so that every piece is exactly the same thickness, so slight variations, above or below the nominal thickness, are acceptable. If your job requires absolute consistency, each piece will have to be measured. That will increase the material cost and may delay delivery.

The slates used on homes are usually about 3/16-inch thick and one of the standard lengths and widths in Figure 8-1. Slates thicker than ½ inch are manufactured in lengths up to 30 inches. Random width and length slates are also available. Sizes ranging from 10 x 6 inch through 14 x 10 inch are the most plentiful. If you insist on non-standard sizes, you'll pay more and wait longer for delivery.

Slate trade names are based on color. They're known as black, blue black, gray, blue gray, green, mottled green, purple, mottled purple, purple variegated and red. Other colors are considered "specials."

The trade name usually has the word "unfading" or "weathering" before it. Slates from some quarries weather. The color of weathering slate "mellows" after it's exposed to the elements for a few months. Unfading slate keeps its original natural shade.

Based on grade, the trade names of slates are No. 1 Clear, Medium Clear, No. 1 Ribbon and No. 2 Ribbon. The architectural classifications for slate roofs include standard slate, textural slate, graduated slate and flat slate.

Course	Slate thickness	Length	Exposure
Under-eaves course	⅜" thick	14" long	No exposure
First course	¾" thick	24" long	10½" exposure
1 course	¾" thick	24" long	10½" exposure
2 courses	½" thick	22" long	9½" exposure
2 courses	½" thick	20" long	8½" exposure
2 courses	⅜" thick	20" long	8½" exposure
4 courses	⅜" thick	18" long	7½" exposure
5 courses	¼" thick	16" long	6½" exposure
3 courses	¼" thick	14" long	5½" exposure
3 courses	3/16" thick	14" long	5½" exposure
8 courses	3/16" thick	12" long	4½" exposure

Courtesy of Vermont Structural Slate Co.

Figure 8-2 Coursing method for graduated slate roof

Textural slate has a rougher surface texture than standard slate, often with uneven butts and variations in shingle thickness, size, and exposure. Slates may be the same in length but vary in thickness, or the thickness may be the same and the length (and exposure) varies. If the thickness varies, it'll usually be between ¼ to ¾ inch, although thicker slates may be installed at random to give the roof a rough textured appearance.

Slates usually come from the quarry with the holes already punched. Some contractors, however, order their slate unpunched. They've invested in slate punching machines so they can keep their crews busy on idle days or during bad weather. Most slates are punched with two holes, but be sure you punch four nail holes through any slates thicker than ¾ inch, or longer than 20 inches. While most specifications call for slates to be machine punched, you can hand punch with a punch and maul for slates cut to fit at hips.

When slate blocks are split to make roofing slates, the thicker slates tend to have rougher faces and edges than the thinner slates. The thinner slates split more cleanly so they tend to have more even surfaces. The thinner slates may make the roof look flat when used beside the rougher, heavier slates near the eaves. This becomes especially important close to the ridge because the ridge slates are farther from the eye and will already look smoother than the ones lower down on the roof. You may want to special-order thinner slates with a rougher surface.

Slating a Graduated Roof

A graduated roof is slated with the longest and thickest slates at the eaves. Use slates that are shorter and thinner as you move up to the ridge. Graduated roofs combine textural slate with a wide variety of slate sizes, thicknesses, exposures and color. Figure 8-2 shows one method of coursing for a graduated roof. For example, a graduated roof might use slates from 12 to 24 inches long and 3/16 to 1½ inches thick. The head lap should be a constant 3 inches for all slates, regardless of the length.

Be sure to lay random slates on a graduated roof so the vertical joints are broken and offset from those in the courses above and below them. See Figure 8-3.

Felt Underlayment

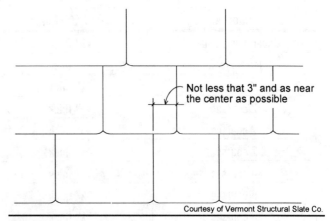

The weight of felt you use for underlayment under slate roofing will depend on the architectural classification of the roof as well as the thickness of slate you install. Standard slate on a residence can be installed

Figure 8-3 Proper jointing

over 15-pound saturated felt. For a textural roof, use 30-pound felt. For slates up to ¾ inch thick on a graduated roof, use one layer of 30-pound felt. For thicker slates, use 45-, 55- or 65-pound roll roofing.

Slate roofing doesn't require felt underlayment to be watertight. Remember that you have to drive nails through the felt to install each piece of slate. The underlayment is just extra insulation and cushioning under the slate surface.

Installation on a Flat Roof

If the roof slope is less than 4 in 12, slate has to be installed with no lap, in a setting bed of asphalt, pitch, or concrete. You can install 3/16-inch thick slate on a flat roof. But if the roof will carry foot traffic, use slates that are at least ¼- to 3/8-inch thick. The 3/16-inch slates have enough wearing surface, but 3/8-inch slates go into the bedding better and stay more securely in place under foot traffic. Make sure you order unpunched slate.

Slate for flat roofs comes in standard sizes of 6 x 6, 6 x 8, 6 x 9, 10 x 6, 10 x 7, 10 x 8, 12 x 6, 12 x 7 and 12 x 8. Set slates that are less than ¾ inch thick in asphalt or pitch. Set slates ¾ inch and thicker over a ¾-inch setting bed of concrete that you've installed over waterproofing felt. Figure 8-4 shows weights of flat slate roofs.

Installation on a Sloping Roof

The weight of slate on a sloping roof varies tremendously, depending on the thickness of the slate, the head lap, and the weight of the slate itself, which varies. For instance, the weight of standard slate varies from 700 to 800 pounds per square, but thicker slates vary from 800 to 3600 pounds per square. See Figure 8-5. Rafters that are strong enough to hold wood shingles or shakes are usually strong enough to hold 3/16-inch-thick commercial standard slate when you lay it with a 3-inch head lap.

Slate Roofing

Weights of flat slate roof without concrete bedding slab			
Materials	Weight of materials per sq. (100 sq. ft.)	Total weight per sq. (100 sq. ft.)	Total weight pounds per sq. ft.
Waterproofing (Weight varies, assumed here to be 150 lbs.)	150		
3/16" slate	250	400	4.0
1/4" slate	335	485	4.85
3/8" slate	500	650	6.5
1/2" slate	675	825	8.25
3/4" slate	1,000	1,150	11.5
1" slate	1,330	1,480	14.8
Weight of flat slate roof with concrete bedding slab			
Materials	Weight of materials per sq. (100 sq. ft.)	Total weight per sq. (100 sq. ft.)	Total weight pounds per sq. ft.
Waterproofing (Weight varies, assumed here to be 150 lbs.)	150		
3/4" concrete bed	750		
3/16" slate	250	1,150	11.50
1/4" slate	335	1,235	12.35
3/8" slate	500	1,400	14.00
1/2" slate	675	1,575	15.75
3/4" slate	1,000	1,900	19.00
1" slate	1,330	2,230	22.30

Courtesy of Vermont Structural Slate Co.

Figure 8-4 Weights of flat slate roofing

Thickness (inch)	Weight (lbs/square)
3/16	700 - 800
1/4	900 - 1000
3/8	1300 - 1400
1/2	1700 - 1800
3/4	2500 - 2800
1	3400 - 3600

Courtesy of Vermont Structural Slate Co.

Figure 8-5 Weights of slate roofing per square with 3-inch head lap

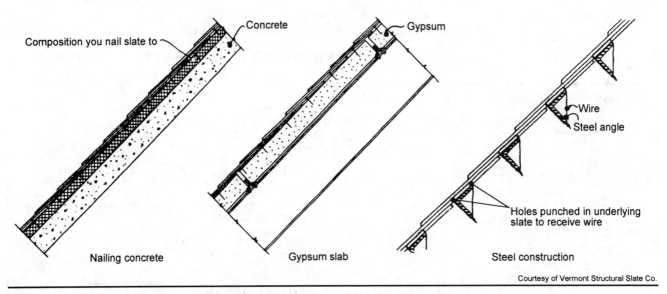

Figure 8-6 Slates on various types of fireproof construction

In residential construction, you usually install slate over solid wood sheathing or plywood. For fireproof commercial work, you can install slates over a variety of substrates, including concrete, gypsum slabs and steel purlins, as shown in Figure 8-6. Over steel purlins, install slates punched with four holes and run wire through each pair of holes and around the steel angle. Then twist the ends of the wire to draw the slate down tightly to the angle.

After you've put in the underlayment, drip edge and valley flashing material, install a starter course (under-eaves slates) at the eaves of the roof. This course is usually applied over a wood cant strip. Allow a 2-inch overhang at the eaves and a ½-inch overhang at the rakes.

To figure out how long the under-eaves slates should be, add 3 inches to the exposure you use on slates in the main part of the roof. For example, if you use a 6½-inch exposure at the center of the roof, use 9½-inch slates at the eaves. You can either lay half slates, or use full slates with the long edge laid parallel to the eaves. You can use under-eaves slates that are the same, or only half as thick, as the field slates.

Install the first slate course over the under-eaves slates with the butts of both courses set flush and the vertical joints offset.

The head lap you use to lay slates will depend on the slope of the roof. Use a 4-inch head lap over slopes between 4 in 12 and 8 in 12. Reduce the head lap to 3 inches on slopes between 8 in 12 and 20 in 12. You can cut the head lap to 2 inches on roofs steeper than 20 in 12, or on vertical walls. Figure 8-7 shows recommended head laps.

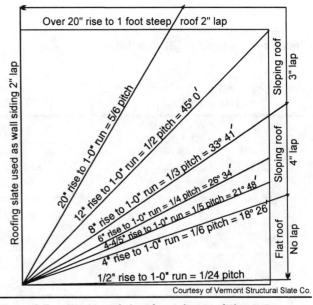

Figure 8-7 Head lap of slate for various roof slopes

Length of slate (inches)	Exposure at slope 8" to 20" per foot, 3" lap (in inches)
24	10½
22	9½
20	8½
18	7½
16	6½
14	5½
12	4½
10	3½

Courtesy of Vermont Structural Slate Co.

Figure 8-8 Exposure for sloping roofs

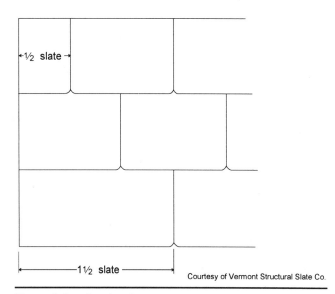

Courtesy of Vermont Structural Slate Co.

Figure 8-9 Starting slates

The part of each slate exposed depends on the head lap. Here's how you figure the exposure:

$$\text{Exposure} = \frac{L - HL}{2}$$

where: L = Length of shingle
HL = Head lap

So, the exposure of a 24-inch slate laid up with a 3-inch head lap is:

$$\text{Exposure} = \frac{24 \text{ inch} - 3 \text{ inch}}{2} = 10\frac{1}{2}"$$

Figure 8-8 shows exposures required for various lengths of slate installed with a 3-inch head lap. The exposure used on a graduated slate roof will often vary, as shown in Figure 8-2. It shows a typical variation in exposure.

Install slates so the joints in each course are offset from the joints in courses above and below. If you're installing random-width slates, lay the slates so that a joint of the underlying slate is located as close to the center of the overlying slate as possible. Offset joints in adjacent courses at least 3 inches. Refer back to Figure 8-3.

When you use slates that are all the same width, proper jointing is taken care of automatically by starting every other course with a half slate. For a more interesting and random jointing pattern, install slates as shown in Figure 8-9.

Install slates with staggered joints all the way up the roof. Adjust the exposure of courses below the ridge so the ridge course of slates doesn't have to be trimmed. The top course should be both nailed and set in spots of roofing cement to keep the wind from lifting them up. Figure 8-10 shows this.

Roofing Construction & Estimating

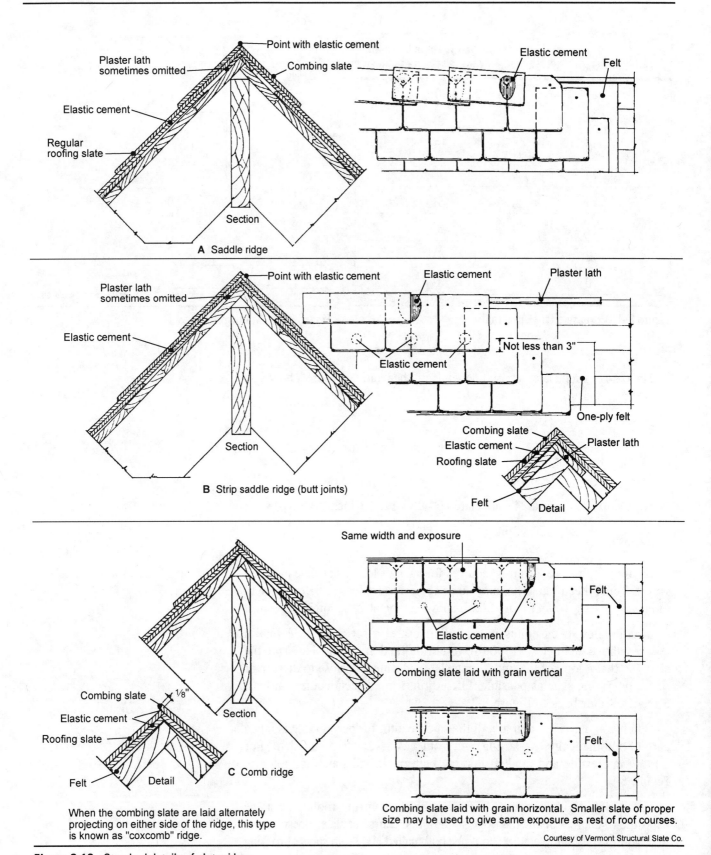

Figure 8-10 Standard details of slate ridges

Length of slate (inches)	Spacing of lath (inches)
24	10½
22	9½
20	8½
18	7½
16	6½
14	5½
12	4½

Courtesy of Vermont Structural Slate Co.

Figure 8-11 Spacing of nailing strips

Normally, you nail slates directly to the sheathing. However, in climates without wind-driven rain or snow, you can omit the roofing felt and nail the slates directly on 1 x 2 or 1 x 3 nailing strips installed parallel to the eaves. Space the nailing strips so they're appropriate for the length of slate and head lap you're using. If you use a 3-inch head lap, space the nailing strips as shown in Figure 8-11.

Ridges

There are three basic types of slate ridges: the saddle ridge, the strip saddle ridge and the comb ridge. Look again at Figure 8-10.

To create any of these ridges, you lay slate right up both sides of the roof so slates from both sides butt flush together at the roof top. Then cover the ridge with *combing slate*. A comb is the ridge of a roof, so combing slate is the topmost row of slates, which projects above the ridge line. Usually the grain of this slate should run horizontally. Nail the combing slate through joints in the underlying slates, not through the slates themselves. Use combing slate that's wide enough so you can make the exposure of the uppermost slate course about the same as the exposure on the rest of the roof. Also, be sure the joints of combing slates are offset from the joints of slates underneath them.

Apply elastic roofing cement along the joint at the peak of the roof. Cover all nail heads with cement. Apply more cement under the unnailed ends of combing slates.

■ **Saddle Ridge** Install combing slate from both sides of the ridge with butts flush. Install each combing slate with two nails and overlap adjacent slates so the nails are covered. See Figure 8-10 A.

■ **Strip Saddle Ridge** Install each combing slate with four nails and butt the end joints over a bed of elastic cement. See Figure 8-10 B.

■ **Comb Ridge** Install about the same way as a strip saddle ridge except you extend the combing slate on one side of the ridge about 1/8 inch beyond the top of the combing slate from the other side. See Figure 8-10 C.

The *coxcomb ridge* is a variation of the comb ridge. You lay the combing slates so they alternately project on either side of the ridge.

Hips

There are five basic types of slate hips. These are the saddle hip, the strip saddle hip, the mitered hip, the Boston hip and the fantail hip. They're shown in Figure 8-12.

On all of these hips, drive nails through the joints of the underlying slates. Apply elastic cement along the joint at the centerline of the hips, over all nail heads and at the lower ends of the hip slates. In addition to nailing, set the slates underlying the hip slates into spots of roofing cement. This keeps slates from blowing off in high winds.

■ **Saddle Hip** Nail each hip slate over a lath strip or 3½-inch cant strip. Install slates that are the same width as the exposure you're using on the main roof. Install each slate with four nails and overlap adjacent slates so the butts of the hip slates line up with the butts of the slate courses on the main roof. See Figure 8-12 A.

■ **Strip Saddle Hip** Install the same way as a saddle hip, except you lay the narrower hip slates up with butted joints that don't necessarily line up with butts of the slate courses on the main roof.

■ **Mitered Hip** Cut the hip slates into triangular pieces so they fit together, forming a tight joint over the hip. The slate roof courses and hip slates are all in the same plane. I recommend that you install metal flashing under mitered hip slates. Why chance a leak in such an expensive roof? See Figure 8-12 B.

■ **Boston Hip** Install the same way as a saddle hip, except cut the hip slates into trapezoidal-shaped pieces. See Figure 8-12 C.

■ **Fantail Hip** Install like a mitered hip, except you cut the bottom corner along the hip at an angle to form the fantail. See Figure 8-12 D.

Slate Roofing

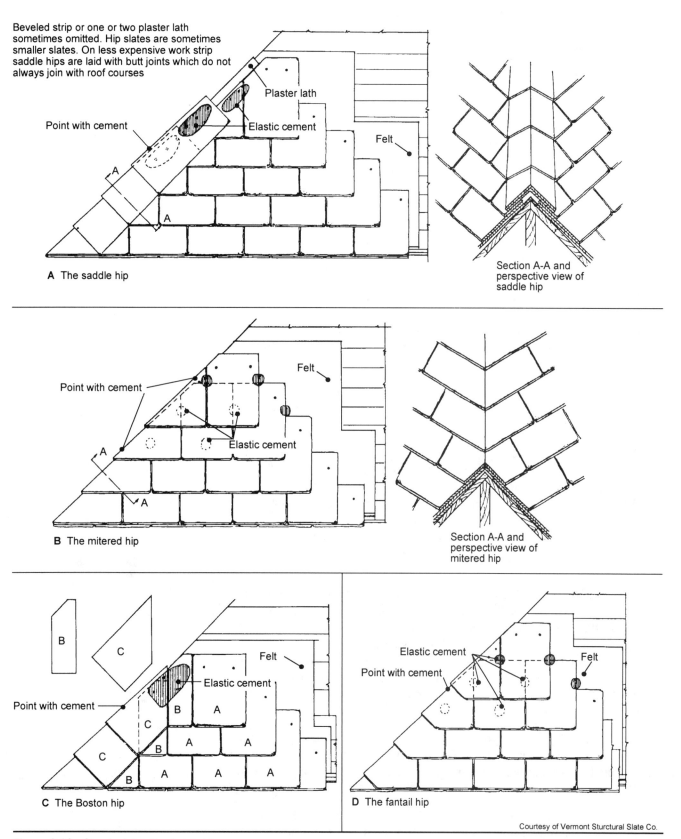

Figure 8-12 Standard details of slate hips

Valleys

There are four basic types of slate valleys. These are the open valley, the closed valley, the round valley and the canoe valley. See Figure 8-13.

■ **Open Valley** It's the easiest to install. Install slates at the upper end of the valley to within 2 inches of each side of the centerline of the valley. Widen this distance about $1/8$ inch per foot going down the valley. See Figure 8-13 A. Flash the valley with metal that's wide enough to extend at least 4 inches under the slate. Six to 8 inches is even better. Use 15-inch wide, 16-ounce copper as valley flashing on slate roofs. Fabricate the valley flashing with a center crimp and water guards. Use cleats spaced on 8- to 12-inch centers to secure the flashing to the sheathing. See Figure 8-13 B. You can also use zinc-coated metal no thinner than 0.0179 inch.

■ **Closed Valley** Trim and work the slates tightly into the valley centerline. Usually you flash a closed valley with sheets of copper fabricated with a center crimp but no water guards. See Figure 8-14. Nail the flashing to the sheathing each 18 inches along the edges of the flashing. Be sure that the nails you use to secure the slates don't go through the flashing.

Fabricate copper valley sheets no longer than 8 feet. If the valley is more than 8 feet long, use two sheets and lap the upper sheet over the lower by at least 4 inches.

You can also flash a closed valley by placing copper pieces under each course of slate entering the valley. See Figure 8-13 C. Make each piece long enough so it extends 2 inches above the top of the underlying slate, and overlaps the underlying piece of flashing by 3 inches. Install each piece flush with the butt of the overlying slate. Make each piece wide enough to extend at least 4 inches beyond each side of the centerline of the valley.

■ **Round Valley** This makes an attractive transition between intersecting slopes when installed on a graduated or textural roof. See Figure 8-13 D. To make the valley slates fit, they must be tapered toward the bottom. To make them lie flat, use valley slates 4 inches longer than the slates you use on the corresponding courses on the roof, and set the top edge of the slate in a mortar bed. This is called *shouldering* the slates. Usually you flash the valley with metal or mineral-surfaced roll roofing cut to the proper radius.

The type of foundation you use to support a round valley will depend on the shape of the valley. In a valley with a slight curve, cut tapered edges on 1 x 12s and nail them into the angle formed by the intersecting roofs.

For a round valley of substantial curvature, cut 3-inch blocks to fit the valley angle. Saw them to the proper radius, then nail them horizontally over the sheathing. Space the blocks vertically at about the same distance

Slate Roofing

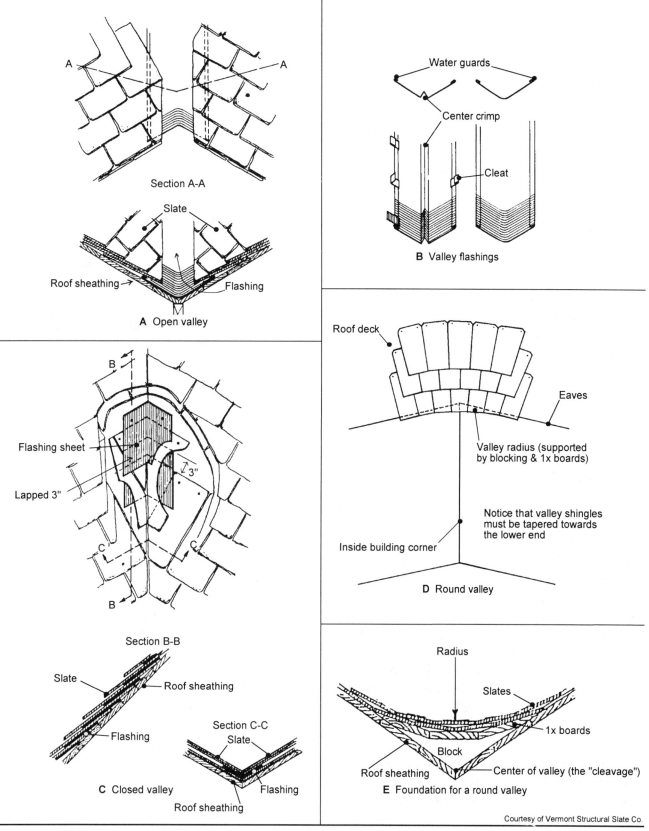

Figure 8-13 Standard details of slate valleys

243

as the exposure of the slates so the blocks form nailing strips under the slate. You'll have to use smaller blocks as the valley gets smaller where it approaches the ridge.

Another method used to form the foundation is a combination of the two methods previously discussed. In this case, you cut 3-inch blocks to fit the curve of the valley. Space them vertically on 20- to 30-inch centers up the valley. Then nail tapered 1 x 2s or 1 x 3s lengthwise up the valley, over the blocks. See Figure 8-13 E.

The radius of a round valley is greatest at the eaves and goes to practically zero at the ridge. The minimum radius allowed at the valley is 26 inches. If roof conditions don't permit this radius, you can use a variation of the round valley called a *canoe valley*. Install a canoe valley the same way as a round valley except that the radius at the eaves and ridge will be practically zero. Then you gradually increase the radius until it reaches its maximum halfway between the eaves and ridge. The result is that the rounded part of the valley is widest halfway up the valley, thinning toward the ridge and eaves. That forms a surface that's canoe-shaped when you see it from above.

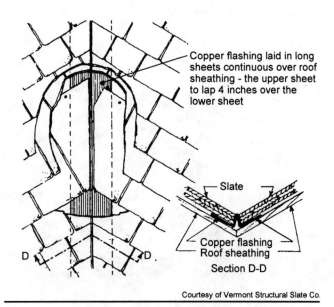

Figure 8-14 Flashing a closed valley using long sheets underneath slate

Fasteners

Slate is a high-quality, durable roofing material. But it has to be installed correctly to perform as expected for many years. That means you have to punch nail holes properly, nail the slates carefully, and use the correct nails.

Drive nails with the heads *just touching* the top of the slate. The slate should hang on the nail, as shown in Figure 8-15 A. If you drive the nail too deep, the nail head will shatter the slate around the nail hole. Eventually the slate will "ride" up the nail and blow off in a heavy wind. Figure 8-15 B shows this condition. If you don't drive the nail deep enough, the head will probably crack the slate above when someone walks on it. Figure 8-15 C shows this.

The shaft of a roofing nail isn't strong enough to withstand the shearing stress of the slates, and its head isn't large enough to keep the slates from being lifted. It's best to use cut copper slating nails on a slate roof.

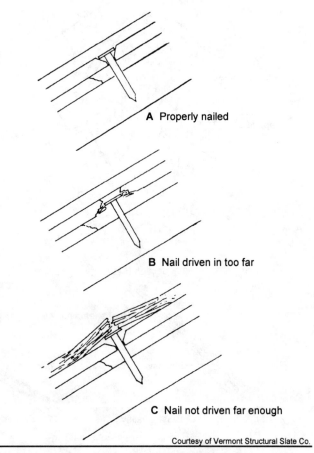

Figure 8-15 Proper nailing for slate roofing

But you could also use cut brass, cut yellow-metal or cut zinc slating nails to fasten slates to the sheathing. In dry climates, hot-dipped galvanized nails are acceptable.

Use 3d nails for commercial standard slates up to 18 inches long, and 4d nails for longer slates. Use 6d nails on hips and ridges. Thicker slates need longer and heavier-gauge nails. As a rule of thumb, use nails 1 inch longer than twice the thickness of the slate. If you're using ¾-inch slate, you'd use 2½-inch nails.

Install each slate with two nails, except for slates thicker than ¾ inch or longer than 20 inches, which need four nails.

Flashing

Install flashing on a slate roof as explained in Chapter 4. Figures 8-16 through 8-18 show basic flashing methods for slate roofs. Use top quality flashing that will last as long as the slate roof itself.

When you use copper flashings, install at least 16-ounce soft copper (16 ounces per square foot). Lock and solder the joints of all base flashings (Figure 8-17). You don't need to solder cap flashings. When soldering copper seams, be sure both sides of each sheet to be soldered are tinned 1½ inches along the edge. Tinning (precoating) means coating a metal with solder or another tin alloy before soldering or brazing to help the solder flow into the joint. Then use a rosin-flux solder, and make sure each joint is thoroughly soldered. Otherwise, the joints may come loose when the metal expands and contracts.

Electrolysis (electrical current passing between dissimilar metals) can dissolve copper. So insulate the joint between copper and iron or steel. Place lead strips between copper and iron or steel, or else heavily tin the iron.

When you use tin flashing, install at least 28-gauge (12-ounce) tin, and lock and solder all base flashing joints. Always paint the underside of tin flashing with metal or bituminous paint.

When you use lead flashing, use 3-pound hard lead for all flashing except for cap flashing, which can be 2½-pound. Install lead so that it can expand and contract. Never drive nails through lead. Instead, use cleats of 16-ounce soft copper or 3-pound hard lead. Fasten the cleats to the sheathing with hard copper wire flat-head nails at least ¾ inch long, or with brass screws equipped with lead shields. Join lead sheets with locked seams beaten with wooden hammers. Don't solder the seams.

Roofing Construction & Estimating

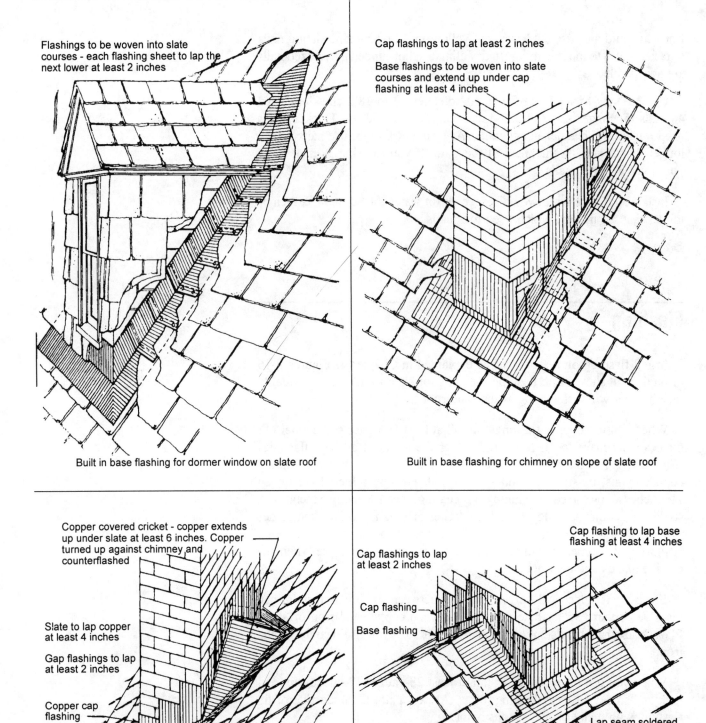

Figure 8-16 Standard flashing details

Slate Roofing

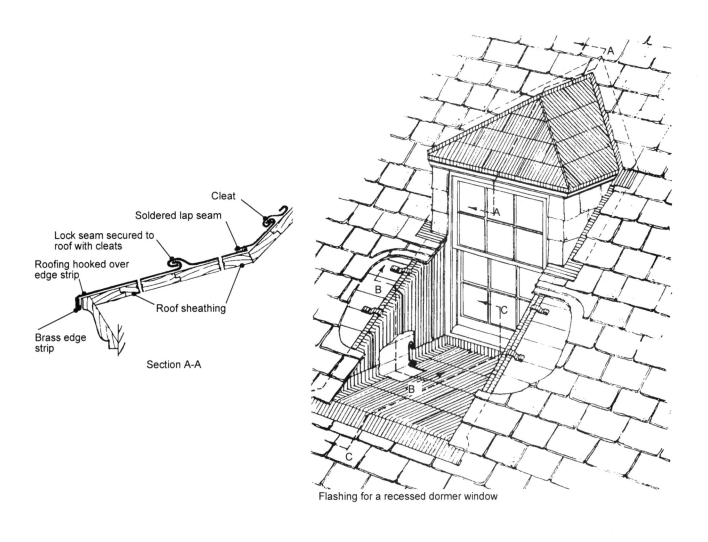

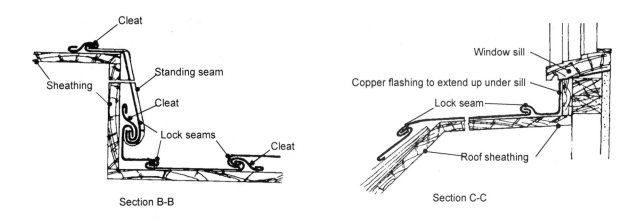

Figure 8-17 Flashing a dormer

Courtesy of Vermont Structural Slate Co.

Roofing Construction & Estimating

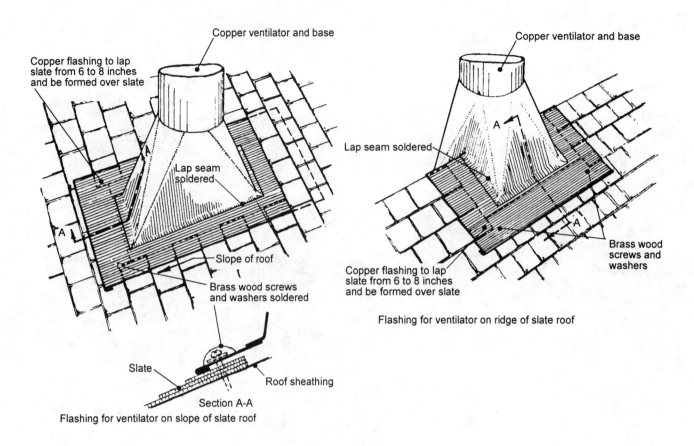

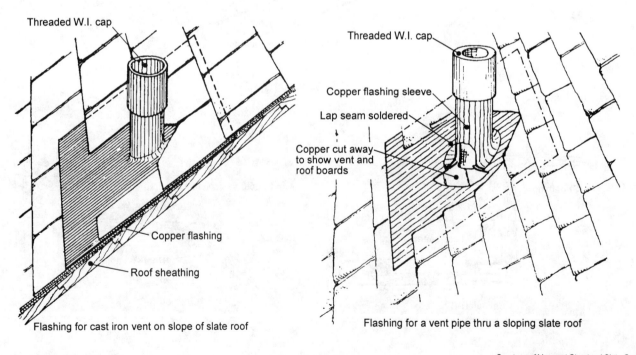

Figure 8-18 Vent flashings

Slate Roofing

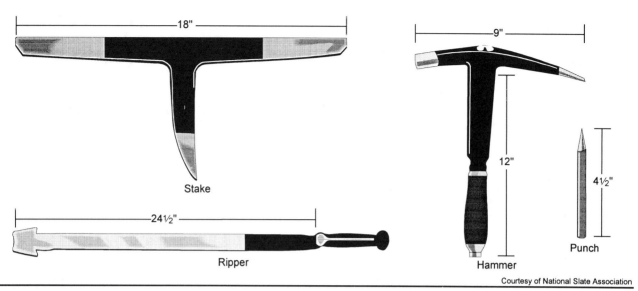

Figure 8-19 Slater's tools

When you use zinc flashing, install at least 11-zinc-gauge (0.024 inch thick) rolled zinc. Don't drive nails through zinc. Use zinc clips or cleats instead (see Figures 8-13 B and 8-17). Use an acid-flux solder to solder all joints.

Slater's Tools

Slate roofers use specialized tools. These include the punch, hammer, ripper and stake shown in Figure 8-19. The punch is 4½ inches long and is used for punching countersunk nail holes in the slates. The hammer also has a sharp point for punching holes in slate. It has a claw at the center of the head for removing nails. One edge of the shank is sharp, for cutting slate. The ripper is about 24 inches long and is used to remove broken slate. The hook at the end of the blade is used to cut and remove slating nails. The stake is 18 inches long and is used as a "work bench" for cutting and punching slates. The short tapered arm is driven into a plank to hold the tool in place. Place slate over the long edge and cut or punch as required.

Estimating Slate Quantities

The first step is to take off slate quantities by the square foot. Then you convert quantities into squares of slates required. Slate prices are based on the number needed to cover a square of roof surface when laid with a 3-inch head lap. Figure 8-20 shows the quantity needed to cover a square of roof surface at each of three different head laps.

Size of slate	Number of slates required per square		
L x W	2" head lap	3" head lap	4" head lap
26 x 14	86	89	94
24 x 14	94	98	103
24 x 12	109	114	120
22 x 12	120	126	133
22 x 11	131	138	146
20 x 14	114	121	129
20 x 12	133	141	150
20 x 11	146	154	164
20 x 10	160	170	180
18 x 12	150	160	171
18 x 11	164	175	187
18 x 10	180	192	206
18 x 9	200	213	229
16 x 12	171	184	200
16 x 10	206	222	240
16 x 9	229	246	267
16 x 8	257	277	300
14 x 12	200	218	240
14 x 10	240	261	288
14 x 9	267	291	320
14 x 8	300	327	360
12 x 10	288	320	360
12 x 9	320	355	400
12 x 8	360	400	450
12 x 7	411	457	514
12 x 6	480	533	600
10 x 8	450	515	600
10 x 7	514	588	686
10 x 6	600	686	800

Figure 8-20 Number of slates required at various head laps

Here's how to find the quantity required if you're not using a 3-inch head lap:

Squares Ordered per Square Covered equal:

$$\frac{\text{Slates per 100 SF at your head lap}}{\text{Slates per 100 SF at 3-inch head lap}} \times \text{Roof area (in squares)}$$

Equation 8-1

Here's an example. Suppose you have a roof area of 500 square feet and you're going to use 12 x 10-inch slates with a head lap of 2 inches, 3 inches, or 4 inches:

First, divide the total roof area by 100 to find the area in squares:

Squares of roof area = Roof Area ÷ 100
= 500 ÷ 100
= 5 squares

Using the values from Figure 8-20 for 12 x 10-inch slates with a 2-inch head lap, you need:

$$\frac{288}{320} \times 5 = 4.5 \text{ squares}$$

So, if you're using a 2-inch head lap, you have to order only 4.5 squares of slate to cover the area.

With a 3-inch head lap, you need:

$$\frac{320}{320} \times 5 = 5 \text{ squares}$$

With a 4-inch head lap, you need:

$$\frac{360}{320} \times 5 = 5.63 \text{ squares}$$

Under-eaves Slates

Slate is installed with a double starter course at the eaves. The formula to figure how much under-eaves starter-course material you need is from Chapter 4:

Starter Course (SF) = Eaves (LF) x Exposure Area (SF/LF)

Equation 4-10

For example, material needed for 80 LF of eaves with a 6-inch exposure is:

Starter-course material (in square feet) = 80 LF of eaves x 0.5 SF/LF
= 40 SF

A more practical formula, since you usually order under-eaves slates by the piece, is:

$$\text{Number of Eaves Shingles} = \frac{\text{Eaves Length (ft.)}}{\text{Slate Width (ft.)}}$$

Equation 4-12

For example, the number of 8-inch-wide under-eaves slates needed for 80 linear feet of eaves (regardless of the exposure) is:

$$\text{Number of under-eaves slates} = \frac{80 \text{ LF}}{0.67 \text{ LF}} = 120 \text{ slates}$$

Cutting Waste at Hips and Valleys

You can account for most of the cutting waste and breakage on a slate roof by adding 1 square foot of area to be covered for each linear foot of hip or valley. Waste beyond this will depend on the complexity of the roof and the skill of the roofing crew.

Hip and Ridge Slates

All slate roofs require *ridge* slates. However, the Boston and saddle hips are the only hips that require hip slates. You don't necessarily install hip and ridge slates at the same exposure as field slates, and they aren't necessarily the same size as the field slates. Figure the number of hip and ridge slates you need using the formula adapted from Chapter 4:

$$\text{Courses} = \frac{\text{Dimension of Structure}}{\text{Exposure}}$$

Equation 4-3

Since each side requires a slate, we need:

$$\text{Hip and ridge slates} = 2 \times \frac{\text{LF of ridge and hip}}{\text{Exposure}}$$

Equation 8-2

Here's an example using the roof shown in Figure 8-21. Assume a roof slope of 5 in 12 and a 2-inch overhang at the eaves. Suppose you use 16 x 8-inch field slates and 16 x 9-inch saddle hip and ridge slates. Also assume an exposure of 10 inches for the saddle hip and ridge slates.

First, the total length of the eaves is 2 x (40' + 20'), or 120 linear feet. Since the roof slope is 5 in 12, you'll install the slates with a 4-inch head lap (according to Figure 8-7). Using a 4-inch head lap, the exposure is 6 inches (material length less the head lap, divided by 2).

Since the under-eaves slates must be only 3 inches longer than the exposure, you can use 12 x 9-inch slates laid horizontally at the eaves. The number of 12 x 9-inch under-eaves slates required is:

$$\text{Number of under-eaves slates} = \frac{120 \text{ feet}}{1 \text{ foot}} = 120 \text{ slates}$$

Using the roof-slope factor from Appendix A, the net roof area, including the 2-inch overhang at the eaves, is:

40'4" x 20'4" x 1.083 = 889 square feet

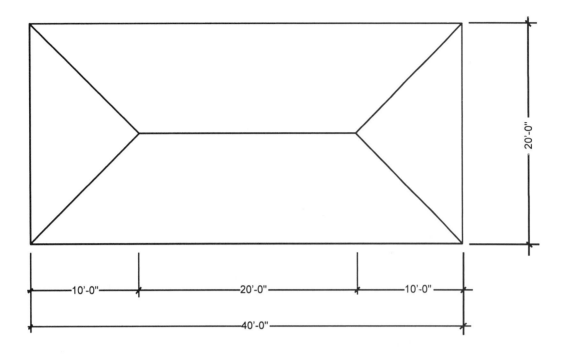

Figure 8-21 Hip roof example

So you need 8.9 squares of field slates. (Use this quantity to determine labor costs.)

The length of any hip is 14.74 linear feet (from Column 3, Appendix A). The total length of all hips is 4 times that, or 59 linear feet. Therefore, include 59 square feet of slate in the estimate for cutting waste based on 1 square foot per linear foot of hip.

Now, the total gross area covered by the field slates, including waste, is:

889 + 59 = 948 SF

Divide that by 100, for 9.5 squares. Now, figure the number of 16 x 8-inch slates for the field (using a 4-inch head lap)

9.5 squares x 300 units per square (from Figure 8-20) = 2,850 slates

But remember, you have to order slate based on coverage using a 3-inch head lap. So to convert, use the formula in Equation 10-1:

Squares (16 x 8 slates) = $\dfrac{300}{277}$ x 9.5 squares = 10.3 squares

That's the number of *squares* to order to get the number of *slates* you require:

10.3 squares x 277 slates per square = 2,853 slates

The total length of hips and ridge is 79 linear feet (59' + 20').

The total number of 16 x 9-inch hip and ridge slates (using Equation 8-2) required is:

$$2 \times \frac{79 \text{ feet}}{0.83 \text{ feet}} = 190 \text{ slates}$$

Notice we converted the 10-inch exposure to feet for this calculation.

Estimating Slate Roofing Costs

Estimating the material cost for a slate roof is tricky business. Always get a current price quote and information about availability before you bid a job. Your dealer will need the following information:

1) The type, color and texture required
2) The thickness, such as commercial, standard, ¼ inch, ⅜ inch, etc.
3) The roof classification, such as standard, textural, graduated or flat
4) Number of nail holes required per slate
5) Types of valleys and flashings required
6) Types of hips and ridges required
7) Required delivery date
8) Job location
9) Total quantity of field slate
10) Total linear feet of hips and ridges

Keep in mind that freight from the quarry is a fixed charge based on either carload or less-than-carload lots. The latter carries a freight charge about double that of carload quantities. With this information in hand, your dealer can quote you a realistic price for your materials.

For example, let's assume that you're covering the roof we discussed in the last section. There's a net roof area of 8.9 squares. But to actually cover the roof, you need 10.3 squares of 16 x 8-inch field slates, 120 each 12 x 9-inch under-eaves slates and 190 each 16 x 9-inch hip and ridge slates. Let's assume your dealer quotes a material price of $500 per square, including freight and sales tax, for all the slate you'll need. So the material will cost:

10.3 squares x $500 per square = $5,150

According to Craftsman's *National Construction Estimator,* a crew of one roofer and one laborer can install a square of slate in 11.3 manhours. If the roofer makes $25 per hour and the laborer makes $20, including labor burden, then the average manhour cost is $22.50 per hour (25 + 20 ÷ 2 = 22.50). So the labor cost for installing the roof is:

8.9 squares x $22.50 per hour x 11.3 hours per square = $2,263

The labor cost is $2,263. Assume a 30 percent markup:

(5150 + 2263) x 1.3 = 9,637

You'll bid the project at $9,637.

9 Metal Roofing and Siding

▶ Sheet metal used in roofing is made of galvanized steel, stainless steel, monel metal, terne metal, copper, lead, zinc, aluminum, or a combination of metal alloys. Any metal sheet heavier than 30 gauge is called *sheet metal*.

You can expect a metal roof to last 25 to 30 years. Although some types can rust, you can restore them with a special paint that contains a plastic fiber which covers the metal and helps prevent further deterioration. Metal panels are rigid and highly resistant to wind uplift. But if even a corner works loose, a strong wind can blow the entire panel off.

Metal panels usually have a low fire rating because they are such good conductors of heat. The only way to mitigate this is to surround the panels with fireproofing material. And unless there's insulation between the roof and the building interior, metal panel roofs are very noisy during rain or hailstorms.

Sheet metal roofing and siding are generally made from cold-rolled metals. Sheet metals are either cold-rolled or hot-rolled. Several properties of the sheet metal depend upon the metal ingot's temperature during rolling. Cold-rolled sheet metal has a lower carbon content. While it's more malleable than hot-rolled sheet metal, cold-rolled sheet metal has the disadvantage of being weaker.

Some materials are factory-coated with other metals or paint to improve the base material's corrosion resistance and appearance. Metal coatings are usually applied with the hot-dip process.

Figure 9-1   Modern ribbed metal panel roof

Before you paint a new galvanized metal roof, clean it with muriatic acid. If the roof has already weathered for at least a year, clean it only with water before you paint.

Modern Metal Panel Systems

There's an endless variety of modern decorative sheet metal systems available for roofing and siding. Figure 9-1 is an example of a ribbed panel metal roof. There are also systems for mansard roofs, fascia and soffit systems. Some systems are designed to be applied over an existing roof. Others are made of a sandwich of metal faces with a foam insulation core. These systems are made of different metals, and they have different surface shapes and finishes. Some systems can be installed on roof slopes as low as ¼ in 12.

Decking Requirements for Metal Roofing Panels

Structurally-sound panels such as corrugated steel can be placed over wood purlins, as in Figure 9-2, or over steel purlins. The installation method depends on the type of metal roofing. Wood purlins are commonly 1 x 4s or 2 x 4s spaced at 16- to 48-inch centers, depending on the roof slope and the anticipated snow load. Steel purlins are normally spaced on 4-foot centers.

Metal Roofing & Siding

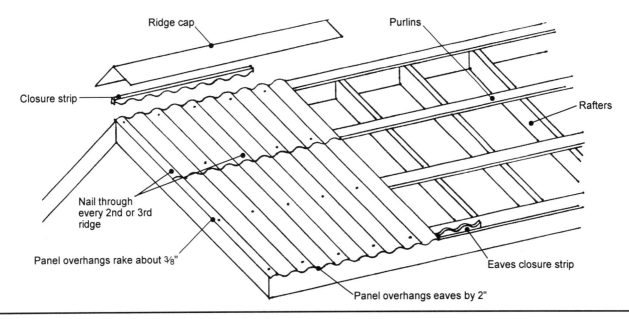

Figure 9-2 Corrugated panels installed over wood purlins

Most other metal roofing panels, including copper, lead, monel metal, terne metal and zinc panels, must be installed over solid sheathing. This is because they have very little structural integrity and bend under foot traffic. Terne-coated stainless steel can be installed over a fluted metal deck as in Figure 9-3.

Installing Metal Roofing Panels

To apply corrugated metal roofing panels, install a closure strip like the one in Figure 9-2 along the eaves of the roof. Then install the first row of panels along the eaves with the ribs parallel to the slope. Allow a 2-inch overhang at the eaves and a 3/8-inch overhang at the rakes. Nail or screw the panels through every other panel ridge into the underlying purlins. Always drive fasteners through the ridges of corrugated panels rather than in the valleys where water flows.

Use self-tapping cap screws or stove bolts to fasten panels to a metal frame. Overlap the panel sides by two corrugations. Install the second row of panels overlapping the ends over the first row by at least 6 inches (on roof slopes 4 in 12 and steeper), or 8 inches (on roof slopes of 3 in 12 to 4 in 12). Seal the end laps with caulking compound on roof slopes less than 4 in 12.

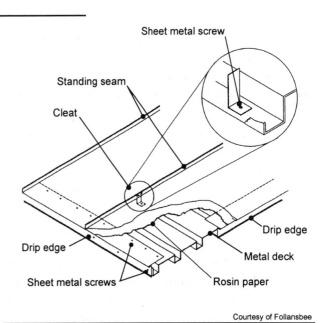

Figure 9-3 Installation over a fluted metal deck

257

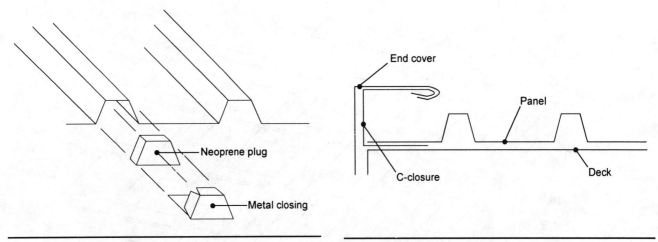

Figure 9-4 Eaves sealed with closure caps

Figure 9-5 Rakes sealed with an end cover

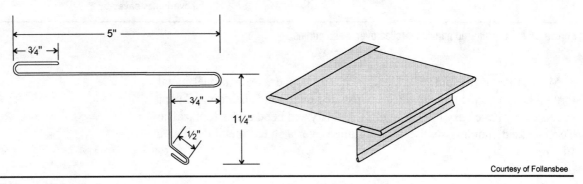

Figure 9-6 Preformed drip edge

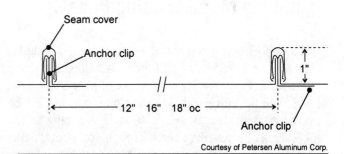

Figure 9-7 Panel ribs with snap-on standing seams

Other metal panel roofs installed without a closure strip may be sealed along the eaves with neoprene plugs covered with metal caps, as in Figure 9-4, and along the rakes with an end cover as in Figure 9-5.

Some metal roofing systems such as terne metal are anchored over a preformed drip edge along the eaves and rakes. This drip edge is 5 inches wide with a ¾-inch hemmed back edge to serve as a water dam. That's shown in Figure 9-6. The drip edge is nailed to wood sheathing or screwed to a metal deck, as in Figure 9-3.

There are many ways to anchor modern metal roofing panels to a deck. Some systems use anchor clips (sometimes called *panel clips*) attached to the deck, or to purlins. The panel rib is attached to the anchor clip and then covered with a snap-on standing seam cover. That's shown in Figure 9-7. Some anchor clips are slotted so the panel can move as it expands and contracts. These clips, called *expansion clips* or *cleats*, are shown in Figure 9-8. Panel laps must be located over purlins since that's where the clips are located.

Metal Roofing & Siding

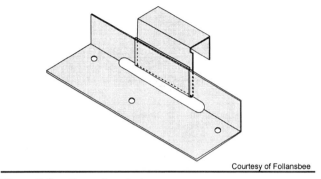

Figure 9-8 Expansion cleat

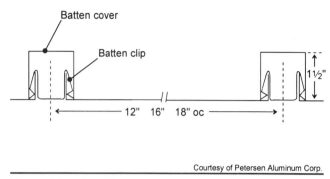

Figure 9-9 Panel ribs with snap-on batten seams

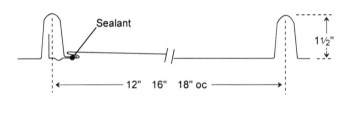

Figure 9-10 Integral standing seams covered by the adjacent panel

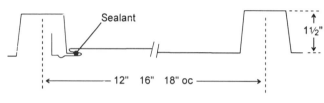

Figure 9-11 Integral battens

Some systems use a batten clip nailed or screwed through the rib and into the deck or purlin. Later, you conceal the clip with a snap-on batten seam cover as shown in Figure 9-9. Other systems are nailed or screwed to the deck or purlins and later covered by an adjacent panel (Figures 9-10 and 9-11) which has an integral standing or batten seam. Integral seams are often made with a factory-applied sealant within the seam, as shown in both Figures 9-10 and 9-11. Still other systems use a standing seam that's sealed by crimping the connection with a seaming machine like the one in Figure 9-12.

Fitting Hips and Valleys

Modern metal panel systems are usually precut at the factory, numbered and delivered with a layout plan showing the locations of the panels. If on-site fabricating is required, flat panels can be cut by a break machine. Ribbed panels can be cut with a portable power saw equipped with a Carborundum blade. Don't worry about making perfect cuts. They'll be concealed by a ridge cap

Figure 9-12 Seaming machine used to crimp standing seams

259

Roofing Construction & Estimating

over the ridge and hips as shown in Figures 9-2, 9-13 or 9-14. Caps may be nailed or screwed into the ridge board and hip rafters, or bolted into steel. Then install the lower edge of each side of the ridge cap over neoprene closure strips as in Figure 9-2, or metal "Z" closure strips as in Figure 9-13. Lap the ridge-cap end joints at least 12 inches.

Ridge caps over standing-seam roofs are often installed over 2 x 2 wood battens spaced to provide ridge venting. Look at Figure 9-14. For exact spacing requirements, check your local building code. For discussion of alternative venting systems for many types of roofs, look ahead to Chapter 13.

The ridge of some standing-seam terne metal roof systems is sealed with a seam like the one in Figure 9-15.

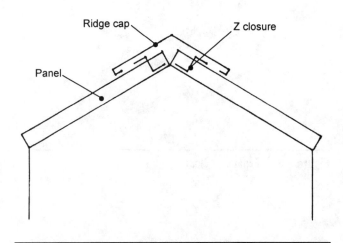

Figure 9-13 Ridge cap with "Z" closure

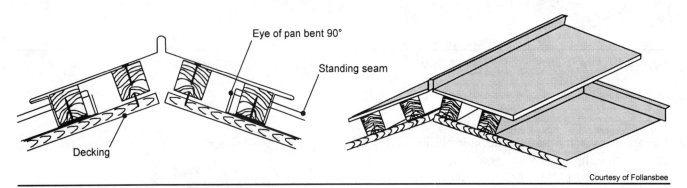

Figure 9-14 Ridge cap over standing-seam roof

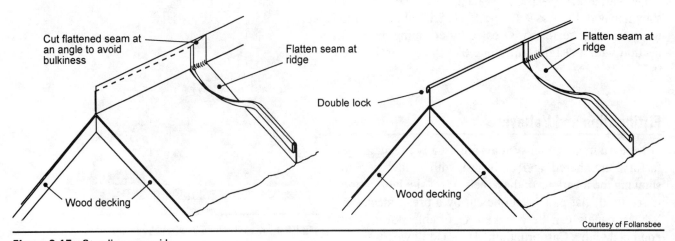

Figure 9-15 Standing-seam ridge

Metal Roofing & Siding

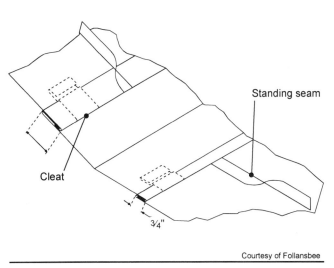

Figure 9-16 Valley flashing installed with cleats

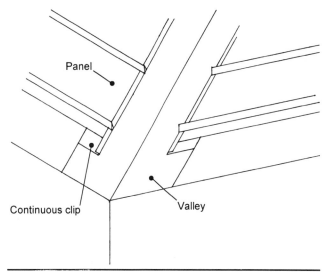

Figure 9-17 Valley flashing installed with a continuous clip

Valley Flashings

You can flash valleys of metal roofs with preformed galvanized iron flashing, or use the same type of metal as on the main roof. Prefabricate the valley flashing with 1-inch water guards along the edges. For even more protection use a valley similar to the one shown in Figure 3-42 back in Chapter 3. This type of valley has a splash diverter running down its center and water guards on both sides as well. The combination provides metal roofs with double protection by preventing water from splashing under the panels adjacent to the valley. Secure a metal valley to the sheathing or the purlins with metal cleats as seen in Figure 9-16, or with a continuous clip, see Figure 9-17. Don't install fasteners into the valley flashing.

If you need more than one piece of flashing over the length of the valley, lap the upper piece at least 6 inches over the lower and seal the lap with roofing cement.

Vent Flashing

On flat-panel metal roofs, you can install rubber vent pipe flashing like that in Figure 9-18. On corrugated roofs, you can't use vent flashing. Instead, pack the roof opening around the pipe with roofing cement. Spread the cement to form a cone that extends about 3 inches up the vent pipe.

Where the roof intersects a vertical surface such as a wall or chimney, install continuous flashing instead of step flashing. That's shown in Figures 9-19 and 9-20.

Figure 9-18 Rubber vent pipe flashing

Roofing Construction & Estimating

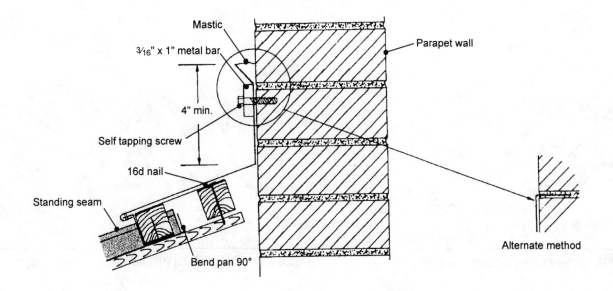

Figure 9-19 Flashing at a wall

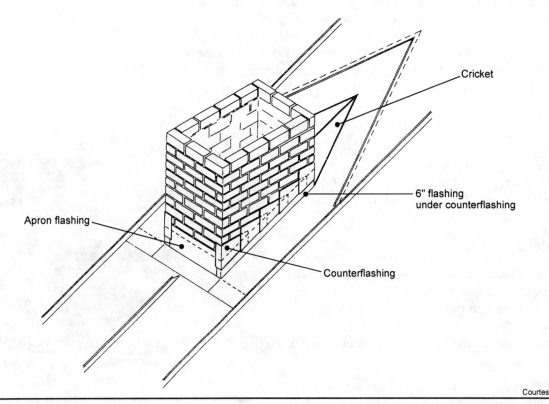

Figure 9-20 Chimney flashing

Metal Roofing & Siding

Fasteners

Use nails, screws and cleats made of the same metal as the roof. The exception is for tin roofs, where you can use steel nails. Use barbed, ringed or screw-shank nails to fasten panels to wood. Use sheet metal screws and rivets to fasten sheet metal to sheet metal. Install neoprene washers under the heads of all exposed screws. In some cases, you can use seams to connect sheets, rather than fasteners.

Closure Strips

Compressible closed-cell neoprene closure strips like the ones in Figure 9-21 seal metal roofing and siding components. Fabricators of metal roofing and siding produce made-to-match closure strips for their product lines. Typically, these closure strips are readily available through suppliers of metal roofing and siding. In exceptional cases the closure strips could be custom-made from wood.

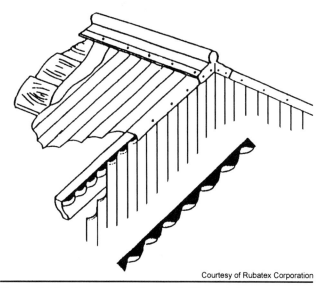
Courtesy of Rubatex Corporation

Figure 9-21 Closure strips for metal siding panels

Job-Fabricated Seams

Sheet metal panels such as copper and terne metal are often joined together by job-fabricated seams. The simplest job-fabricated seam (and the strongest) is the riveted soldered seam shown in Figure 9-22. Use this type

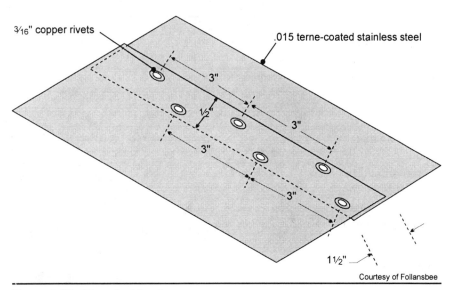
Courtesy of Follansbee

Figure 9-22 Riveted soldered seam

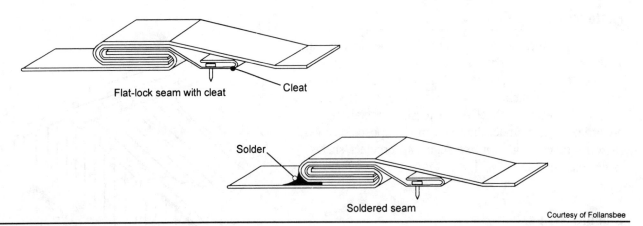

Figure 9-23 Flat-locked seam

of seam for any metal roof that must withstand high thermal stress, heavy snow loading, or a lot of foot traffic. Riveted soldered seams allow a metal roof to expand and contract as a single unit. The result is a very strong roof.

Flat and Flat-Locked Seams

A flat-locked seam is shown in Figure 9-23. Most flat-lock seam metal roofing is factory formed. However, it is possible to site-fabricate flat-lock seams on copper and terne metal roofing. You can use a flat-locked seam on roof slopes as low as ¼ in 12. You can also use this type of metal roofing on a curved surface, like the dome shown in Figure 9-24.

To form flat seams on long panels, the long edges are turned back about ¾ inch. While this is usually done at the factory, you can site-fabricate flat-lock seams with special bending tools. Turn one edge up and the opposite edge down, as in Figure 9-25.

Figure 9-25 shows a flat seam and an expansion batten for thermal expansion. An engineer should determine the battens' size, spacing, and what material should be used for the battens. The proper batten depends on the local climate conditions, the kinds and amounts of load expected, and the expansion coefficient of the metal roofing.

To form flat seams in small panels (normally 14" x 20" or 20" x 28"), notch the corners and turn in all four edges of the sheets about ¾ inch. Turn the top and one

Figure 9-24 Copper dome

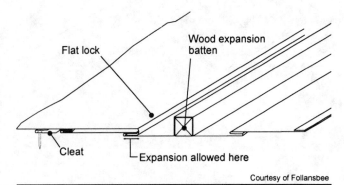

Figure 9-25 Expansion batten

Metal Roofing & Siding

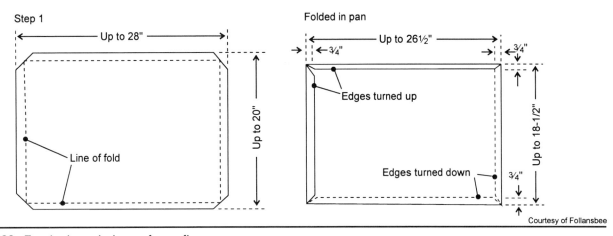

Figure 9-26 Turn back panel edges to form a flat seam

adjacent edge of each sheet up, and the bottom and remaining adjacent edge under to form a flat seam along each panel edge as in Figure 9-26.

On roofs that slope more than ½ in 12, seal the joints of flat seams of most metals with caulking compound or white lead. On roofs with slopes less than ½ in 12, mallet the joints and sweat them full of solder (Figure 9-23).

Fasten each panel to the roof with cleats locked into the seams on each edge. Fasten the cleats with nails but don't drive the nails through the panels. Place the panels (pans) starting at the low point of the roof. Work up-slope, placing the higher pan over the upper edge of the lower pan. Stagger all joints as shown in Figure 9-27.

Standing Seams

The job-fabricated standing seam shown in Figure 9-28 is another way to interlock metal sheets. If you use job-fabricated standing seams, choose panels with pre-formed edges (preformed pans) as in Figure 9-29. Use a seaming machine like the one in Figure 9-12 to crimp the seams. But don't install a standing seam on roof slopes less than 3 in 12.

Use cleats to fasten job-fabricated standing seams to the deck. Cleats for terne metal panels are 2 inches wide, spaced on 12-inch centers. Nail each cleat with two ⅞-inch (minimum) flat-head galvanized roofing nails. The cleat becomes an integral part of the seam as the seam is folded. You can see that in Figure 9-28.

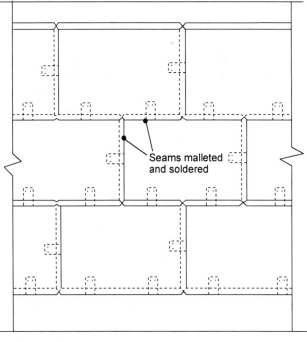

Figure 9-27 Cleats for fastening a flat-seam roof

265

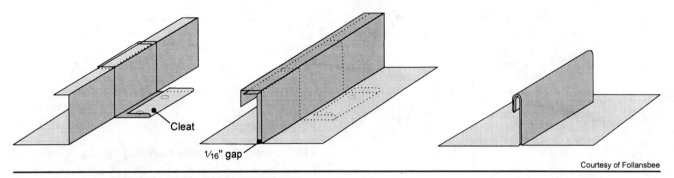

Figure 9-28 Job-fabricated standing seam

Allow 1/16 inch between job-fabricated standing-seam panels so the metal can expand and contract. The maximum recommended distance between standing seams of terne metal is 21 inches. For panels more than 20 feet long, the maximum recommended seam center is 17 inches. For panels more than 30 feet long, install expansion cleats like the ones in Figure 9-8 to allow for thermal movement.

Batten Seams

Another common type of job-fabricated seam is the *batten seam*. Some batten seams are installed over wood batten strips, as shown in Figure 9-30. You can install batten seams on roof slopes of 3 in 12 and steeper. Batten seams are recommended when you're using the heavier sheet metals, or where the coefficient of expansion of the roofing material is significant. (Look ahead to Figure 9-41 for the coefficients of expansion for various metals.) It's a good idea to use batten seams for copper, terne metal, zinc, or aluminum materials. Install 2-inch-wide cleats on 12-inch centers. Nail the cleats into the wood batten, where they become an integral part of the batten as the seam is folded as shown in Figure 9-30.

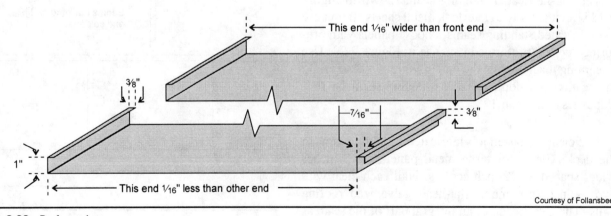

Figure 9-29 Preformed pans

Metal Roofing & Siding

1. Batten seam

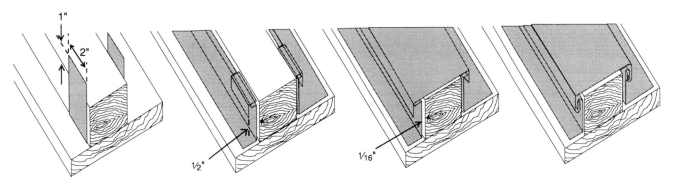

2. Finishing batten end

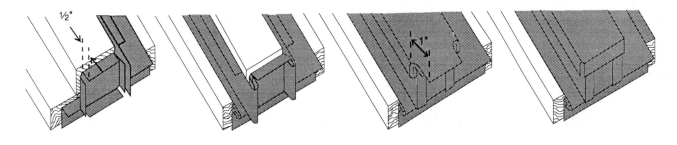

3. Ridge

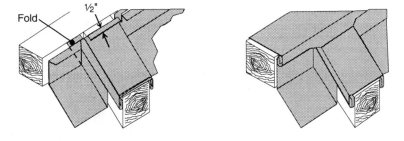

Courtesy of Follansbee

Figure 9-30 Job-fabricated batten seam

Cross Seams

You can get panel systems with the ends of the panels (the cross seams) preformed for a flat-seam lap joint, no matter what kind of side seams are used. Some cross seams are prefabricated (already shaped and with the clip installed) as shown in Figure 9-31. All you have to do is hook the next panel into them.

Other cross seams (like terne metal systems) are job-fabricated, as in Figures 9-32 and 9-33. The type of cross seam used on a terne metal roof depends on the slope of the roof. On roof slopes between 3 in 12 and 6 in 12, install a solder strip 4 inches below the upper edge of the down-slope panel. A solder strip is a strip of metal that is soldered across a panel that forms an attachment point for an overlying panel, as in Figure 9-32. On slopes of 6 in 12 and steeper, you don't need a solder strip at the seam. Figure 9-33 shows this seam. In either case, fasten the cross seam to the roof structure with two cleats and seal the seam with white lead or butyl rubber sealant.

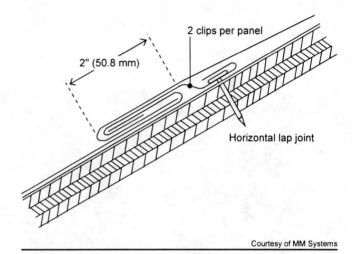

Figure 9-31 Prefabricated cross seam

Sealing Job-Fabricated Joints

The most common type of solder used on sheet metal roofing is 50-50 lead-tin alloy. Terne metal tarnishes quickly and requires surface cleaning with a rosin flux prior to soldering. You can use 50-50 solder on most metals, including lead. For stainless steel and nickel alloys, use high-tin (60 to 70 percent) solder. There are special solders for aluminum, but they're seldom used in sheet metal work. Use caulk instead.

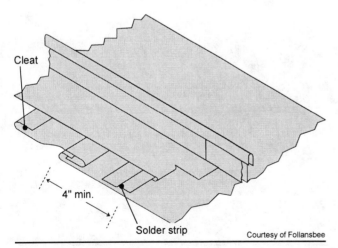

Figure 9-32 Job-fabricated cross seam (low slope)

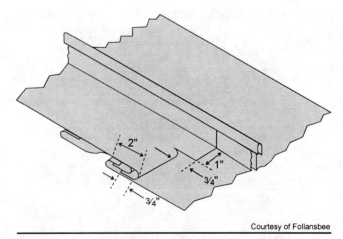

Figure 9-33 Job-fabricated cross seam (steep slope)

Use soft solders only to conceal joints and to make them watertight. They aren't meant to resist stresses.

You can weld joints instead of soldering to make them waterproof. The advantage of welding is that the resulting joints don't require mechanical fasteners or locked seams to resist stresses. The disadvantage is that some metals are hard to weld, and the heat tends to burn off the protective metal coating. I don't recommend that you weld panels to a steel frame when you're going to cover the panels with insulation, or a built-up roof system. You want the roof covering and the panels to expand and contract as a single unit. That's not possible if you've used welded joints.

Brazing isn't practical due to the expense, and contact between dissimilar metals can cause galvanic corrosion.

When you can't solder or weld, you can still produce waterproof joints. You do so by simply applying a sealant between the two metal sheets you're joining. Then you finish the joint by using mechanical fasteners, or by forming locked seams (any seam except the riveted soldered seam in Figure 9-22). Noncuring butyl-based caulk is a good sealant. Give edge-lapped material a bead of caulk along the full length of the joint. You can also buy panels with an integral standing seam that contains a continuous bead of sealant material. (Refer back to Figures 9-10 and 9-11.)

Estimating Metal Roofing and Siding

Take off metal roofing and siding by the square, based on the total area to be covered. As a rule of thumb, you deduct only half of any opening smaller than 100 square feet, but you deduct 100 percent for openings larger than 100 square feet.

You must also account for flashing and accessories, including ridge and hip covers and closures, eaves closures or drip edge, rake end covers, valley flashing, continuous wall flashing and gutters. Other accessories you'll sometimes need include snap-on seams, concealed panel clips and cleats, solder strips, roof curbs and ridge ventilators.

Due to side laps and end laps (cross seams), or because of material lost in making seams, the net area that some metal panels will cover is less than the total area the panel is sold by. However, many modern ribbed metal panels are sold according to the net length and width (exposure) of the panel. If necessary, you can calculate the waste factor with the following formula:

$$\text{Waste Factor} = \frac{\text{Gross Panel Area}}{\text{Net Panel Area}}$$

Equation 9-1

Find the number of panels required per square using:

$$\text{Panels per Square} = \frac{100 \text{ SF}}{\text{Net Panel Area (SF)}}$$

Equation 9-2

If the panel dimensions are given in inches, use:

$$\text{Panels per Square} = \frac{14{,}400 \text{ sq. in.}}{\text{Net Panel Area (sq. in.)}}$$

Equation 9-3

▼ **Example 9-1:** Find the waste factor and the number of panels required per square for 28" x 120" panels installed with flat-seam construction. Assume that you lose 1½ inches in both panel dimensions due to the lap of flat seams.

The waste factor from Equation 9-1 is:

$$\text{Waste Factor} = \frac{28 \text{ in.} \times 120 \text{ in.}}{26.5 \text{ in.} \times 118.5 \text{ in.}}$$

= 1.07, or 7 percent waste due to seams

The number of panels required per square (using Equation 9-3) is:

$$\text{Panels/Square} = \frac{14{,}400 \text{ sq. in.}}{26.5 \text{ in.} \times 118.5 \text{ in.}}$$

= 14,400 ÷ 3,140.25
= 4.6 panels

The charts in this chapter contain waste factors for only the most commonly-produced metal panels. They don't include waste due to cutting off the uppermost panel so its end terminates at the ridge. Consult the manufacturer's figures when you estimate a specific metal roof panel system.

You can order many panels factory-cut to the lengths required for your job. Depending on the manufacturer's longest available length, some panels are long enough to run continuously from the ridge to the eaves. Modern panel lengths vary from 10 to 40 feet. Longer panels can be special ordered, but it will require special truck transport permits. Unless the roof is designed in increments of the net panel width, you'll have to cut the last panel on the job, or special order the width of the last panel.

Steel Roofing and Siding Quantities

Steel roofing and siding sheets are coated with zinc, tin, nickel, lead, aluminum, or a combination of materials. Steel coated with zinc is called *galvanized* steel. Steel coated with tin is called *bright plate* and steel coated with lead and tin is called *terne plate*.

Galvanized Sheet Metal

Galvanized steel is relatively inexpensive and corrosion-resistant. It can be lead-soldered and used in direct contact with concrete, mortar, lead, tin, aluminum and wood (except redwood and red cedar).

Galvanized sheet metal is manufactured in many gauges and in a number of coating classes, depending on the amount of zinc applied to the sheet. The common galvanized sheet metal used for construction purposes has 0.625 ounces of zinc applied per square foot on one side of the sheet. Galvanized sheet metal comes in 6- to 10-foot lengths (in 1-foot increments) and in 24- to 48-inch widths (in 2-inch increments).

When produced with a flat seam, the edges of the sheet are each turned in about half an inch. This means the sheet covers about 1½ inches less than its size in each dimension. Use Figure 9-34 to determine waste factors and the number of metal sheets required per square for flat-seam galvanized steel roofing.

You'll have more waste when you install a standing seam roof than when you use the flat seam. A standing seams loses 2¾ inches of width at each standing seam, and 1⅛ inches of length at each cross seam. The waste factors and coverage for galvanized steel sheets with a standing seam are given in Figure 9-35. You can install a standing seam roof on roofs with slopes of 2 in 12 and steeper.

Galvanized metal also comes in rolls 14, 20, 24 and 28 inches wide. You can use Figure 9-36 to determine waste factors and coverage for galvanized steel rolls with a standing seam.

Ribbed Metal Panel Quantities

There are many types of ribbed metal roofing and siding panels available. A selection of styles offered by one manufacturer appears in Figure 9-37. We'll take a closer look at one of them, corrugated steel panels.

Corrugated Steel Roofing and Siding

Corrugated steel comes in many corrugation sizes. Sheets with 2⅔-inch pitch (nominally, a 2½-inch pitch) and 3-inch pitch are the most common. See Figure 9-37.

Corrugated roofing and siding sheets come in lengths of 5 to 12 feet, in 1-foot increments, in both standard and wide widths.

- Standard sheet widths are 26 inches for siding and 27½ inches for roofing (2⅔- or 3-inch pitch).

- Wide sheet widths are 31¼ inches (2⅔-inch pitch) and 32 inches (3-inch pitch) when used for siding material.

- When used for roofing, wide sheet widths are 32¾ inches (2⅔-inch pitch) and 33½ inches (3-inch pitch).

Corrugated steel sheets come in a variety of finishes, including uncoated (black), painted, galvanized, vinyl and ceramic.

Panel size (inches)	Waste factor	Panels required per square
24 x 96	1.08	6.8
24 x 120	1.08	5.4
26 x 96	1.08	6.2
26 x 120	1.08	5.0
28 x 72	1.08	7.7
28 x 84	1.08	6.6
28 x 96	1.07	5.8
28 x 108	1.07	5.1
28 x 120	1.07	4.6
30 x 96	1.07	5.4
30 x 120	1.07	4.3
36 x 96	1.06	4.4
36 x 120	1.06	3.5
42 x 96	1.05	3.8
42 x 120	1.05	3.0
48 x 96	1.05	3.3
48 x 120	1.05	2.6

These figures allow for a loss of 1½" in panel length and width for making the flat seams.

Figure 9-34 Waste factors and coverage for galvanized steel sheet metal panels (flat-seam method)

Panel size (inches)	Waste factor	Panels required per square
24 x 96	1.14	7.1
24 x 120	1.14	5.7
26 x 96	1.13	6.5
26 x 120	1.13	5.2
28 x 72	1.13	8.1
28 x 84	1.12	6.9
28 x 96	1.12	6.0
28 x 108	1.12	5.3
28 x 120	1.12	4.8
30 x 96	1.11	5.6
30 x 120	1.11	4.5
36 x 96	1.10	4.6
36 x 120	1.09	3.6
42 x 96	1.08	3.9
42 x 120	1.08	3.1
48 x 96	1.07	3.4
48 x 120	1.07	2.7

These figures allow for loss of 2¾" of panel width for making the standing edge seam and 1⅛" in panel length for making the flat cross seam.

Figure 9-35 Waste factors and coverage for galvanized steel sheet metal panels (standing-seam method)

Roll width (inches)	Waste factor	Linear feet required per square
14	1.24	107
20	1.16	70
24	1.13	57
28	1.11	48

This table allows for loss of 2¾" in panel width for making the standing edge seam.

Figure 9-36 Waste factors and coverage for galvanized steel sheet metal rolls (standing-seam method)

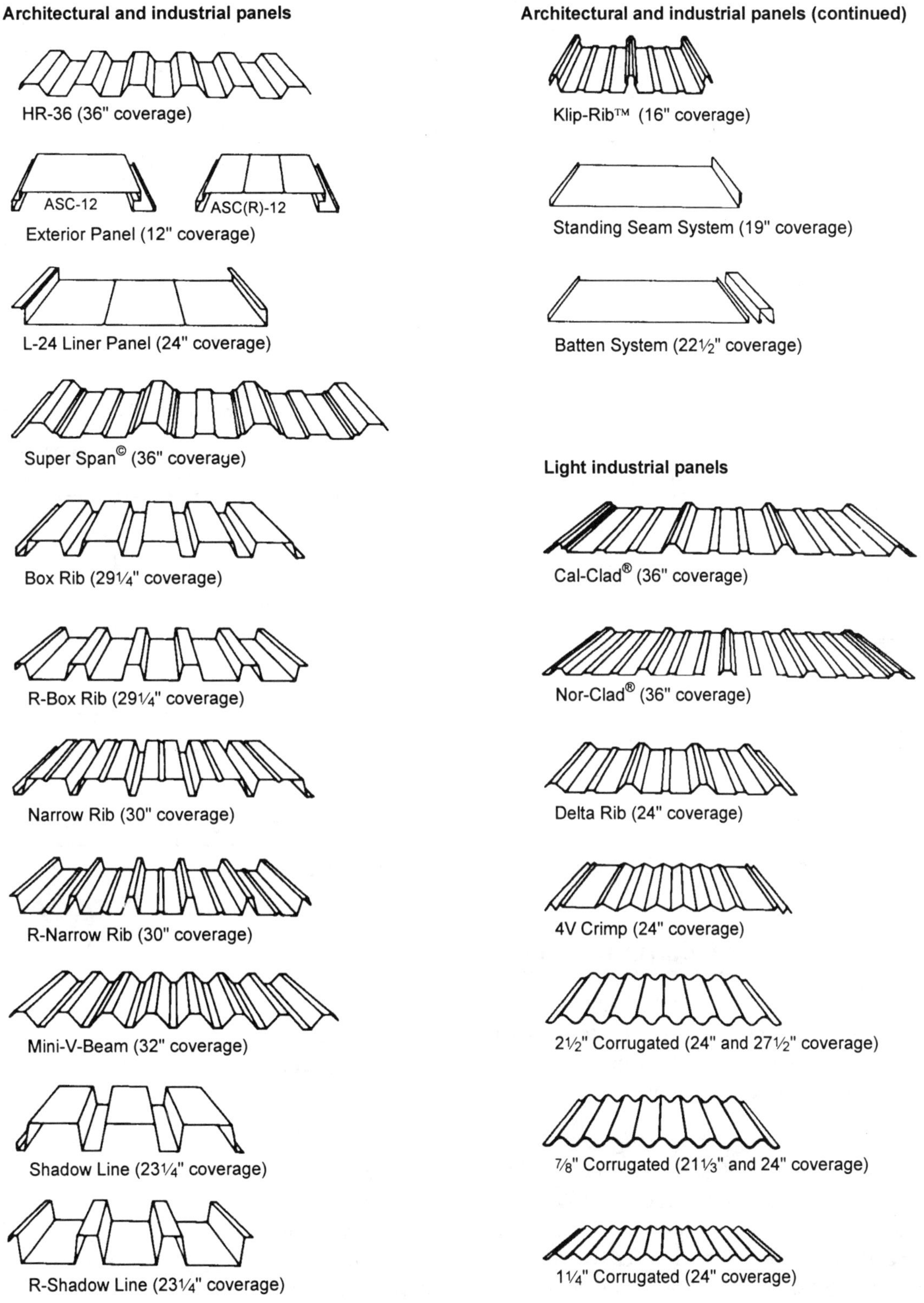

Metal Roofing & Siding

Architectural and industrial panels

HR-36 (36" coverage)

ASC-12 / ASC(R)-12
Exterior Panel (12" coverage)

L-24 Liner Panel (24" coverage)

Super Span© (36" coverage)

Box Rib (29¼" coverage)

R-Box Rib (29¼" coverage)

Narrow Rib (30" coverage)

R-Narrow Rib (30" coverage)

Mini-V-Beam (32" coverage)

Shadow Line (23¼" coverage)

R-Shadow Line (23¼" coverage)

Architectural and industrial panels (continued)

Klip-Rib™ (16" coverage)

Standing Seam System (19" coverage)

Batten System (22½" coverage)

Light industrial panels

Cal-Clad® (36" coverage)

Nor-Clad® (36" coverage)

Delta Rib (24" coverage)

4V Crimp (24" coverage)

2½" Corrugated (24" and 27½" coverage)

⅞" Corrugated (21⅓" and 24" coverage)

1¼" Corrugated (24" coverage)

Courtesy of ASC Pacific, Inc.

Figure 9-37 Various ribbed roofing and siding panels

Length of sheets (feet)	26-inch wide (2½-inch pitch)		26-inch wide (3-inch pitch)	
	Waste factor	Sheets required per square	Waste factor	Sheets required per square
5	1.16	10.7	1.14	10.5
6	1.15	8.8	1.12	8.6
7	1.14	7.5	1.11	7.4
8	1.13	6.5	1.11	6.4
9	1.13	5.8	1.10	5.7
10	1.12	5.2	1.10	5.1
11	1.12	4.7	1.09	4.6
12	1.11	4.3	1.09	4.2

Length of sheets (feet)	31¼-inch wide (2⅔-inch pitch)		32-inch wide (3-inch pitch)	
	Waste factor	Sheets required per square	Waste factor	Sheets required per square
5	1.12	8.6	1.18	8.9
6	1.11	7.1	1.17	7.3
7	1.10	6.0	1.16	6.2
8	1.09	5.2	1.15	5.4
9	1.09	4.6	1.15	4.8
10	1.08	4.2	1.14	4.3
11	1.08	3.8	1.14	3.9
12	1.07	3.4	1.14	3.6

This table allows for a one-corrugation side lap and a 4" end lap

Figure 9-38 Waste factors and coverage for corrugated steel sheets used for siding

Installation Allowance for Corrugated Siding

For siding, fasten side laps on 18-inch centers. For roofing over solid sheathing, fasten side laps on 12-inch centers. When installing panels over purlins, fasten side laps over each purlin. Fasten end laps over each purlin at every second or third corrugation.

When installed as siding, one-corrugation side lap and a minimum of a 4-inch end lap is enough protection. For roofing, provide a two-corrugation side lap and a 6-inch end lap (on roof slopes of 4 in 12 and steeper), or an 8-inch end lap (on roof slopes less than 4 in 12). Figure 9-38 shows waste factors and coverage for various-sized corrugated steel sheets used for siding. Figure 9-39 shows the same information for sheets used for roofing.

Figure 9-40 summarizes corrugated steel's physical properties in a compact form.

Length of sheets (feet)	27½-inch wide (2⅔-inch pitch)		27½-inch wide (3-inch pitch)	
	Waste factor	Sheets required per square	Waste factor	Sheets required per square
5	1.23	10.7	1.25	10.9
6	1.21	8.8	1.23	8.9
7	1.19	7.4	1.21	7.5
8	1.18	6.4	1.20	6.5
9	1.17	5.7	1.19	5.8
10	1.17	5.1	1.18	5.2
11	1.16	4.6	1.18	4.7
12	1.16	4.2	1.17	4.3
Length of sheets (feet)	32¾-inch wide (2⅔-inch pitch)		33½-inch wide (3-inch pitch)	
	Waste factor	Sheets required per square	Waste factor	Sheets required per square
5	1.21	8.9	1.21	8.7
6	1.19	7.3	1.20	7.2
7	1.17	6.1	1.18	6.1
8	1.16	5.3	1.17	5.3
9	1.15	4.7	1.16	4.6
10	1.15	4.2	1.16	4.1
11	1.14	3.8	1.15	3.8
12	1.14	3.5	1.15	3.4

This table allows for a two-corrugation side lap and a 6" end lap.

Figure 9-39 Waste factors and coverage for corrugated steel sheets used for roofing

Miscellaneous Metal Roofing Quantities

You can use corrugated plastic sheets along with corrugated sheet metal panels to let light enter through the roof. You can also use these panels alone as the entire roof covering. Both colored and translucent sheets are available.

Plastic corrugated sheets are made from acrylics. Sheets are commonly 50½ inches wide and 8, 10, 12, 16 or 20 feet long. The sheets are usually installed with a one-corrugation side lap and an 8-inch end lap.

Reinforced corrugated fiberglass sheets are made from glass fiber-reinforced acrylics. Corrugated sheets come 26, 34, 35, 40 or 42 inches wide and 8, 9, 10, 11 or 12 feet long. Sheets are installed with a one- or two-corrugation side lap and an 8-inch end lap. All plastic sheets (reinforced or not) weigh about 0.4 pounds per square foot.

U.S. Mfr's Gauge	Thickness (inches)	Uncoated (black)		
		Weight (lbs per square foot)		
		Flat	Corrugated	
		--	2⅔ x ½	3 x ¾
12	0.1046	4.38	4.77	5.05
14	0.0747	3.13	3.41	3.61
16	0.0598	2.50	2.73	2.89
18	0.0478	2.00	2.18	2.31
20	0.0359	1.50	1.64	1.73
22	0.0299	1.25	1.36	1.44
24	0.0239	1.00	1.09	1.15
26	0.0179	0.75	0.82	0.87
28	0.0149	0.63	0.68	0.72
29	0.0135	0.56	0.60	0.65

U.S. Mfr's Gauge	Thickness (inches)	Galvanized		
		Weight (lbs per square foot)		
		Flat	Corrugated	
		--	2⅔ x ½	3 x ¾
12	0.1084	4.53	4.94	5.23
14	0.0785	3.28	3.58	3.79
16	0.0635	2.66	2.90	3.07
18	0.0516	2.16	2.35	2.49
20	0.0396	1.66	1.81	1.91
22	0.0336	1.41	1.53	1.62
24	0.0276	1.16	1.26	1.33
26	0.0217	0.91	0.99	1.05
28	0.0187	0.78	0.85	0.90
29	0.0172	0.72	0.78	0.83

Figure 9-40 Gauges, thickness and weights of corrugated steel

Corrugated structural glass sheets come 47½ inches wide and 8, 10 or 12 feet long. Sheets are ⅜ inch thick and weigh about 6.3 pounds per square foot. Sheets are installed with a two-corrugation side lap and a 6-inch end lap.

You can get prefabricated membrane roof and wall panels made of a foam polyisocyanurate insulation core sandwiched between a 22-, 24- or 26-gauge galvanized steel (or aluminum) outer face and a 26-gauge prepainted steel interior face. The outer face comes with a laminated coating of PVC membrane for protection and a more attractive appearance. They may have interlocking edge joints, or be sealed with heat-welded batten strips. Panels are 2 to 6 inches thick and a standard width of 42 inches. You can get them in any desired length.

Aluminum Roofing and Siding

Aluminum panels are alloyed with copper, zinc, manganese, magnesium, chromium or nickel. The core metal is designed for strength and the corrosion-resistant finish (normally an aluminum alloy or commercially pure aluminum) is fused to the core metal under heat and pressure during a hot-rolling manufacturing process.

Aluminum is light, non-corrosive, rigid and durable. But it also has a very high coefficient of thermal expansion, as you can see in Figure 9-41.

Materials (metals and alloys)	Linear expansion per 100 degrees	
	Centigrade	Fahrenheit
Aluminum, wrought	0.00231	0.00128
Brass	0.00188	0.00104
Bronze	0.00181	0.00101
Copper	0.00168	0.00093
Iron, cast, gray	0.00106	0.00059
Iron, wrought	0.00120	0.00067
Iron, wire	0.00124	0.00069
Lead	0.00286	0.00159
Magnesium, various alloys	0.00290	0.00160
Nickel	0.00126	0.00070
Steel, mild	0.00117	0.00065
Steel, stainless, 18-8	0.00178	0.00099
Zinc, rolled	0.00311	0.00173

Figure 9-41 Coefficients of expansion of various metals

Gauge	Nominal thickness (inches)	Weight (pounds per SF)
26	0.016	0.226
24	0.020	0.285
22	0.025	0.355
20	0.032	0.455
18	0.040	0.568
16	0.051	0.725
14	0.064	0.910
12	0.081	1.152

Figure 9-42 Aluminum sheet data

Aluminum is produced in sheets and coils, and in a variety of gauges, with properties as shown in Figure 9-42.

You can install corrugated aluminum sheets on roof slopes of 3 in 12 and steeper, and V-beam sheets on slopes as low as 2 in 12. When using concealed clip panels, the minimum slope allowed is 4 in 12 unless a single panel will cover the total span (ridge to eaves). Then the single panel can be applied over roof slopes as low as ½ in 12.

Terne Metal Roofing and Siding

Terne metal (terne plate) roofing was first introduced in Wales in 1720. It's also called *valley tin* or *roofer's tin*. It was originally made by applying tin over an iron core metal. Terne metal roofs are very durable and come with a 20-year warranty. Some terne metal roofs have survived over 100 years. The Faculty Club of Yale University boasts an original terne metal roof that's over 230 years old!

Terne metal is one of the strongest roofing materials known. It has a tensile strength of 45,000 PSI, 1½ times that of copper, 1¼ times that of zinc across grain, almost double that of zinc with the grain and almost 25 times that of lead. Because of its great strength and its compatibility with solder, terne metal is also used to flash copings, chimneys, and valleys.

Terne metal has the lowest coefficient of expansion of any metal roof, less than half that of zinc or lead, little more than half that of aluminum and ¾ that of copper. As a result, terne metal roofing needs no expansion seams:

- on runs less than 30 feet when both ends are free to move (when covered with built-up roofing, for example)
- on runs less than 15 when the ends are securely fastened

Metal Roofing & Siding

Because of its strength and low coefficient of expansion, terne metal is often used for window and door flashing — including fireproof doors because it's fireproof. It's also used for termite shields.

Another advantage of terne is its light weight. Thirty-gauge terne weighs only 62 pounds per square and can be supported with only ½-inch plywood. This is half the weight of 16-ounce copper (125 pounds per square), less than a third of 20-gauge corrugated galvanized steel (225 pounds) and about 6 percent that of lead (1000 pounds).

Paint adheres extremely well to terne metal. This material isn't affected by the most damaging industrial, chemical and marine environments. Terne-coated stainless steel is *anodic,* which means that the terne coating will sacrifice itself to protect the core metal. Terne metal can be formed into very intricate shapes so it's often used for gutters and downspouts.

Traditional terne metal roofing and siding materials are now produced by applying a lead-tin alloy over a core metal sheet of copper-bearing carbon steel by the hot-dip process. Terne metal comes in varying thicknesses and in three coating weights of tin-lead coating (0.29, 0.73 and 1.47 ounces per square foot).

Gauge	Thickness (inches)	Weight (pounds per SF)
28	0.0155	0.71
26	0.0180	0.82
24	0.0240	1.07

Figure 9-43 Physical properties of terne metal coating over stainless steel

Gauge	Thickness (inches)	Weight (pounds per SF)
30	0.0120	0.54
28	0.0150	0.65
26	0.0180	0.78

Figure 9-44 Physical properties of terne metal coating over copper-bearing steel

It comes in two grades: short terne and long terne. *Short terne* is made by dipping a copper-bearing steel alloy in a hot lead-tin alloy that's about 85 percent lead and 15 percent tin. Short terne comes in 14-, 20-, 24- and 28-inch standard widths and in various lengths. It's also available in 50- and 100-foot rolls.

Long terne has a coating alloy whose lead and tin content ranges from 80 to 87.5 percent, and from 12.5 to 20 percent, respectively. Long terne comes in sheets up to 4 feet by 10 feet.

The main factor affecting the choice between short and long terne is whether you need to work with rolls or sheets.

Terne-Coated Stainless Steel (TCS)

Terne-coated stainless steel (TCS) is also made by coating Type 304 stainless steel on both sides with terne alloy (20 percent tin, 80 percent lead). Terne-coated stainless steel comes in three gauges described in Figure 9-43.

Terne-coated copper-bearing steel comes in three gauges as described in Figure 9-44.

	Suggested gauge		Suggested gauge
Coping	.018	**Roofs**	
Flashing		Standing seam	.015
Chimney	.015	Batten seam	.015
Base	.015	Flat lock	.015
Counter	.015	Bermuda	.015
Valley	.015	Flat lock decks	.015
Hip	.015	**Perimeters**	.015 or .018
Ridge	.015	**Mansards**	.015 or .018
Ledge	.015	**Fascia**	.015 or .018
Dormer	.015	**Gravel stop fascia**	.015 or .018
Sill	.015	**Wall covering**	.015 or .018
Curtain wall	.015	**Drainage**	
Thru wall	.015	Gutter 5"	.015
Lintel	.015	Gutter 6" or more	.024 or .018
Spandrel	.015	Conductor pipe	.015
New wall to old	.015	Built-in gutter	.018 or .015
Termite shield	.015	Valley	.015

Courtesy of Follansbee

Figure 9-45 Design factors for terne-coated steel

Figure 9-45 lists uses for TCS with the suggested material gauge for each application. You can use 30-gauge material when the initial sheet of terne-coated copper-bearing steel is 20 inches wide or less, or 28- or 26-gauge metal if the initial sheet is wider than 20 inches.

You can install flat-seam terne metal roofing over roof slopes as low as ¼ in 12. Use batten- and standing-seam terne metal roofing on slopes of 2½ in 12 and steeper.

Nominal plate sizes for flat seam roofing are 14" x 20", or 20" x 28". There's a loss of 1½ inches in both panel dimensions due to forming the flat seams so the plates have a net covering capacity of 12½" x 18½" and 18½" x 26½", respectively. Waste factors and coverage for terne metal plates used in flat-seam roofing are given in Figure 9-46.

Terne metal roofing material also comes in rolls 14, 18, 20, 24, 28, and 36 inches wide. Rolls are manufactured in widths of 4, 6, 8, 10, and 12 inches to use for flashings. When used for flat-seam roofing, 1½ inches of the covering width of a roll will be lost due to turning the edges of the roll. Use Figures 9-47 and 9-48 for the waste factors and coverage for 50-foot and 100-foot rolls of terne metal with various seam patterns.

Panel dimensions (inches)	Waste factor	Panels required per square
14 x 20	1.21	62.3
20 x 28	1.14	29.4

This table allows for the loss of 1½" for both panel dimensions for making the seams.

Figure 9-46 Waste factors and coverage for terne metal roofing panels (flat-seam method)

Flat-seam method

Roll width (inches)	Waste factor	Rolls required per square
14	1.13	1.9
18	1.10	1.5
20	1.09	1.3
24	1.08	1.1
28	1.07	0.9
36	1.05	0.7

This table allows for the loss of 1½" of roll width for making the flat edge seams, and 6" in roll length for making the cross seams.

Standing-seam method

Roll width (inches)	Waste factor	Rolls required per square
14	1.32	2.3
18	1.23	1.6
20	1.21	1.5
24	1.17	1.2
28	1.14	1.0
36	1.11	0.8

This table allows for the loss of 3¼" of roll width for making the standing edge seams, and 6" in roll length for making the cross seams.

2" x 2" batten-seam method

Roll width (inches)	Waste factor	Rolls required per square
14	1.57	2.7
18	1.40	1.9
20	1.35	1.6
24	1.28	1.3
28	1.23	1.1
36	1.17	0.8

This table allows for the loss of 5" of roll width for making the batten edge seams, and 6" in roll length for making the cross seams.

4" x 4" batten-seam method

Roll width (inches)	Waste factor	Rolls required per square
14**	--	--
18	2.02	2.7
20	1.84	2.2
24	1.62	1.6
28	1.49	1.3
36	1.35	0.9

**A 14" roll is too narrow to accommodate a 4" x 4" batten seam.

This table allows for the loss of 9" of roll width for making the batten-edge seams, and 6" in roll length for making the cross seams.

Bermuda roof, 2-inch riser

Roll width (inches)	Waste factor	Rolls required per square
14	1.35	2.3
18	1.25	1.7
20	1.22	1.5
24	1.18	1.2
28	1.15	1.0
36	1.12	0.8

This table allows for the loss of 3½" of roll width for making the riser, and 6" in roll length for making the cross seams.

Figure 9-47 Waste factors and coverage for 50-foot rolls of terne metal roofing

Flat-seam method

Roll width (inches)	Waste factor	Rolls required per square
14	1.13	1.0
18	1.10	0.7
20	1.09	0.7
24	1.07	0.5
28	1.06	0.5
36	1.05	0.4

This table allows for the loss of 1½ inches of roll width for making the flat edge seams, and 6 inches in roll length for making the cross seams.

Standing-seam method

Roll width (inches)	Waste factor	Rolls required per square
14	1.31	1.1
18	1.23	0.8
20	1.20	0.7
24	1.16	0.6
28	1.14	0.5
36	1.11	0.4

This table allows for the loss of 3¼" of roll width for making the standing edge seams, and 6" in roll length for making the cross seams.

2" x 2" batten-seam method

Roll width (inches)	Waste factor	Rolls required per square
14	1.56	1.3
18	1.39	0.9
20	1.34	0.8
24	1.27	0.6
28	1.22	0.5
36	1.17	0.4

This table allows for the loss of 5" of roll width for making the batten edge seams, and 6" in roll length for making the cross seams.

4" x 4" batten-seam method

Roll width (inches)	Waste factor	Rolls required per square
14**	--	--
18	2.01	1.4
20	1.83	1.1
24	1.61	0.8
28	1.48	0.6
36	1.34	0.5

**A 14" roll is too narrow to accommodate a 4" x 4" batten seam.

This table allows for the loss of 9" of roll width for making the batten-edge seams, and 6" in roll length for making the cross seams.

Bermuda roof, 2-inch riser

Roll width (inches)	Waste factor	Rolls required per square
14	1.34	1.2
18	1.25	0.9
20	1.22	0.7
24	1.18	0.6
28	1.15	0.5
36	1.11	0.4

This table allows for the loss of 3½" of roll width for making the riser, and 6" in roll length for making the cross seams.

Figure 9-48 Waste factors and coverage for 100-foot rolls of terne metal roofing

Terne metal roofing comes in lengths of 96 and 120 inches, and widths of 14, 18, 20, 24, 28, and 36 inches. However, 24" x 120" is the largest sheet size recommended for standing- or batten-seam construction. Figure 9-49 gives waste factors and coverage for these sheet sizes.

You don't have to paint terne-coated stainless steel because it weathers to a pleasing slate gray color. However, you should paint terne-coated copper-bearing steel. Apply an iron oxide linseed oil base primer to both sides of each panel. Then use a linseed oil finish coat on the exposed surface. To meet conditions of the warranty, terne must be repainted with a finish coat at least once every eight years.

▼ **Example 9-2:** Assume a roof slope of 5 in 12, then find the quantity of corrugated steel panels required for a gable roof with a plan area of 80 feet by 40 feet. Assume that you'll use 27½-inch-wide panels (3" pitch), 12 feet long, with a two-corrugation side lap and 6-inch end lap.

Solution: The net roof area to be covered is:

Net Roof Area = 80' x 40' x 1.083 (from Column 2 of Appendix A)
= 3466 SF ÷ 100
= 34.66 squares (Use this quantity
to figure labor costs.)

The total length from the eaves to the ridge is:

Total Required Panel Length = 20' x 1.083
= 21.66'

The gross roof area (excluding cutting excess panel length at the ridge) is:

3466 SF x 1.17 (from Figure 9-39) = 4055 SF

Using two 12-foot panels with a 6-inch end lap (23.5 feet), we must increase the end lap, or cut off the top of the second panel by:

Panel loss = 23.5' - 21.66
= 1.84'

The total square feet of excess panel length is:

2 x 80' x 1.84' = 294 SF

The gross roof area is:

Gross Roof Area = 4055 SF + 294 SF = 4349 SF
= 4349 SF ÷ 100
= 43.49 squares

Standing seam

Sheet dimensions (inches)	Waste factor	Sheets required per square
14 x 96	1.39	14.9
14 x 120	1.37	11.8
18 x 96	1.30	10.9
18 x 120	1.29	8.6
20 x 96	1.27	9.6
20 x 120	1.26	7.5
24 x 96	1.23	7.7
24 x 120	1.22	6.1
28 x 96	1.21	6.5
28 x 120	1.19	5.1
36 x 96	1.17	4.9
36 x 120	1.16	3.9

This table allows for the loss of 3¼" of sheet width for making the standing-edge seams, and 6" in sheet length for making the cross seams.

2" x 2" batten seam

Sheet dimensions (inches)	Waste factor	Sheets required per square
14 x 96	1.66	17.8
14 x 120	1.64	14.1
18 x 96	1.48	12.3
18 x 120	1.46	9.7
20 x 96	1.42	10.7
20 x 120	1.40	8.4
24 x 96	1.35	8.4
24 x 120	1.33	6.7
28 x 96	1.30	7.0
28 x 120	1.28	5.5
36 x 96	1.24	5.2
36 x 120	1.22	4.1

This table allows for the loss of 5" of sheet width for making the batten-edge seams, and 6" in sheet length for making the cross seams.

Bermuda roof construction (2-inch riser)

Sheet dimensions (inches)	Waste factor	Sheets required per square
14 x 96	1.42	15.2
14 x 120	1.40	12.1
18 x 96	1.32	11.0
18 x 120	1.31	8.7
20 x 96	1.29	9.7
20 x 120	1.28	7.7
24 x 96	1.25	7.8
24 x 120	1.23	6.2
28 x 96	1.22	6.5
28 x 120	1.20	5.2
36 x 96	1.18	4.9
36 x 120	1.17	3.9

This table allows for the loss of 3½" of sheet width for making the riser, and 6" in sheet length for making the cross seams.

4" x 4" batten seam

Sheet dimensions (inches)	Waste factor	Sheets required per square
20 x 96	1.94	14.6
20 x 120	1.91	11.5
24 x 96	1.71	10.7
24 x 120	1.68	8.4
28 x 96	1.57	8.4
28 x 120	1.55	6.7
36 x 96	1.42	5.9
36 x 120	1.40	4.7

This table allows for the loss of 9" of sheet width for making the batten-edge seams, and 6" in sheet length for making the cross seams.

Figure 9-49 Waste factors and coverage for terne metal sheets

The waste factor is:

Waste Factor $= \dfrac{4349 \text{ SF}}{3466 \text{ SF}}$
$= 1.25$

Other items that will probably be required are:

Eaves closure strip = 2 x 80' = 160 feet
Ridge cap = 80 feet
Rake trim = 4 x 20' x 1.083 = 87 feet

Chrome-Nickel Stainless Steel Roofing

Stainless steel, which is highly resistant to corrosion, has chromium and nickel added during the steel manufacturing process. There are six types of stainless steel commonly used as building materials, all with 14 to 20 percent chromium and 6 to 14 percent nickel. The type of stainless steel most widely used for exterior work is Type 302 (Grade 18-8). It has 17 to 19 percent chromium and 8 to 10 percent nickel. Type 430 contains no nickel and is used mostly indoors. Type 316 (Grade 18-12) is used in corrosive marine environments. Some stainless steel is manufactured with a coating of terne metal.

Gauge	Weight (pounds per SF)
16	2.88
17	2.59
18	2.30
19	2.02
20	1.73
21	1.58
22	1.44
23	1.30
24	1.15
25	1.01
26	0.86

Figure 9-50 Weights of monel metal roofing

Because of its high strength, stainless steel is produced in relatively thin sheets. But it's so hard that you have to do all fabricating in the shop. You can install stainless steel with a standing seam and a continuous weld over roof slopes as low as ¼ in 12.

Monel Metal

Monel metal is a trade name for a white metal that looks like stainless steel. The principal alloys in monel are nickel (about 66 percent) and copper (about 30 percent) with small amounts of iron, manganese, silicone and other elements. Monel metal is expensive and hard, requiring shop fabrication.

Monel is manufactured in 16 to 26 gauge weights. The 26 gauge (20-inch seam centers) or 25 gauge (24-inch seam centers) metal is used most often for roofing material. Standard widths are 24, 30 and 36 inches. Standard lengths are 96 and 120 inches. Weights of monel metal are given in Figure 9-50.

You can install monel roofing with a flat, standing or batten seam on roofs with slopes of 3 in 12 and steeper. Install monel roofing over solid sheathing, using monel clips and cleats nailed or screwed to the structure.

Lead Roofing

Sheet lead is very pliable so you can draw and stretch it to cover warped surfaces. That makes it a popular flashing material for tile and slate roofs. Sheet lead is sold by its weight in pounds per square foot.

Sheet lead is usually hard lead, weighing 2.5 pounds per square foot and containing 6 to 7.5 percent antimony. Sheet lead is also available in weights from 1 to 8 pounds per square foot. It's normally sold in 24" x 48" panels. Some steel or copper sheets are coated with lead, producing a material with a lightweight core whose surface has the color and provides the protection of lead.

Lead roofing is often installed with standing and batten seams. You can install batten-seam, standing-seam and flat-seam lead roofing on roof slopes as low as ¼ in 12.

Gauge	Thickness (inches)	Weight (pounds per SF)
9	0.018	0.67
10	0.020	0.75
11	0.024	0.90
12	0.028	1.05
13	0.032	1.20
14	0.036	1.35
15	0.040	1.50
16	0.045	1.68

Figure 9-51 Physical properties of zinc roofing

Zinc Roofing

Zinc is lighter and stiffer than lead, but it's more susceptible to damage from acids and has a high coefficient of expansion. Rolled sheet zinc is used for roofing and flashing, and zinc is often used as a surface coating for sheet steel. Zinc roofing comes in thicknesses of 9 through 16 gauge, as shown in Figure 9-51.

Avoid using zinc flat-seam construction except over surface areas less than 200 square feet (10 gauge is recommended). Install standing seams on slopes of 2 in 12 and steeper. I recommend 10 gauge zinc for flat-seam and standing-seam installations. Use batten seams on slopes between 3 in 12 and 6 in 12. The gauge used in batten-seam construction depends on the on-center batten spacing:

- For spacing of 18" oc, use 11 or 12 gauge.
- For spacing of 30" oc, use 12 or 13 gauge.
- For spacing of 40" oc, use 13 or 14 gauge.

Solder cross seams on roof slopes less than 4 in 12. Stock sheet sizes are 20, 30, 36 and 40 inches wide, and 7 or 8 feet long.

Copper Roofing

Copper used in sheet metal work is called *tough pitch copper*. It comes in hot-rolled (soft or dead soft) or cold-rolled (hard or cornice temper) forms. Soft-rolled copper is easier to work with because it's more pliable.

Cold-rolled copper, however, is stronger, harder, and less pliable, but also less expensive. Most flashing work is done using cold-rolled copper. Sheet metal copper comes in weights of 16, 20, 24 and 32 ounces per square foot.

Roofing sheets of copper are 20, 24, 30 and 36 inches wide and 8 and 10 feet wide. Strip copper is 10 and 20 inches wide and 8 and 10 feet long. Rolled copper comes in rolls 6 and 20 inches wide, 50 to 100 feet long, and weighing between 80 and 100 pounds per square.

Install 20-ounce cold-rolled copper plates no larger than 16" x 18" with ¾-inch flat, locked and soldered (or white-leaded) joints. Use 2-inch-wide copper cleats to install the panels on the deck. You can use flat seams on roof slopes as low as ¼ in 12.

After folding the seams, a 16" x 18" sheet will cover only 14½" x 16½". That produces a waste factor of 1.20, so you need about 60 sheets per square. You can install standing seams on roof slopes of 2½ in 12 and steeper. Waste factors and coverage for standing-seam construction are given in Figure 9-52. You can install batten seams on roof slopes as low as 3 in 12.

Sheet size (inches)	Waste factor	Sheets required per square
20 x 96	1.25	9.4
24 x 96	1.20	7.5
30 x 96	1.16	5.8
36 x 96	1.13	4.7
20 x 120	1.25	7.5
24 x 120	1.20	6.0
30 x 120	1.16	4.6
36 x 120	1.13	3.8

This table allows for a loss of 3¾" of panel width for making the standing edge seams, and 1½" in panel length for making the flat cross seams.

Figure 9-52 Waste factors and coverage for copper roofing using a standing seam

Install 16-ounce copper when seam widths are on 20-inch centers, or less. Use 20-ounce copper when seam centers are farther apart. Don't rivet or solder standing seams. You don't need to solder cross seams when the roof slope is greater than 3 in 12.

Copper-Bearing Steel Roofing

Copper-bearing steel roofing comes in two sizes of 24-gauge sheets:

- 26½ inches by 50 feet when installed by the standing seam method

- 25 inches by 12 feet when installed by the pressed standing-seam method

You can install standing-seam rolled panels on slopes as low as 2 in 12 and pressed standing-seam panels on slopes as low as ¼ in 12. Use cleats and nails to install this type of steel roofing.

Titanium-copper-zinc alloy (T-C-Z) roofing usually comes in 20- or 24-inch widths, and in 10-foot lengths. It's usually installed using the batten- or standing-seam methods. The weight of T-C-Z roofing varies from 125 to 150 pounds per square.

Metal Shingles

Metal shingles come in a wide variety of shapes, styles, colors and textures. Finishes include aluminum, galvanized steel, copper, and enamel- or porcelain-coated surfaces. Metal shingles are available as single units or multiple strips (Figure 9-53). Metal shingles can be used safely for all single coverage applications.

The most popular metal shingles are made from 0.02-inch-thick aluminum, finished in a wood shake pattern. The strips are 36 or 48 inches long, and 10 inches (nominal) wide. The strips are designed to interlock at the edges and ends. Weights of aluminum shingles vary from about 40 to 60 pounds per square. A big problem with an all-aluminum shingle is that it's easily dented by hail.

Figure 9-53 Metal strip-shingle roof

You can get porcelain enamel on iron shingles in 10" x 10" giant individual shingles. These expensive shingles weigh about 225 pounds per square.

A variety of shingles are made with a core of 0.02-inch-thick aluminum or 22- through 28-gauge galvanized steel, coated with a siliconized polyester finish. They come in a wide variety of colors. Styles include mission tile, Spanish tile, and shakes. These shingles are relatively lightweight. Aluminum-core shingles weigh from 40 to 65 pounds per square and steel-core shingles weigh from 90 to 250 pounds per square. Don't install this type of shingle over roof slopes less than 3 in 12 unless you first apply a minimum 30-pound felt underlayment. Aluminum or galvanized nails are often used to fasten these shingles. However, some metal shingle panels are fastened with self-drilling nylon-headed steel screws with bonded aluminum or neoprene washers.

Another variety of metal shingle is the interlocking copper shingle. It's made from rolled 10-ounce copper, producing a lightweight shingle. The underlayment is either one layer of 30-pound felt or two layers of 15-pound felt. You can use copper shingles as a re-roofing material over cedar or asphalt shingles, or over built-up roofing. Don't install copper shingles over roof slopes less than 3 in 12.

Metal shingles and panels are usually installed with special accessories such as sidewall and end-wall flashing, rake and eaves trim, and hip and ridge caps. Figure 9-54 shows some of the accessories you need to install metal shingles.

Metal Roofing & Siding

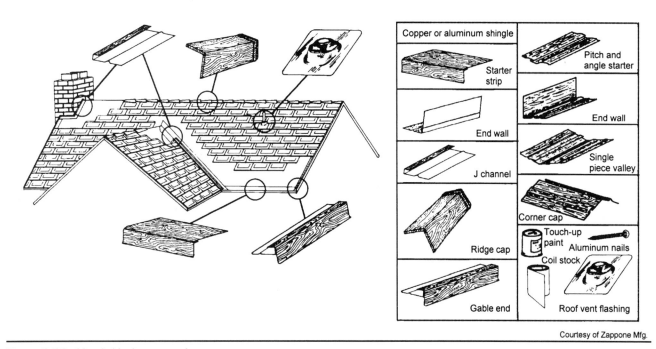

Figure 9-54 Metal shingle accessories

Labor for Installing Metal Roofing

Let's assume you're going to cover the roof described in Example 9-2 with corrugated roofing that costs $2.00 per square foot, including sales tax. From Example 9-2, you must purchase 4349 square feet of panels. So the total material cost is:

4349 SF x $2.00/SF = $8698.00

According to the *National Construction Estimator*, one sheet metal worker can install a square foot in 0.026 hours. If we assume that the worker makes $20.00 per hour, including labor burden, then the labor for installing the roof will be:

3466 SF x 0.026 hours/SF x $20.00/hour = $1802.00

The total cost is:

$8698.00 + $1802.00 = $10,500.00

10 Built-up Roofing

▶ A built-up roof (BUR) is membrane roofing, made up of asphalt- or tar-saturated felts, coated felts, fabrics or mats with alternate layers of asphalt or pitch bitumen. The broad membrane roofing classification also includes elastomeric roofing, which we'll cover in detail in the next chapter.

There are three types of built-up roofing:

- Aggregate-surfaced roofing, which uses a layer of aggregate embedded in a flood coat of bitumen

- Smooth-surface roofing, which uses a glaze coat of bituminous or fibrated aluminum materials

- Mineral-surfaced roofing, which uses a granule-surfaced sheet

Bitumen makes a BUR waterproof. Felts make it strong and flexible so the roof can expand and contract. If properly installed and regularly maintained, the life expectancy of a BUR is 20 years or more. Besides long life, they have another advantage: BUR systems are extremely resistant to wind uplift.

The way you install a BUR system depends on several things. Before we discuss the mechanics of applying a built-up roof, let's see how to figure out the best installation method.

Roof Slopes

The maximum slope of a BUR depends on what type interply and surface bitumen and roofing surface you use. Don't install pitch bitumen over roof slopes more than $1/4$ in 12. Don't apply surface aggregate over slopes greater than 3 in 12. You can install smooth-surfaced or mineral-surfaced roofing (using asphalt bitumen) on slopes up to 6 in 12. On non-nailable decks with roof slopes that are more than 1 in 12 (using asphalt bitumen) or $1/2$ in 12 (using pitch bitumen), you must install wood nailers to hold the roofing felts or insulation in place. We'll discuss that in detail later.

Substrate Design

The substrate (insulation, or deck) must be compatible with the BUR system that covers it. Decking for BUR systems must be strong enough to resist deflecting under normal live loads like snow, standing water and foot traffic. Decking that does deflect excessively under a load causes cracks in the overlying BUR.

Report any defects you find in the deck (to the general contractor in writing) and make sure they are corrected before you install a BUR. Look for improper or inadequate fastening of the deck, weak spots or holes, warped boards or improper drainage. If the deck is plywood, look for unsupported edges or inadequate thickness.

Remember, if the roof you installed fails due to an inadequate or defective deck, there's no way you're going to convince the customer that it's not your fault. You should also check that the roof deck is designed to adequately shed and drain water. For good drainage, the roof must have a minimum slope of $1/8$ inch per foot; however, a slope of $1/4$ inch per foot is better. Make sure there are plenty of roof drains, that leaders and gutters are large enough to handle the flow of water, and that the roof deck is sloped to carry water to drainage areas.

Don't install metered roof drains (or any drain designed to restrict water flow) unless the roof is also designed to retain water. Check that any ponded water areas can evaporate within 24 hours after a rain. Roofs designed to pond water may not be warranted, depending on the manufacturer. Water is heavy: water just 1 inch deep weighs about 5 pounds per square foot. That weight is a primary cause of deck deflection. Sometimes the deflection is permanent. If this happens too often, the roof structure could eventually collapse.

A properly-designed roof slopes toward one or more roof edges to carry drainage water off. You can contour sprayed-on foam insulation to slope toward roof drains, or install tapered rigid insulation to provide good drainage. A roof deck with a proper slope will show no signs of standing water 48 hours after a rain.

Metal Decks

If you use a metal deck for a BUR substrate, be sure it's 22 gauge (minimum) with 1½-inch-deep ribs. The ribs are designed with a maximum deflection of $1/240$ of the span between supporting members under a 300-pound load placed on a 1-square-foot area at mid-span.

Install the deck so there's no more than a $1/16$-inch horizontal deviation across any three adjacent top flanges. Use a straightedge to check the deviation. If the deck fails this flatness tolerance, you have to use mechanical fasteners to secure insulation to the deck. Over steel decking, use fasteners long enough to penetrate the deck. Figure 10-1 shows several types of fasteners, and the substrate materials you use them with.

The metal deck beneath built-up roofing must be galvanized or painted. Stagger the deck end laps and weld, crimp or button-punch the side laps.

The type of metal deck and rib determine both the type and the thickness of the insulation you install. Use a minimum ¾-inch-thick insulation over a metal deck. Due to potential deflection and/or irregular surfaces, some manufacturers recommend that you install insulation over a metal deck using one mechanical fastener for each 2 to 4 square feet, or 12 to 15 pounds of hot steep asphalt per square (hot bitumens are described in detail on page 303). Manufacturers recommend using mechanical fasteners.

If you use a vapor retarder base sheet beneath the insulation, install it in hot steep asphalt applied in continuous ½-inch-wide ribbons placed on 6-inch centers parallel with the ribs. You can also apply bitumen over the entire deck in contact with the vapor barrier. You can then install the overlying insulation with mechanical fasteners, or in a solid mopping of hot steep asphalt.

If you're using more than one layer of insulation, you can install both layers in one operation with mechanical fasteners. Or you can install the second layer in 12 to 25 pounds of hot steep asphalt per square. Always install the upper layer of insulation with end and side joints staggered over the joints of the bottom layer.

Lightweight Insulating Decks

Lightweight insulating decks (wet fill decks) include lightweight aggregate concrete (20 to 40 pounds per cubic foot dry density), foamed concrete, gypsum concrete and thermosetting asphaltic decks.

Over lightweight insulating concrete fill or poured gypsum concrete decks, you install a coated or vented base sheet using mechanical fasteners. Don't mop hot bitumen over lightweight concrete or gypsum concrete. If you don't use insulation board, install a ¾-inch protection board into a solid mopping of hot bitumen between the base sheet and the subsequent roofing

Deck type	Special instructions	Fastener recommendations
Wood, tongue & groove & plywood		1, 3, 4, 7, 8, 11
Lightweight insulating concrete	See note "1"	9, 10
Existing gypsum decks (reroofing)	See note "2"	6, 9, 10, 12
Structural wood fiber	See note "3"	2, 5, 6
Fiberboard roof insulation		1, 3, 4, 7, 8, 12

Note "1" Hydro-Stop™ vapor barrier applied with Celotex® Anchor Bond® LWC™ fasteners must be used as the first ply over new lightweight insulating concrete decks.

Note "2" Due to variation in hardness of existing gypsum decks, fastener selection should be field tested to determine penetration and holding power.

Note "3" A Vaporbar™ base sheet must be used as the first ply or embedment course over structural wood fiber decks if roof insulation is to be hot mopped to the deck.

Use Celotex® Anchor Bond® roof insulation fasteners for attachment of roof insulation to decks.

Detailed description of fasteners & sources of supply

1 	2 	3 	4 
Roofing nail 11 or 12 ga. 3/8"-1/16" diam. head National Nail Co.	Insuldeck Loc-Nail E.G. Building Fasteners Corp.	Roofing nail annular thread 11 ga. 3/8" diam. head Independent Nail Co.	Roofing nail spiral thread 11 ga. 3/8" diam. head W.H. Maze Co.
5 	6 	7 	8 
Capped Es-Nail 1" cap ES-Products	Tube-Loc nail 1" diam. cap Simplex Nail & Mfg. Co.	Squarehead cap nail annular thread 1" diam. cap Simplex Nail & Mfg. Co.	Squarehead cap nail spiral thread 1" diam. cap Independent Nail Co.
9 	10 	11 	12 
Nail-Tite type A 1¼" diam. cap ES-Products	Anchor Bond LWC fasteners Zonolite or Nail-Tite Mark III ES-Products Olympic lightweight concrete fasteners	Roofing staple for power driven application only Bostitch Spotnails	Do-All nail hardened E.G. Building Fasteners Corp.

Figure 10-1 Mechanical fasteners

Courtesy of Celotex

membranes. For insulation over lightweight concrete or gypsum concrete, mechanically fasten a vapor retarder sheet between the fill and the insulation board. Also, vent the fill with roof relief vents like those in Figure 10-2.

When top venting is required, install pressure relief vents. As a rule of thumb, install one vent per every 900 to 1000 SF of roof area. Some manufacturers recommend that you install 4-inch-diameter (minimum) hooded pressure relief vents 20 feet from the roof edges and at 40-foot centers thereafter. Make sure the vent openings penetrate the roof fill a minimum of 2 inches. I recommend filling the penetration with loose fiberglass insulation, as shown in Figure 10-2, to prevent condensation from forming on the roof deck.

Freshly poured lightweight insulating concrete decks must cure before you install a BUR system. As a rule of thumb, allow at least ten days of dry weather to pass between the pour and installing a BUR system. This rule of thumb makes one assumption: the deck that the concrete was poured on must have underside venting, like a perforated steel deck. Unfortunately, this isn't always the case. If the insulating deck is poured over a nonporous structure, concrete takes longer to cure. As a rule of thumb, allow a full 30 precipitation-free days curing time to pass before you install a BUR system.

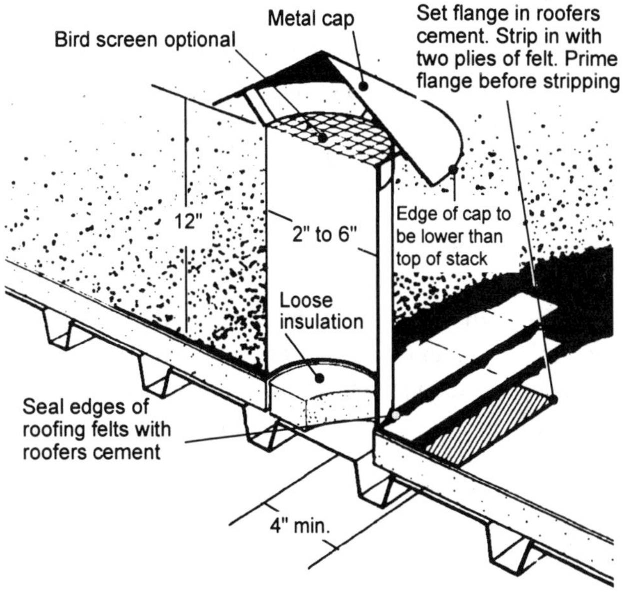

Note: This detail is used to minimize vapor pressure from insulation. The moisture may have entered due to leaks, faulty vapor retarders or during construction. The spacing is determined by the type of insulation used and the amount of moisture to be relieved. It is sometimes used for new roofs when vapor retarders are used and a venting system is desired.

Figure 10-2 Roof relief vent

Install gypsum concrete decks a minimum of 2 inches thick and pour the deck over cement-fiber or glass-fiber form boards that vent the underside. You can use slotted corrugated panels instead of form boards if their slots provide at least 1.5 percent open area. Never install lightweight insulating concrete to receive a BUR system over a vapor retarder, or over unvented metal decks, poured concrete decks, existing built-up roofs or any substrate that doesn't let the fill dry from below. Never install lightweight concrete decks on roofs with slopes greater than 1 in 12. Also be aware that some manufacturers of BUR products categorically refuse to warranty systems installed over lightweight insulating concrete decks or lightweight aggregate-asphaltic compacted fills.

Seal joints between precast gypsum panels with an 8-inch-wide felt strip embedded in roofing cement. Then you can install a vapor retarder sheet with mechanical fasteners. Over this, install insulation board or felts in a solid mopping of hot bitumen.

Over thermosetting insulating fill, install the base sheet in a hot mopping of bitumen. If you use additional insulation, install a vapor retarder between the fill and the insulation board, and vent the fill.

Mechanical fasteners like the ones in Figure 10-1 (used over gypsum and insulating concrete decks, or over the insulation board or other nailable decks) are made by various manufacturers for use with specific materials. Make sure you choose the right fastener for the material you're using.

Precast and Prestressed Concrete Decks

Over a precast or prestressed concrete deck, prime the deck and install insulation board into a hot mopping of bitumen at 30 pounds per square. If joints between the concrete members aren't sealed, don't mop within 4 inches of the joints to prevent the bitumen from dripping into the joints.

Never install a membrane directly over a deck. Many manufacturers recommend that you install a minimum of 2-inch-thick, 2500 PSI concrete fill or a ¾-inch protection board over prestressed concrete roof decks before you apply roofing. That's to compensate for surface irregularities due to slab misalignment and camber. If you pour a concrete deck, don't fill the joints between concrete members because it keeps the fill from drying from below. If you seal the joints, you must vent the fill from above.

Reinforced Concrete Decks

Prime reinforced concrete decks with 7½ pounds of asphalt primer per square. Use a special thin liquid asphalt primer that you can use over non-nailable decks. This comes in both 5-gallon pails and 55-gallon drums. After the primer, apply hot bitumen at 20 to 30 pounds per square. Install the first roofing ply into the hot bitumen.

If you use a base sheet, spot-mop it in place over the deck. Spot-mopping allows venting of any remaining moisture in the concrete. However, be aware that some manufacturers' warranties only apply when the base ply is solid-mopped to the deck.

Wood and Plywood Decks

Use mechanical fasteners to install the base sheet over wood board or plywood decks. Choose 11-gauge galvanized nails with 7/16- to 5/8-inch heads, or drive nails through caps with an equivalent diameter. Use large-head annular ring-shank nails over plywood decks. The large heads help prevent the wind uplifting the roofing felts. Use nails that are long enough to go at least 5/8 inch into the wood board, and at least ½ inch into the plywood. One advantage of nailing is that you can install the base sheet to the deck and still allow horizontal passage of water vapor trapped between the deck and the roof system.

Structural Wood Fiber Decks

Use structural wood fiber decks manufactured with a factory-applied felt surface. Strip all joints in the deck with 6-inch-wide felt embedded in roofing cement. Nail a vapor retarder base sheet over structural wood fiber decks. Then install at least 1 inch of rigid insulation into a solid mopping of hot steep asphalt. If you use decking without the felt surface, install rosin paper between the deck and the base ply. Install only as much decking as you can cover with roofing that same day.

Back Nailing

If a built-up roof slopes more than 1 in 12 (using asphalt bitumen) or more than ½ in 12 (using pitch bitumen), you'll have to install roofing felts with mechanical fasteners in addition to the interply bitumen mopping. That's to stop the membranes from slipping. This is called *back nailing*. Some manufacturers recommend back nailing only on slopes exceeding 2 in 12 with asphalt bitumen. Don't apply coal-tar pitch over slopes exceeding 1 in 12.

Nailable decks include wood, plywood, gypsum (poured or precast plank), lightweight insulating concrete and structural wood fiber. Over those decks, back nail each felt with the appropriate fasteners as specified in Figure 10-1. Apply fasteners on 12-inch centers along a row 1 inch below the top edge of each felt.

Over non-nailable decks, back nail each ply into high-density fiberboard insulation at 12-inch centers along a row located 1 inch below the top edge of each felt, or back nail into treated wood nailers 1 inch (or about 10 and 12 inches) below the top edge of each felt. Some manufacturers recommend that you install nailers parallel with the slope and space the nailers face-to-face at a maximum of 48 inches, as shown in Figure 10-3.

Install the membranes perpendicular to the wood nailers and be sure that at least two roofing plies cover each nail head. Also install nailers at ridges, eaves, gable ends and around openings. Where there's no insulation, install the tops of the nailers flush with the top of the deck. Over insulated decks, install the tops of the nailers (now called insulation stops) flush with the top of the insulation boards. Use insulation stops that are at least 2 inches wide (some recommend 4 inches) and as thick as the insulation. Install 6-inch-wide nailers at the bottom of the slope. That's also shown in Figure 10-3.

Other manufacturers recommend that you install the nailers perpendicular to the slope and the membranes parallel to the slope. This is called the "strapping" method. The spacing you use for the nailers depends on the roof slope and the type of roof surface installed. Figure 10-4 lists nailer locations for various applications.

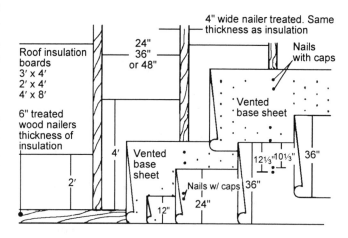

Figure 10-3 Wood nailers

Roof slope	Type of surface	Nailer locations (face-to-face)
1 to 2 in 12	Smooth	None required
	Aggregate or cap sheet	20'
2 to 3 in 12	Smooth	20'
	Aggregate or cap sheet	10'
3 to 4 in 12	Smooth	10'
4 to 6 in 12	Smooth	4'

Figure 10-4 Nailer placement recommendations

Still other manufacturers recommend that you use the strapping method and install the nailers 20 feet face-to-face over slopes through 3 in 12, and 4 feet face-to-face over slopes greater than 3 in 12.

Base Sheets (Vapor Retarders)

The first (and normally the heaviest) membrane you install in a BUR system is the base sheet. In most cases, the base sheet acts as a vapor retarder which protects the insulation or subsequent membranes from damage due to moisture from within the building.

A vapor retarder is normally a coated sheet or a vented membrane. A coated sheet (also called smooth roll-roofing) is a roofing felt (organic or inorganic) coated on both sides with bitumen and finished with a heavy coat of non-sticking mineral powder. You can use a coated sheet as a vapor-retarding base sheet, or as the top ply on a smooth-surfaced built-up roof. Coated sheets are 36 inches wide and 36 to 144 feet long. They weigh 45 to 80 pounds per square.

One type of vented base sheet is made of an asphalt-impregnated glass fiber mat with ¼-inch-diameter holes located on 5-foot centers. These holes let moisture escape at the time of installation. Some systems require that you install fiberboard, perlite board or fibrous glass insulation between the insulation board and the base sheet to cut down on blisters made by air and moisture trapped between the two layers.

Another type of vented base sheet is made of an asphalt-saturated glass fiber mat whose bottom surface is covered with mineral granules embossed to provide channels for moisture vapor to escape. The top surface is smooth to receive the bitumen and overlying roof membranes.

A vented base sheet weighs about 72 pounds per square. This type of sheet is highly recommended over vented fills such as gypsum decks. Manufacturers also recommend it when you're re-roofing over existing roof membranes.

Another kind of base sheet has ⅝-inch perforations spaced on 3-inch centers in both directions. This sheet weighs 60 pounds per square. It's secured by the asphalt that penetrates the vent holes as you install the subsequent roofing felt or insulation board with a solid hot bitumen mopping.

Turn down the edges of vented base sheets over the edge of the roof to allow venting under the metal edge strip, as shown in Figures 10-5, 10-6 and 10-7. Turn the edges up along vertical surfaces and spot-mop the sheets at 2-foot centers to allow venting between the base flashing and metal counterflashing, as shown in Figure 10-5. Provide additional venting with relief vents when any dimension of a roof is more than 60 feet.

Built-up Roofing

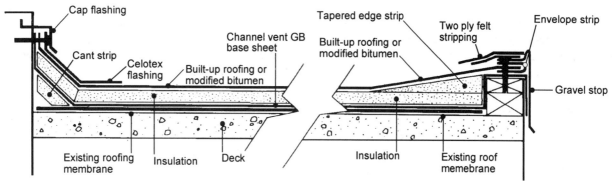

Channel Vent GB Base Sheet is made on a tough, durable fiber glass mat. The sheet is ashalt-impregnated and coated with high quality, mineral-stabilized asphalt coating. The bottom surface of the sheet is covered with mineral granules and then embossed to provide channels for the escape of moisture vapor.

Channel Vent GB Base Sheet is intended to replace Celotex regular base sheets on all roofing systems which require superior venting. Channel Vent GB Base Sheet is required as the first ply for reroofing over existing roofing membranes

Figure 10-5 Cross section of a BUR system

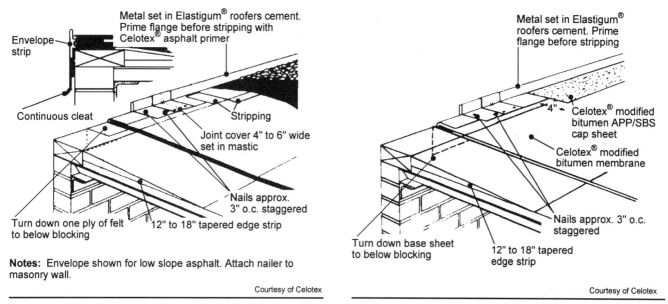

Notes: Envelope shown for low slope asphalt. Attach nailer to masonry wall.

Courtesy of Celotex

Figure 10-6 Light metal roof edge detail stripped with felts

Courtesy of Celotex

Figure 10-7 Light metal roof edge detail stripped with cap sheet material

Install a vapor retardant base sheet whenever these conditions exist:

- The average outside winter temperature is 40°F or less
- You expect unusually high relative humidity generated from within the building (45 percent or greater during winter)
- The attic air space is used for a return-air plenum for heating or air conditioning

Since water vapor tends to go from warm to cooler areas, always install the vapor retarder on the side of the roof assembly that's warmer most of the time.

Install the base sheet with 4 inch side laps and 6 inch end laps. Over nailable decks such as wood, plywood, structural cement-fiber or poured gypsum, nail all laps on 9-inch centers and stagger-nail the center of the sheet in two rows on 18-inch centers. Locate each row 12 inches from the edges of the sheets. Drive about 100 nails per square. Drive nails through flat metal discs at least 1 inch in diameter.

On wood decks, install the base sheet over rosin paper. Prime non-nailable decks using 1 gallon of asphalt primer per square. Embed the base sheet in hot steep asphalt using 23 pounds per square. Don't install vapor retarders with cold-applied adhesives.

Over nailable or non-nailable decks, turn the sheets up parapet walls, curbs and other vertical surfaces at least 4 inches above the top of the insulation. Use roofing cement or steep asphalt to adhere the sheet to a vertical surface. At pipe penetrations, flash the base sheet with two plies of asphalt-saturated felt and roofing cement.

When you expect extreme humidity inside the building (50 percent at 70°F or greater), install two base sheets. Embed the second base sheet in a solid mopping using 23 to 25 pounds of asphalt per square.

Roofing Membranes

Membranes used in built-up roofing include roofing felts, fabrics, coated sheets, rosin-sized paper and mineral-surfaced sheets. Most manufacturers recommend that you apply roll material with its long dimension perpendicular to the slope of the roof.

Rosin-sized Paper

In places where hot bitumen might drip through cracks or spaces in the deck, nail rosin-sized sheathing paper or unsaturated felts over decks (especially wood) before you apply the bitumen. Install rosin paper between a wood roof deck and the roofing membrane to prevent damaging chemical reactions between wood resins and the membrane.

You can also install rosin paper over a deck as a slip sheet, to prevent adhesion of the base sheet to the deck. This helps relieve expansion-contraction problems. Nail the base sheet over the rosin paper. Lap rosin paper sheets 2 inches and secure them with enough nails to prevent slippage and wind uplift. Glass fiber felts don't require rosin paper, so you can nail them directly to the deck. Rosin paper, or some other separating layer, is required over plywood decks, and you must install blocking beneath the end joints of the plywood. Rosin paper weighs about 5 pounds per square.

Roofing Felts

Roofing felts make a BUR system strong. They also help protect the lower layers of bitumen from debris, dirt, water, air, and the drying and weathering effects of the sun. Roofing felts are absorbent, so they help prevent bituminous materials from melting and flowing away during hot weather. Felts are perforated so trapped air can escape from under the sheets when you apply them. Glass fiber felts don't have perforations because they're already porous.

Some felts come with a coating of fire-resistant materials. Felts are often produced with a coating of fine mineral powder on one or both sides to keep them from sticking to each other when they're packaged in a roll. If you're using uncoated felts, keep them cool so they don't stick together.

Apply 25 pounds of asphalt or pitch per square between felts with an application tolerance of plus or minus 15 percent. Too thin a layer of bitumen weakens the lap, and too thick reduces friction and allows the felts to slip on sloped roofs.

Porous substrates such as low-density decks and insulation tend to absorb bitumen. In this case you may need more than 25 pounds of bitumen per square to make sure the felt adheres. You've used enough if there's fluid bitumen remaining of the surface after absorption.

Control the number of plies by varying the widths of the side laps. Apply felts in a shingle-like fashion, starting at the low end of the roof. Felt exposures vary, depending on the number of plies you install (Figure 10-8). You can determine the felt exposure required for any BUR system with the following formula (the 34 inches is the 36-inch felt width minus a 2-inch edge lap):

$$\text{Exposure} = \frac{34 \text{ inches}}{\text{No. of Plies}}$$

Equation 10-1

▼ **Example 10-1:** To find the exposure of each ply of a BUR system when two plies of felts are applied in a shingle-like pattern:

$$\text{Exposure} = \frac{34 \text{ inches}}{2}$$
$$= 17 \text{ inches}$$

Using the same formula, three plies require an exposure of $11\frac{1}{3}$ inches, and four plies require an exposure of $8\frac{1}{2}$ inches.

Roofing felts are either organic or inorganic. The most commonly used organic felt weighs 15 pounds per square. An equivalent glass fiber (inorganic) felt weighs $7\frac{1}{2}$ pounds per square. Roofing felts are usually typed by their weight. For example, Type 8 means a $7\frac{1}{2}$-pound glass fiber felt and Type 30 means a 30-pound felt. Full information about roofing felts is in Chapter 3.

Roofing Construction & Estimating

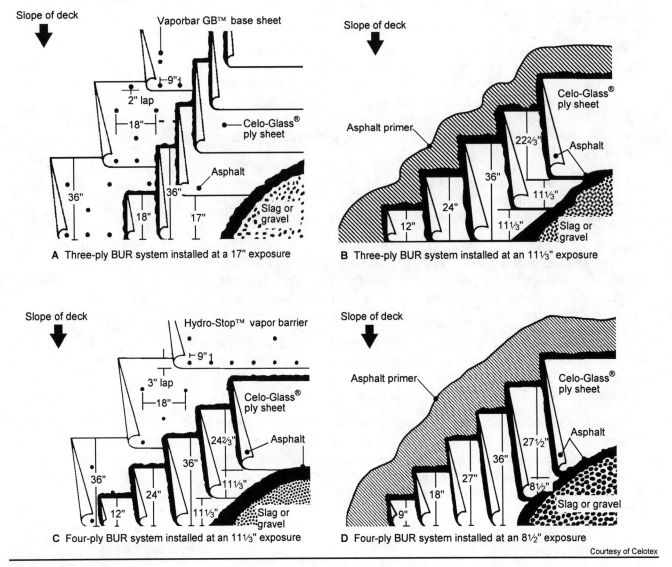

Figure 10-8 Felt exposures

Some roofing felts are coal-tar saturated and you install them with coal-tar bitumen (pitch). Others are asphalt-saturated and you install them with asphalt bitumen.

Three-ply BUR System

A three-ply BUR system normally consists of:

- Base sheet (nailed or mopped) followed by two layers of roofing felts laid like shingles with a 17-inch exposure, as shown in Figure 10-8 A; or,

- Three plies of felts laid like shingles with an 11⅓-inch exposure, as shown in Figure 10-8 B.

Four-ply BUR System

A four-ply BUR system normally consists of:

- A base sheet (nailed or mopped) followed by three layers of roofing felts laid with a 11⅓-inch exposure, as shown in Figure 10-8 C; or

- Four plies of roofing felts laid with an 8½-inch exposure, as shown in Figure 10-8 D.

Fabrics

Fabric membranes made of bitumen-saturated cotton, glass fiber or jute are sometimes used for roofing membranes. For more detailed information about these products see Chapter 12. Uncoated polyester filaments bonded under heat and pressure produce a roofing fabric that's lightweight (0.9 pounds per square foot) and elastic. Polyester fabric is usually glued and surfaced with cutback asphalt.

Hot Bitumens

Bitumen is defined as an amorphous, semi-solid, organic hydrocarbon mixture. Asphalt and coal-tar pitch are the most widely-used bitumens in the roofing industry. Asphalt occurs naturally, and it's also made during petroleum processing. It weighs an average of 8.7 pounds per gallon.

Coal-tar bitumen (coal-tar pitch) is a product of the distillation of coal tar in the manufacture of coke or gas. It is softer than asphalt. It has excellent self-healing properties after it's cooled and is virtually unaffected by water. It weighs about 10.6 pounds per gallon.

Coal-tar pitch is more expensive than asphalt. But despite its higher cost, it's recommended on dead-level roofs and low-slope roofs with slow drainage and occasional ponding. Coal-tar bitumen is required on roofs designed to retain water. It's also recommended in highly corrosive environments.

Asphalt and coal-tar pitch are generally incompatible, with a couple of exceptions:

- You can install asphalt flashing material with coal-tar pitch.

- You can use hot Type II asphalt over coal-tar saturated felts.

When you apply glaze coats of bitumen, it produces a fusion between plies because of the partial melting of asphalt within the plies. Bitumens behave like fluids when they're heated: during warm weather, they're elastic; during cold weather, they harden.

Bitumens are normally applied with a mop and the membranes are pushed into the hot bitumen with a push broom. This is called *brooming-in*. To avoid displacement of the membrane, you drag the broom across the membrane from the unmopped side rather than pushing from behind the roll. It's important that you broom-in the membrane while the bitumen is hot to avoid blisters caused by vapors trapped beneath a membrane. If fishmouths (wrinkles at the edge of the felt) or buckles develop during the broom-in, immediately cut and repair them.

Four types of asphalt are available, with varying softening-point temperatures:

- Type I (dead-level) asphalt softens at 135° to 151°F. This type of bitumen is self-healing, and you can install it over roof slopes up to $1/4$ in 12.

- Type II (flat) asphalt softens at 158° to 176°F and you can use it on slopes up to $1/2$ in 12.

- Type III (steep) asphalt is the most commonly used. It softens at 180° to 205°F and you can apply it on roof slopes up to 3 in 12. Due to the intensity and length of the summer season, parts of some southern states such as Texas, Florida, New Mexico, Arizona and California require Type III asphalt on roof slopes ranging from dead level to $1/2$ in 12.

- Type IV (special steep) asphalt softens at 205° to 225°F. Use this type of asphalt on roof slopes up to 6 in 12 where there is a potential during hot weather for the asphalt to melt and run, or for the roofing felts to slide away. You can use Type IV asphalt on smooth roof surfaces, but it lacks the self-healing properties of the low-melt asphalts.

Using the correct type of asphalt will result in a roof that softens and self-heals, yet doesn't sweat melted bitumen during the heat of the day. The temperature of the lower roof plies can exceed 175°F on a 100° day. In winter, the temperature of a built-up roof over insulation will be close to the outdoor temperature.

The flash point of asphalt is the lowest temperature at which it gives off enough vapor to form an ignitable concentration. Most asphalts have a flash point ranging from 437° to 500°F. You can safely heat dead level asphalt or pitch to 400°F and flat and steep asphalts to 450° and 475°F. Don't heat asphalt to within 25°F of the flash point.

Bitumens don't waterproof as well if you overheat them or heat them too long. If you won't use a bitumen for four hours or more, shut off the kettle or reduce the temperature to about 325° to 350°F. Don't apply Type I and II asphalts or coal-tar bitumen at a temperature lower than 350°F, and Types III and IV at lower than 400°F.

Bitumen congeals quickly in cold weather. But don't overheat the bitumen to compensate for the conditions. Instead, keep the roofing roll close behind the mop (approximately 5 feet) and broom-in the membrane as quickly as possible.

The equiviscous temperature (EVT) of a bitumen is the optimum temperature and viscosity to apply the material so it will stick properly and be waterproof. This is the temperature at which 125 centipoise is attained. Apply roofing bitumen at a temperature within 25°F of the EVT.

Store and protect bituminous materials from rain and snow. If moisture gets into them, they foam up in the kettle and you'll have problems applying them.

Cold-applied Bitumens

There are many cold-applied bituminous materials available. Asphalt mastic consists of asphalt and graded mineral aggregate. Bituminous grout is made of bitumen and fine sand. You can pour either product when heated, but they must be troweled when cool.

Asphalt emulsions are made of fine droplets of water dispersed in asphalt with the aid of an emulsifier such as bentonite clay. Some emulsions also contain glass fibers or other materials. Asphalt emulsions are so thin that you usually spray them on. Use emulsions for surfacing bitumen on slopes up to 6 in 12. Apply surface emulsions at the rate of 3 gallons per square. Since emulsions contain water, don't let them freeze.

Cutback bitumens are thinned with organic solvents and light oils to make them flow freely. Cutbacks are used mainly for interply applications and less frequently as a surfacing material. When used as a flood coat for embedding aggregate, apply cutback asphalt at the rate of 6 to 7 gallons per square. Cutbacks contain flammable solvents; don't expose them to an open flame. Liquid emulsions and cutbacks solidify when the water or solvent in them evaporates, leaving only the bitumen behind.

The most common system used in cold-process roofing uses three layers of 53-pound cold-process felts saturated with a cold asphalt emulsion. Secure the felts with nails and asphalt adhesive applied at the rate of 2½ gallons per square. Then cover the surface with a layer of cold-applied asphalt-fibrated emulsion at the rate of 3 to 4 gallons per square. This roof system is sometimes referred to as a Type 5 roof.

Flashing Sealers

Asphalt cements are thick pure asphalts with no fillers and only enough solvent to permit application. Asphalt cements are used mainly for sealing the laps of flashing materials.

Plastic cements, flashing cements, all-purpose roofing cements and cutback cements are trowelable mixtures of asphalt or coal tar with fillers such as glass fibers, powdered aluminum, rubbers and solvent. These products are used for securing flashing material.

Store asphalt cement and other related products in a warm place until you use them. If you need to warm the cement, place the unopened container in hot water until it's pliable. Never heat the container directly over a flame.

Cold-applied Roof Coatings

Many cold-applied asphalt roof coatings are available. While their composition varies, the main ingredients are always emulsified asphalt and mineral colloids. Acrylic- and neoprene-based adhesives are other popular cold-applied products. These materials are very durable and elastic, and can be applied by brush or spray at a rate of approximately 2 to 6 gallons per square, depending on the specific product.

Roof coatings consisting of asphalt or coal tar, fiber and nonvolatile penetrating oils are sometimes applied over an existing built-up roof to rejuvenate the old asphalt (or tar) to its original condition. This type of coating is called a *resaturant*.

Asbestos

Many roofing products, including some roof coatings, contain asbestos fibers. Evidence to date indicates that the asbestos in these materials is contained, so it doesn't present the health hazards of free asbestos. Nevertheless, many building owners won't allow these products to be used in their buildings due to potential exposure to liability. Also, many designers refuse to specify asbestos products of any kind. As a result, new asbestos-free roofing materials have been developed and are widely available.

Surface Aggregate

Aggregate is often applied over a hot flood coat of bitumen on the surface of a BUR system. The flood coat is normally applied at the rate of 60 to 70 pounds per square (asphalt), or 70 to 75 pounds per square (coal-tar pitch). In high-wind areas (70 mph and more), an additional 80 pounds of asphalt or 90 pounds of coal-tar pitch will sometimes be specified per square.

On water-retaining roofs, apply aggregate at the rate of 400 pounds per square (300 pounds for slag) into a flood coat of pitch applied at 70 pounds per square. Then sweep off all loose gravel or slag. If the roof doesn't have a controlled flow drainage system, apply a second coat of pitch at 85 pounds per square with aggregate at 300 pounds per square (200 pounds for slag). Again, sweep away any loose aggregate or slag and roll lightly to ensure the remaining aggregate is embedded in the pitch.

Some manufacturers maintain that felt plies can be left uncoated for up to six months before aggregate is installed. But when you apply aggregate, it's very important that you do so while the flood coat of bitumen is still hot. If you can't apply the aggregate soon after installing the membranes, or choose not to because other workers must walk on the roof, apply a glaze coat of hot bitumen to the surface at the rate of 8 to 10 pounds per square. When you can finally place the aggregate, sweep the glaze coat clean before you install the flood coat for the aggregate.

Aggregate does several important things:

- It protects the underlying roof system from the heating, drying and weathering effects of the elements.

- It prevents damage from foot traffic.

- It serves as a ballast to resist wind uplift and improves the roof's fire resistance.

- It reduces the roof surface temperature. (Due to high reflectivity, white aggregate such as marble chips (dolomite) can save up to 12 percent in energy costs when compared to other types of roofing aggregates.)

A properly installed aggregate-surfaced BUR system has a life expectancy of 20 years or more. The most common aggregates are rounded gravel, slag, or crushed rock. Less common are marble chips, scoria, pumice and crushed tile, brick and limestone. Grade gravel aggregate from ¼ to ½ inches, and slag from ¼ to ⅝ inches.

Make sure the aggregate is clean and dry so that it will stick well to the bitumen. The maximum acceptable moisture content (by weight) is 0.5 percent for crushed stone or gravel, and 5.0 percent for crushed slag.

Apply gravel or crushed stone embedded in a flood coat of bitumen at a rate of 400 pounds per square, slag at 300 pounds per square and marble chips at 400 to 500 pounds per square. Adhere approximately 50 percent of any aggregate to the bitumen. In high-wind areas (70 mph and more), apply an additional 300 pounds of gravel or 200 pounds of slag per square.

When embedded in cutback bitumen, apply gravel, crushed stone and slag at the rates of 400, 450 and 350 pounds per square, respectively. Don't install aggregate over asphalt-emulsion coatings. Avoid using low-density aggregates like bauxite, kaolin, gypsum, caliche and shale. They're too easily blown and washed off of the roof.

Choose slag for roof slopes greater than 2 in 12 because it stays embedded better than gravel. I wouldn't install aggregate on roof slopes steeper than 3 in 12.

Mineral granules of the type used to make composition shingles can also be used for roof surfacing over cutback asphalt or mastic. If mastic is used, apply it at the rate of 2½ to 3 gallons per square and deposit the granules at the rate of 50 to 60 pounds per square. You can install a granule surface on roof slopes up to 6 in 12.

Smooth-surface Roofing

You can install smooth-surfaced (black) roofing on roof slopes as steep as 9 in 12. On nailable decks, nail the top edge of each membrane on 12-inch centers for slopes greater than ½ in 12. Install a 43-pound asphalt-coated base sheet, then three plies of 15-pound asphalt-saturated felt solidly mopped with bitumen applied at the rate of 25 pounds per square.

Over non-nailable decks install four plies of 15-pound felts into solid moppings of asphalt applied at the rate of 25 pounds per square. On inclines greater than ½ in 12, install nailers for fastening the top edge of each membrane, driving nails at 12-inch centers. Surface the roof with a glaze coating of 20 pounds of hot asphalt per square, followed by 3 gallons of cold-applied asphalt emulsion per square. Instead of an asphalt coating, you can cover the roof surface with 1.2 gallons of fibrated aluminum coating per square. Use a roller to apply this coating.

Smooth-surfaced roofs are lightweight and easy to inspect and repair. Re-roofing can often be done without removing the old smooth-surface membrane. However, all smooth-surfaced roofs are more susceptible to damage and leaking due to hail impact than aggregate-surfaced roofs. The life expectancy of a smooth-surfaced roof is only six years.

Cap Sheets

You can install heavy-duty cap sheets instead of an aggregate surface. Cap sheets are made from a variety of materials, but mineral-surfaced sheets and reinforced polyester sheets are the most common.

Refer back to Chapter 5 for information about the physical features of mineral-surfaced asphalt sheets. Mineral-surfaced sheets are inexpensive, but not very durable. The life expectancy of a mineral-surface roof is only six years. Mineral-surfaced sheets are usually installed on roof slopes between ½ in 12 and 6 in 12.

When you apply a standard roll-roofing BUR system to a nailable deck, nail a sheet of rosin paper to the deck. Cover the rosin sheet with two plies of 15-pound felt nailed and mopped with hot asphalt bitumen applied at the rate of 25 pounds per square.

Built-up Roofing

Unless you're concerned about bitumen dripping into the building interior, no rosin sheet is required over a non-nailable deck or over insulation. Cover the deck with two plies of 15-pound felt solidly mopped with hot asphalt bitumen applied at the rate of 25 pounds per square.

Over either type of deck, install an 80-pound cap sheet over roof slopes of ½ in 12 to 6 in 12. Mop a 55-pound cap sheet over roof slopes of 3 in 12 to 9 in 12. Precut cap sheets into 12-foot lengths before you install them. Lap the sides 2 inches and the ends 6 inches. Offset adjacent end laps at least 3 feet. Over nailable decks, nail end laps over slopes greater than 2 in 12.

A heavier type of roll-roofing BUR system is composed of one layer of rosin paper (over wood decks), followed by a nailed 15-pound asphalt-saturated felt. Following this, mop two layers of 15-pound felt followed by two layers of 120-pound slate-surfaced felt.

Another type, the three-ply glass fiber cap-sheet membrane system, consists of a base sheet, followed by one ply of felt and a cap sheet, as shown in Figure 10-9 A. A four-ply system normally consists of a base sheet, two plies of felt and a cap sheet, as shown in Figure 10-9 B, or three plies of felt overlaid with a cap sheet, as shown in Figure 10-9 C.

You can also use cap sheets made from a blend of synthetic and glass fibers impregnated with modified (rubberized) asphalt, surfaced with mineral granules. It's installed over a fiberglass base sheet mechanically fastened (over nailable decks) or spot-mopped (over primed non-nailable decks or an existing aggregate surface), followed by the cap sheet solidly mopped onto the base sheet. This sort of cap sheet is approximately 0.1 inch thick and weighs about 73 pounds per square.

Cap sheets can also be used for flashing at parapet walls, roof edges and roof penetrations. Look ahead to the illustrations in the flashing section later in this chapter (starting on page 315).

Aluminum Roof Coatings

Aluminum roof coatings consist of selected asphalts, asphalt emulsions, reinforced mineral fibers, fillers, mineral spirits and pure aluminum pigment flakes.

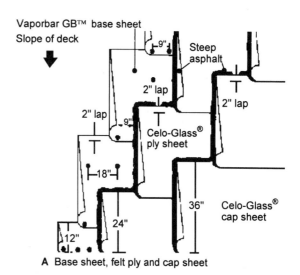
A Base sheet, felt ply and cap sheet

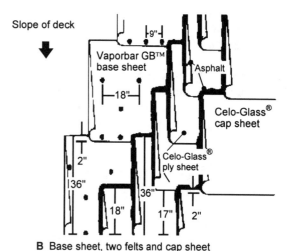
B Base sheet, two felts and cap sheet

C Three felts and cap sheet

Courtesy of Celotex

Figure 10-9 Three- or four-ply glass fiber cap-sheet membrane systems

An aluminum coating is usually applied to roofs where the outside temperature gets extremely high. The coating reduces expansion and contraction of roofing materials due to heat, which stabilizes roof membranes and vertical flashing surfaces. It prolongs the life of the roof by reflecting heat and ultraviolet rays and slows oxidation. It makes the building more energy-efficient by reducing the heat load by 45 percent compared to dark-surfaced roofing. Because the roof insulation is less efficient at high temperatures, this reduced heat also makes the insulation work better, so you need less. And there's a decreased load on the air conditioning system.

This product comes in 1-gallon cans or 5-gallon pails. You spray or brush the coating on the roof surface at the rate of about 1 to 2 gallons per square.

Aluminum roof coating is ideal for application over metal roof decks. There's a specialized aluminum coating available for insulating and protecting aluminum roofs. This product comes in 5-gallon pails. Instead of aluminum roofing coatings, a white acrylic emulsion is sometimes applied over an asphalt surface.

Before you apply an aluminum roof coating, prime the roof surface with asphalt-emulsion at the rate of ¾ gallon per square. Allow the newly-coated roof to cure for a month before you apply an aluminum coating. Allow a three-month curing period for new asphalt- or solvent-coated roofs. Because roofing membranes might shift, some manufacturers recommend that you let the roof surface weather through at least one summer before you apply an aluminum roof coating. Don't apply an aluminum surface coating over areas subject to water ponding — the water will eventually degrade it.

Fibrated aluminum roof coatings are often applied over roof flashing for extra protection. You'll see it in the figures accompanying the section on roof flashing a little later in this chapter.

Phasing

The practice of installing only a portion of a BUR and allowing an unfinished area to remain exposed to weather for a period of time is called *phasing*. But I consider this to be poor roofing practice. Plan the BUR installation so that you don't install more insulation than you can completely cover with membranes and aggregate by the day's end.

If you can't finish applying the gravel, cover all installed insulation with the membranes and apply an asphalt glaze coat with a squeegee at the rate of 10 pounds per square (for organic felts) or 20 pounds per square (for glass fiber felts). You can omit the glaze coat when the bitumen is coal-tar pitch or dead-level asphalt. Then place the surface aggregate no later than one week after you install organic felts, or 30 days after you install glass fiber membranes.

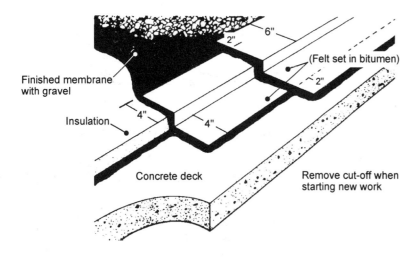

Figure 10-10 Water cut-off

Water Cut-offs

When you must phase roof construction, install temporary water cut-offs as shown in Figure 10-10 to protect the insulation at the end of the day. A cut-off is just a felt strip mopped over the exposed edge of the insulation to provide temporary protection. When you resume work, completely remove the water cut-offs so that insulation joints can be firmly butted together.

Cant Strips

Install cant strips where the built-up roof intersects vertical surfaces in order to break the sharp angle between the wall and roof deck. You can see the cant strips in Figures 10-11 through 10-14. Install cant strips over roof insulation (or deck, in the absence of insulation) with one edge flush against the vertical surface. Run the cant strip out over the roof at least 3 inches and up the vertical surface at least 5 inches.

Prime masonry surfaces with an asphalt primer before you install a cant strip. Use cant strips that are compatible with the roof membrane and bond well with mopped asphalt to the horizontal and vertical substrates. The vertical substrate you install depends upon whether the roof and wall are designed to move independently as shown in Figure 10-11, or the wall helps support the roof deck as shown in Figures 10-12 through 10-14. Nail cants in place over nailable decks or wood nailers, or embed them into a mopping of hot steep asphalt or roofing cement over insulation.

Roofing Construction & Estimating

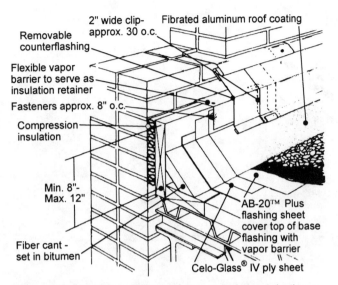

Figure 10-11 Parapet flashing where the deck and wall move independently

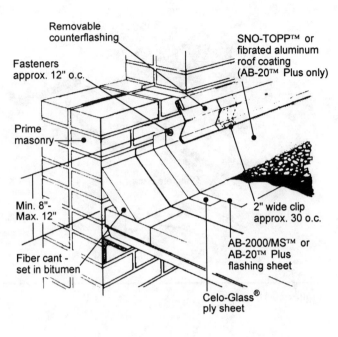

Figure 10-12 Parapet flashing where the deck is supported by the wall

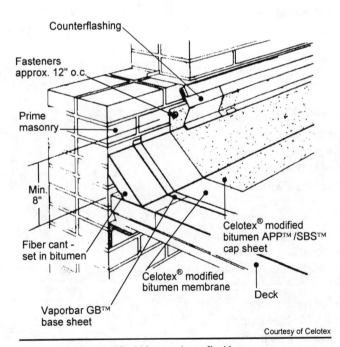

Figure 10-13 Modified bitumen base flashing

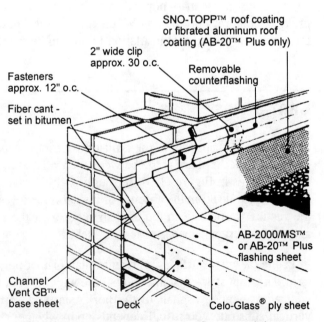

Figure 10-14 Base flashing for vented base sheet

312

Cant strips are made from a variety of materials including preservative-treated wood, rigid fiberboard, perlite board or concrete. Concrete cant strips are cast along with the wall and deck. Don't use metal cant strips because there's too much expansion and contraction.

Tapered Edge Strips

Tapered edge strips are sometimes installed at the inside edge of wood nailing strips to route water away from the roof's edge as shown in Figure 10-5. As with cant strips, nail or mop edge strips into place. Edge and cant strips are made from the same materials. Install edge strips at least 18 inches wide that provide an incline of approximately 1 inch per foot.

Envelope Strips (Bitumen Traps)

Install envelope strips at the edge of a roof and around openings to prevent bitumen from dripping onto surfaces beneath the roof. Make the strips using 12-inch-wide felt or 6-inch-wide soft metal. Allow half of the strip to overhang the edge or opening. Install the envelope strips before you install the roofing. After you install the membranes, fold the strip back flush with the top of the roof and secure it with roofing cement or bitumen as in Figures 10-5 through 10-7.

Temporary Roofs

Install a temporary roof when the building interior has to be kept "in the dry" during construction before the permanent roof is installed.

To install a temporary roof over a non-nailable deck, prime the deck and install a coated base sheet in hot steep asphalt. Over a nailable deck, nail a coated base sheet over rosin paper. In either case, the base sheet remains in place to serve as a vapor barrier. Never install a temporary roof over insulation which will remain as part of the permanent roofing system. The temporary roof goes on before the insulation.

Sometimes you may not realize that you need a temporary roof to protect the contents of a building you're working on. For example, I know of a 12,000-square-foot building that was built with a roof deck of prestressed double tees. Because cranes were used to erect the structure, the slab-on-grade was going to be poured after the tees were erected. Shortly after the last roof tee was installed and before the first slab pour, 4 inches of rain fell within 48 hours. With no temporary roof, all of the rainwater poured through the joints between the tees. The compacted fill (now in the shade) inside the building became a muddy quagmire. Among other nightmarish problems created by the flooding, the project was delayed for weeks because of the time it took the fill to dry.

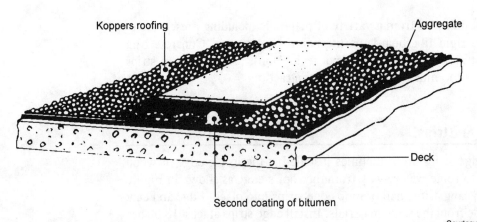

Figure 10-15 Traffic pad

Roof Traffic Pads

Roof traffic pads (roof walkways) as shown in Figure 10-15 provide a protective walkway over built-up or elastomeric roofs. Install traffic pads for paths to areas where security or service personnel need access. As a rule of thumb, install a walkway when more than five people a month will walk over the same path on a roof. Also install pads in areas subject to damage from dropped tools and parts.

Roof walkways for BUR systems are generally made of a uniform core of asphalt, plasticizers and inert fillers bonded by heat and pressure between two saturated and coated sheets of organic felt. The bottom is smooth while the top surface is faced with non-slip mineral or ceramic granules.

Asphalt traffic pads come in 1 x 2-, 3 x 3- or 3 x 6-foot panels, and ½, ¾ and 1 inch thick. Weights of ½-, ¾- and 1-inch asphaltic traffic pads are 3.5, 5.2 and 7.0 pounds per square foot, respectively. Don't install this type of pad in high-traffic areas, especially where there may be walkers wearing high heels.

Rubber walkway pads are also available. These pads resemble a heavy-duty door mat and are manufactured in 3' x 3'8" x ⅜" panels. Colors are normally black or white.

Install traffic pads after you've completed the BUR system. Sweep away all loose aggregate and embed the panels in industrial roofing cement, or into the same material (asphalt or pitch) you used to embed the aggregate. Some manufacturers recommend that you install roof walks before you apply the surface aggregate. Whichever way you do it, make sure it's well secured.

Don't install roof walks on slopes steeper than 2 in 12, or over membranes installed over fiberglass roof insulation. Space traffic pads at least 3 inches apart to allow drainage between them.

You can also use precast rod- or mesh-reinforced paver blocks for roof walkways, with non-skid and exposed-aggregate surfaces if you choose. Sizes vary from 12 x 24 inches to 24 x 36 inches. Install asphalt roof traffic pads, laid dry, under the pavers as a protective layer between the roof and the paver.

Water-retaining Roofs

Some roofs are designed to be cooled during hot weather by water on the surface, either held in ponds or sprayed on. Pond depth is governed by controlled-flow drainage systems installed by the mechanical contractor. Before cold weather begins, the system must be shut down and the water drained off.

Because of the abnormally high dead loads, you must install a water-retaining roof over poured or precast concrete, or other heavy load-bearing decks. The structure must be designed for a minimum 50 PSF dead load with a deflection not exceeding $1/360$ of the span between supporting members.

Use coal-tar bitumen and tar-saturated felts in a water-retaining roof system. Install insulation over two plies of vapor retarder material that's solidly mopped to the deck. Over a non-nailable deck, embed four plies of 15-pound tar-saturated felts in solid moppings of coal-tar bitumen. Install a flood coat of coal-tar bitumen at the rate of 75 pounds per square, followed by either 400 pounds of gravel or 300 pounds of slag per square. Sweep away all loose aggregate. Then apply a second surface flood coat of 85 pounds per square of coal-tar bitumen. Into the second flood coat, embed a second layer of gravel or slag at 300 or 400 pounds per square, respectively. Then sweep away all loose aggregate.

Flashing on Flat Roofs

You need flashing on flat roofs at:

- parapet walls
- the roof edge
- roof penetrations such as skylights, structural members, piping and air conditioning units
- equipment stands
- roof drains
- expansion joints

On any roof, complete all openings and roof penetrations before you install the roof system. Install flashing as roofing progresses to prevent damage to the insulation at roof openings or the deck beneath.

Flashing at Parapet Walls

Water is most likely to get into a flat roof where the roof deck and a vertical surface meet. Install flashing at the intersection of a BUR and a vertical wall to make the connection waterproof.

Install base flashing over a cant strip set at a 45-degree angle. Extend the flashing 8 to 12 inches up the parapet wall and at least 5 inches out over the roofing surface, overlapping the roof membrane. You can use a heavy fiberglass flashing material (Figures 10-11 and 10-12) or a modified bitumen cap sheet (Figure 10-13).

Fiberglass flashing comes in rolls 12, 18 and 36 inches wide, and 72 feet long. It weighs about 38 pounds per 100 square feet. Be sure the flashing material is compatible with the roof membrane and that it has similar expansion and contraction characteristics. Don't use metal base flashings. Install 35 pounds of base flashing per 100 square feet in two layers with roofing cement (or hot steep asphalt). Lap the ends at least 3 inches and cement the laps. When you install felts next to metal, nail or spot-cement them (don't solidly mop) to allow for movement. I recommend that you use roofing cement to attach felt base flashing to metal because it's permanently soft and waterproof.

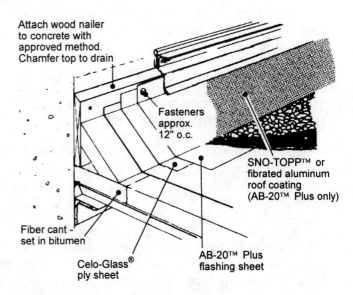

Note: Where deck is supported by and fastened to the concrete wall vertical wood nailer should be secured to the wall with suitable fasteners.

Figure 10-16 Counterflashing for concrete walls or parapets

Nail the top of the second base flashing strip, on 12-inch centers, to the vertical substrate 1 inch below the top edge of the flashing, driving the nails through tin discs. If you're nailing into wood, use 11-gauge, galvanized, barbed roofing nails with 5/8-inch-diameter heads, long enough to penetrate the full depth of the nailing strip. For masonry walls, you use case-hardened nails. Drive them through 28- or 30-gauge tin caps at least 1 3/8 inches in diameter. Or, you can use large-head Simplex nails without tin caps. After you install the base flashing, cover it with 1 to 2 gallons of an aluminum roof coating per 100 square feet. One method for installing base flashing over a wet-fill roof is shown in Figure 10-14.

Install metal counterflashing (cap flashing) over the base flashing. Bend the upper edge and insert it into a reglet (concrete wall) or a 1 1/2-inch deep raggle or slot in the mortar joint (masonry wall). In new construction, install the cap flashing in mortar joints as shown in Figures 10-11 through 10-14. Seal the joint with mortar or roofing cement.

Some counterflashing is designed to be surface-mounted to the parapet wall as shown in Figure 10-16. Install counterflashing about 8 to 12 inches above the surface of the finished roof and extend it down 4 inches over the base flashing. One method of installing counterflashing over a concrete parapet wall is shown in Figure 10-16. Lap counterflashing end joints at

Built-up Roofing

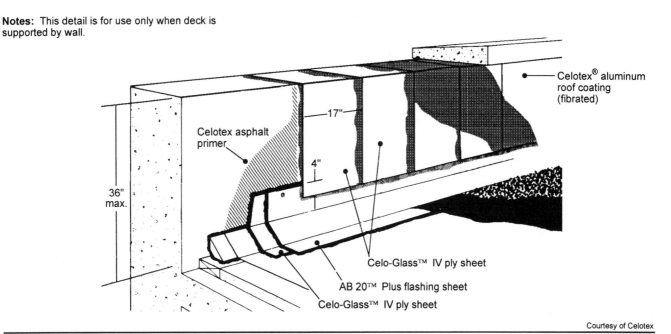

Figure 10-17 An alternative to metal counterflashing

least 3 inches and secure the lap with a clip. Don't solder the joint. Install two-piece counterflashing as shown in Figures 10-11 through 10-14 so the roof can be redone without destroying the original counterflashing.

As an alternative to using metal counterflashing, you can install two plies of glass fiber sheet either horizontally or vertically. Embed each layer in a 1/8-inch-thick troweled-on coat of roofing cement. Coat the second ply with roofing cement as shown in Figure 10-17. Flashing made of plasticized polymeric vinyl or PVC is also used for flashing at parapet walls.

Roof Edge Flashing (Gravel Stops)

Install metal flashing at the edge of a flat roof. The flashing serves as a fascia cover and gravel stop. Make the bead of a gravel stop rise at least 3/4 inch on an aggregate roof, and at least 3/8 inch over a smooth-surfaced roof. As added protection against weather, you can install metal flashing over flexible vinyl or PVC flashing material. On roof slopes less than 1 in 12, prime both sides of the flashing. Then install it into a bed of roofing cement you apply over the top of the completed roof membrane as shown in Figures 10-6 and 10-7. Make the nailing flange of the edge flashing no wider than the nailer (usually 3½ inches). Nail the nailing flange to wood nailers or insulation stops in two staggered rows on 3- to 4-inch centers. Never nail the flange into insulation. Then cover the flange with two strips of 15-pound felt. Make the lower and upper strips 8 and 12 inches wide respectively, as shown in Figure 10-6. You can install cap sheet material instead of felt strips at the roof's edge as shown in Figure 10-7.

On roof slopes of 1 in 12 and steeper, install the roof edge flashing before you apply the roofing material as shown in Figure 10-18. Nail the nailing flange in two staggered rows on 3- to 4-inch centers. Mop all membranes covering the nailing flange up to the lip of the flange.

Flashing at Roof Penetrations

You'll usually find counterflashing built into a skylight unit as shown in Figure 10-19. If it's not, you install flashing here using the same method as you would for a parapet wall.

Framed roof penetrations such as openings for structural members (Figure 10-20), piping (Figure 10-21) and curbs for rooftop air-handling units (Figure 10-22) require you to use shop-fabricated flashing.

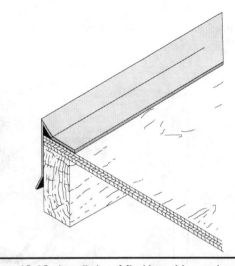

Figure 10-18 Installed roof flashing with gravel stop

You flash small-diameter roof penetrations by setting a primed metal flange over the top ply of membrane into a bed of roofing cement. Follow that with two plies of felt stripping. Set flanges of small vent pipe roof penetrations in mastic over the base ply. On nailable decks, you nail the

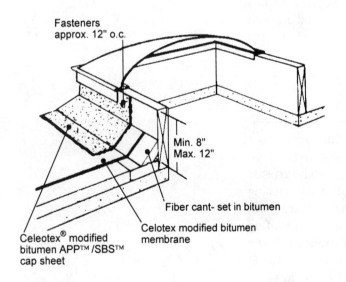

Figure 10-19 Flashing around a skylight

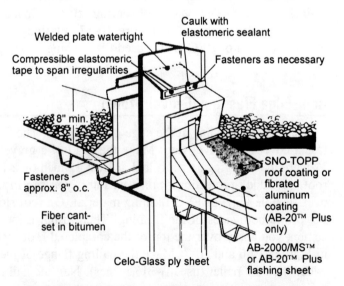

Figure 10-20 Flashing a structural member through roof deck

Built-up Roofing

flange edges on 3-inch centers. Strip the flange with 6-inch glass fiber felt or cap sheet material set in bitumen. Then install subsequent membrane plies over the roof. Don't install penetrations within 15 inches of the lower edge of a cant strip. Typical small-diameter roof penetrations include:

- roof relief vents (Figure 10-2)
- mechanical equipment stands (Figure 10-23)
- hooded pitch pans (Figure 10-24)
- roof drains (Figure 10-25)

Roof Expansion Joints

You install expansion joints in roofs to let the roof expand and contract with changes in temperature. Here's what happens to a BUR roof that doesn't have any expansion joints. High temperatures cause the roof to expand and since there aren't any expansion joints, the felts split. Low temperatures, meanwhile, cause the roof to contract, pulling the felts into ridges in the process. The useful life span of this roof is obviously very short.

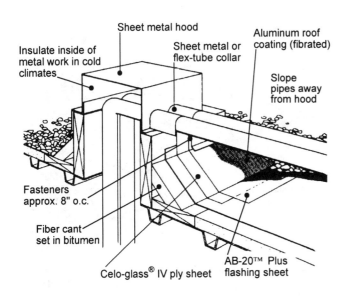

Figure 10-21 Piping through roof deck

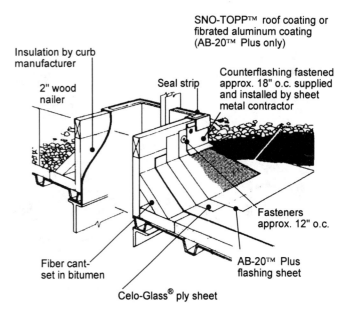

Courtesy of Celotex

Figure 10-22 Curb detail for rooftop air-handling units

Roofing Construction & Estimating

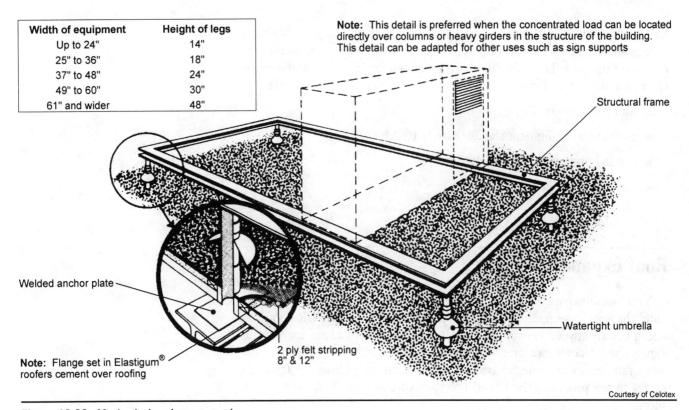

Width of equipment	Height of legs
Up to 24"	14"
25" to 36"	18"
37" to 48"	24"
49" to 60"	30"
61" and wider	48"

Note: This detail is preferred when the concentrated load can be located directly over columns or heavy girders in the structure of the building. This detail can be adapted for other uses such as sign supports

Courtesy of Celotex

Figure 10-23 Mechanical equipment stand

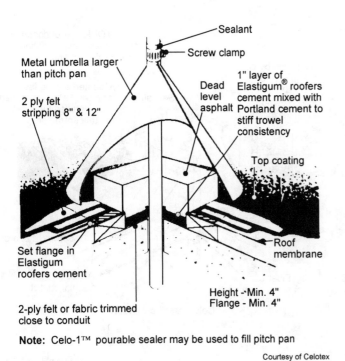

Note: Celo-1™ pourable sealer may be used to fill pitch pan

Courtesy of Celotex

Figure 10-24 Hooded pitch pan

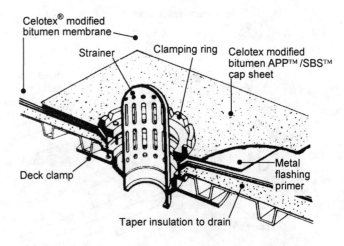

Notes: Min 30" square 2½# lead or 16 oz. soft copper flanges set on finished roof - prime flange before torching modified bitumen sheet.
Membrane plies, metal flashing and cap sheet stripping extend under clamping ring.
Extend Celotex® modified bitumen stripping 4" min. beyond edge of metal flashing

Courtesy of Celotex

Figure 10-25 Roof drain

Align roof expansion joints with the structural expansion joints located throughout the building. Extend each expansion joint to the full length or width of the roof right to the edge of the deck. Never bridge expansion joints with either insulation or roofing membrane. You always elevate expansion joints above the finished roof by at least 4 inches. Never install expansion joints flush with a roof surface. When you design a roof, make sure that water won't flow over or around expansion joints. Finally, never install an expansion joint in a valley.

Here are some obvious places for expansion joints:

- the structural framing or decking changes direction
- the composition of decking material changes
- new construction intersects existing construction
- there's a difference in the elevations of two adjoining decks
- the roof deck intersects a wall
- interior heating or cooling conditions change
- wings where L-, U- or T-shaped buildings intersect
- there may be seismic movement between dissimilar structures
- canopies or exposed overhangs are attached to an air-conditioned building

Expansion joint shields (or covers) insure that the joints are waterproof while still allowing movement throughout the expansion joint. Expansion joint shields are manufactured in three basic configurations as shown in Figure 10-26:

1) the curb flange
2) the straight (low-profile) flange
3) the curb-to-wall flange

Expansion joint covers, regardless of their shape, are essentially a bellows. This bellows is made out of Dacron-reinforced, chlorinated polyethylene laminated to a core layer of closed cell foam. In order to install expansion joint covers, you connect the bellows to a pair of metal (galvanized iron, copper, stainless steel or aluminum) nailing flanges. Expansion joint shields come in 50-foot rolls. Expansion joint accessories include splice covers and connections for crossovers, tees and corners as shown in Figures 10-26 and 10-27.

Coefficient of Linear Expansion

As you've seen over and over throughout the last ten chapters, all roofing materials (metal especially) expand and contract after you install them. You've also seen that if the roof isn't able to do so as a single unit, it won't

Roofing Construction & Estimating

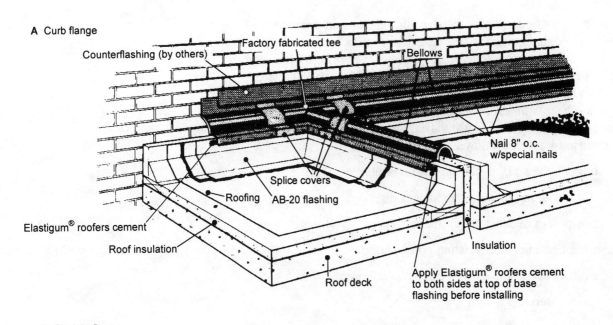

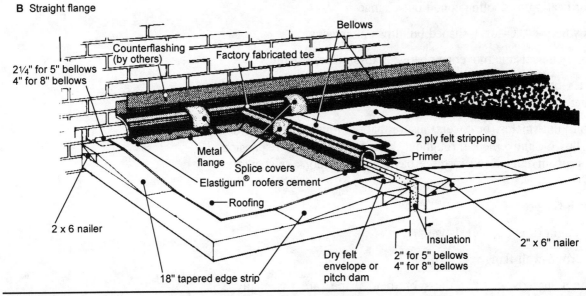

Figure 10-26 Expansion joints

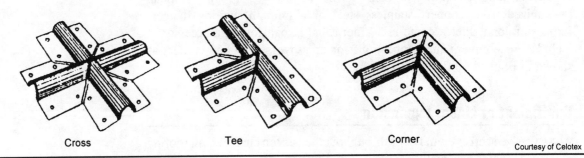

Figure 10-27 Expansion joint accessories

last for very long. Now we'll look at how you go about determining how much expansion room you need to allow for the particular roofing material. To answer that question, you must know the material's coefficient of linear expansion. Simply put, the coefficient of linear expansion is the measured change in length, per unit of length, when the temperature changes by one degree. Assuming that the ends of the material are free to move, the total change in length due to an increase or decrease in temperature is expressed as follows:

Change of Length = E x T x L

| Equation 10-2 |

Where: E = coefficient of linear expansion
             (from Figure 9-41, Chapter 9)
        T = change in temperature
        L = length of member

▼ **Example 10-2:** A continuous roll of zinc flashing is 200 feet long at 60°F. Let's assume the metal is free to move, and find the change in the length of the material when the temperature increases to 95°F (an increase of 35°):

From Figure 9-41, the coefficient of expansion for rolled zinc is 0.00173 per 100°F.

Change in Length = 0.00173 x 35/100 x 200
                 = 0.12 linear feet

Converting to inches, we have:

Change in Length = 0.12' x 12 in./LF
                 = 1.44 inches
                 = 1 7/16 inches

Estimating BUR Systems

When you estimate the cost of BUR systems, note the type of roof, including components, roof deck, and any unusual requirements. Your description, for example, might say:

2" insulation, 4-ply (glass base sheet plus three No. 15) plus gravel

Take off the area of the roof in square feet. Then convert this area into squares. Find the roof area from roof edge to roof edge on roofs without parapet walls. On roofs bordered with parapets, measure the area from outside the parapet walls to allow for material you turn up the wall.

Don't deduct anything for openings less than 100 square feet and deduct only half of the area for openings larger than 100 square feet and less than 500 square feet. Deduct the entire opening area for openings 500 square feet and larger.

Roofing Construction & Estimating

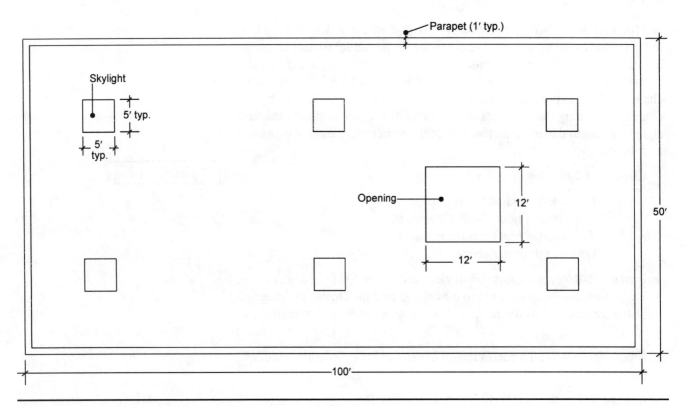

Figure 10-28 Built-up roof example

You must also take off accessories such as fasteners, cant strips, edge strips, envelope strips, flashing, counterflashing, expansion joint cover assemblies, roof relief vents, temporary roofing and glaze coats, water cut-offs, pitch pans, roof drains, skylights and other roof penetrations, primer coats, aluminum coatings, roofing cement, roof traffic pads, additional aggregate, and so forth.

Roof slope, roof size, height of the roof above the ground, and the number and types of roof penetrations will also affect your estimate. For example, the roof slope affects the type of bitumen used. The size of the roof also has an effect. You add more profit for a small job than for a large one. If the roof is high enough to require a crane to raise materials instead of a mobile placer, it's more expensive. And every penetration requires flashing, which adds to the cost.

▼ **Example 10-3:** Find the net area of four-ply built-up roofing required for the roof of the building in Figure 10-28. Assume a wood deck.

Gross Area	= 100' x 50'	= 5,000 SF
Less	½ x 12' x 12'	= (72 SF) (half the opening)
Net Area		= 4,928 SF ÷ 100
		= 49.3 squares

I'd round it up to 50 squares for figuring the total amount of material required for the BUR system.

Specific quantities for such a roof would be:

Item	Weight (pounds/square)
Rosin paper (nailed to wood deck)	5
Coated base sheet (1 ply, nailed)	43
No. 15 felts (3 plies, mopped)	45
Asphalt (3 applications @ 25 pounds each)	75
Surface flood coat	75
Gravel	400
Total weight	**643**

Depending on the job, your take-off items could also include:

```
Insulation    = 98' x 48'    = 4704 SF
   Less: 6 x 5' x 5'         = (150 SF)
   Less: 12' x 12'           = (144 SF)
                               ─────────
Net installed                = 4410 SF ÷ 100 = 44.1 SQ
(10% waste)                    x 1.10
                               ─────────
Purchased                    = 4851 SF ÷ 100 = 48.5 SQ

Base flashing:
   At Parapets     = 2 x (98' + 48')    = 292 LF
   At Small Openings = 6 x 20'          = 120 LF
   At Large Openings = 4 x 12'          =  48 LF
                                          ──────
Net Installed                             460 LF
(5% waste)                              x  1.05
                                          ──────
Purchased                               = 483 LF

Counterflashing (as above)    = 460 LF Installed
                              = 483 LF Purchased

Cant strip (as above)         = 460 LF Installed
                              = 483 LF Purchased
```

To price out this job, let's assume the following material prices:

```
2", R-5.30 perlite insulation board                        = $76.20/SQ
4-ply fiberglass BUR (base sheet, 3 felt plies + gravel)   = $85.35/SQ
Fiber cant strip                                           =  $0.44/LF
Base flashing                                              =  $1.50/LF
Counterflashing                                            =  $2.50/LF
```

The total material costs would be figured as follows:

2" insulation	= 48.5 SQ x $76.20/SQ	= $3,696.00
4-ply BUR	= 50 SQ x $85.35/SQ	= $4,268.00
Cant strip	= 483 LF x $0.44/LF	= $213.00
Base flashing	= 483 LF x $1.50/LF	= $725.00
Counterflashing	= 483 LF x $2.50/LF	= $1,208.00

Total material cost $10,110.00

According to the *National Construction Estimator*, an R-3 crew consisting of two roofers and one laborer can install roofing materials at the following rates:

2" insulation: 1 SQ in 0.82 hours
4 ply BUR: 1 SQ in 1.93 hours
Gravel surface: 1 SQ in 0.718 hours
Cant strip: 1 LF in 0.017 hours

Assuming a roofer makes $20.00 per hour and a laborer makes $15.00 per hour, including labor burden, the average cost per manhour is: [(2 x $20.00) + $15.00] = $55.00 ÷ 3 = $18.33 per manhour. Then the labor costs for the items above would be:

2" insulation   = 44.1 SQ x 0.82 manhour/SQ x $18.33/manhours
                     = $663.00
4 ply BUR   = 49.3 SQ x 1.93 manhours/SQ x $18.33/manhour
                     = $1,744.00
Gravel surface   = 49.3 SQ x 0.718 manhours/SQ x $18.33/manhour
                     = $649.00
Cant strip   = 460 LF x 0.017 manhours/LF x $18.33/manhour
                     = $143.00

An SM crew consisting of one sheet metal worker can install flashing material at the following rates:

Base flashing   = 1 LF in 0.072 hours
Counterflashing = 1 LF in 0.054 hours

Assuming the sheet metal worker makes $20.00 per hour, including labor burden, the labor for installing the flashing would be:

Base flashing   = 460 LF x 0.072 manhours/LF x $20.00/manhour
                     = $662.00
Counterflashing = 460 LF x 0.054 manhours/LF x $20.00/manhour
                     = $497.00

The total labor cost is: $663 + $1,744 + $649 + $143 + $662 + $497
                      = $4,538.00.

The total material and labor cost is: $10,110 + $4,538 = $14,468.

Testing BUR Systems

Sometimes the designer will ask you to make a test cut in the roof to make sure you applied the amount of interply bitumen he specified. Do these cuts before you apply the final surfacing so you can completely repair them.

Cut out an area at least 10 by 42 inches at right angles to the length of the felts to show that the felts have been properly lapped and laid to the correct exposure. To pass the test, the weight of the test-cut components must be within 15 percent, plus or minus, of the specified requirements.

I don't recommend test cuts. Avoid doing them if you can. They're not very accurate and they don't reflect the true condition of the overall roof. Knowing that a test cut is required, a roofing contractor may use too much interply bitumen so that the roof is prone to sliding, even on low slopes. Also, a test cut leaves a patched area that's weaker than the original roof and prone to leaks. Furthermore, making a test cut often voids any warranty. You can avoid having to make a test cut by having an inspector watch the application.

Built-up Roofing Warranties

When a designer specifies a warranted (bonded) roof, an approved roofing contractor must be selected to perform the work. An approved roofing contractor is one whose reputation for good workmanship and financial stability qualifies him for a bond by the material manufacturer furnishing the bond.

When the roof is finished, the roofing contractor sends a "Statement of Compliance" to the manufacturer. The manufacturer then issues a "Limited Service Warranty" to the roofing contractor. The manufacturer keeps a job file during the life of the warranty and sends a representative to make an inspection before the roof is two years old. A limited service warranty usually runs for 10 years and provides a specified dollar amount of coverage for five years and prorated coverage for the remaining five. Some manufacturers offer warranties for 5, 10, and 20 years. The longer the warranty, the more costly. The length depends on how much the owner is willing to spend, since the warranty costs must be included in your bid.

Roofing systems on as many as three separate buildings can sometimes be covered by one warranty if:

- they're for a single owner
- they're all located on the same site
- their completion dates are all within a 12-month period
- all of the roofing systems qualify for the same warranty

A warranty won't cover damages due to contractor errors, omissions or poor workmanship. Damage due to ponded water, natural disasters or building settlement or other structural movement isn't covered. The roof must be capable of shedding all ponded water within 24 hours after a rain. This depends on roof slope, roof deflection and proper roof drain and gutter design and maintenance. Ponded water allows seepage through bare spots, blisters and minute cracks in the roof's surface. Ponded water provides an environment for the growth of bacteria and fungi which can weaken the roof membranes.

The warranty can be voided if the type of building usage is changed from its original intended purpose. Test cuts will often void a roof warranty. It can be difficult, if not impossible, to get a warranty for a building of variable occupancies, residences, apartment buildings, condominiums, nursing homes, or any building where the total area of the roof is less than 5,000 square feet. Buildings this size and smaller often have a high turnover of occupants, each making alterations in the building to suit unique needs. Buildings such as storage silos, soybean or milk-processing plants as well as buildings located outside of the U.S. are also regularly excluded in the terms of the warranty. Roofs over domed structures, heated tanks, dry kilns, car-wash buildings, swimming pools and other structures with abnormally high interior humidity conditions also can't be warranted.

Roof decks which are not properly anchored to the structure, roofs whose insulation is not properly installed, roofs applied over lightweight insulating concrete decks or roof applied directly over prestressed concrete decks usually can't be warranted.

An existing warranty can be voided if improper fasteners or other unapproved construction methods or materials were used in any phase of the roof or substrate construction. These unacceptable conditions include phased roofing or materials not produced by the manufacturer issuing the warranty. The subsequent installation of conduit or pipes over the roof can also void the warranty. Other actions that can void a warranty include failure due to repairs or alterations performed by a non-sanctioned contractor, the installation of roof penetrations, or alterations such as aerials, signs, water towers, antennae, and building expansion or additions.

Some limited service warranties cover only the cost of replacing the membrane. Others cover the cost of replacing the membrane and insulation. The liability covered in the warranty doesn't include damage to the building interior and contents, roof deck or lost profits and/or rents.

The warranty automatically expires at the end of the specified time period or when the dollar amount of the warranty is used up.

Built-up Roofing Repairs and Re-roofing

As soon as you install a built-up roof, it begins to degrade. Loose aggregate gradually and continually washes away, bit by bit. Over time, the roof develops bare spots and the exposed bitumen oxidizes, exposing the underlying cap felt layer to ultraviolet radiation. As a result, the entire roof surface gradually hardens. It's easier for moisture to penetrate the roof, and the alligatoring of felts soon follows. As preventive maintenance, apply a resaturant to the roof surface to lower the surface bitumen softening point and to restore flexibility to felt membranes. The resaturant also provides a base for new aggregate to be embedded.

Before you apply a resaturant, remove all loose aggregate and sweep the roof surface clean. If you're going to reuse the aggregate, clean it with water and let it dry. After you've made minor roof surface repairs, apply about 7 gallons of the resaturant per square. Replace the aggregate immediately.

You can generally resurface and rejuvenate a roof that delaminates when it's flexed and shows blistering, and has insulation that's not damaged or wet.

Moisture trapped within voids between felt plies can expand and make blisters. The voids happen if:

- You didn't broom the felts tightly.
- The bitumen was too hot when you applied it.
- You didn't use enough interply bitumen.
- You didn't store the materials properly and they absorbed moisture.

Large bubbles caused by these problems are often the source of roof leaks, especially if they're located in a low spot. To repair a large bubble, remove aggregate at the problem area and cut an "X" in the bubble. Embed the flaps of the cut material into roofing cement and nail a patch of heavy membrane material over the hole. Drive nails on 1-inch centers along the edges of the patch, then cover the patch with roofing cement.

A roof will usually split if it moves excessively because it doesn't have enough roof expansion joints or they aren't in the correct places. They also occur if the roof membrane or insulation isn't properly attached. Use a spud to remove aggregate adjacent to the split and cut off the curled edges of the membrane. To allow for expansion and contraction, install a loose 4-inch glass fiber strip covered with a 16-inch Tedlar sheet cemented at the edges. Cover the edges of the Tedlar sheet with fully cemented 8- and 10-inch glass fiber sheets. Then cover the sheets with roofing cement or hot asphalt as shown in Figure 10-29.

Alligatoring is the result of a drying and deteriorating roof surface. Apply asphalt emulsions to help eliminate this problem.

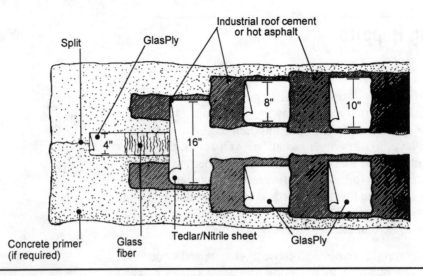

Figure 10-29 Repairing a split

Fishmouths are caused by inadequate brooming of felt edges or improper alignment of the felts. To repair fishmouths, cut the loose material and embed it into roofing cement or hot asphalt.

To repair splits, blisters, alligatoring and fishmouths, embed a patch of heavy membrane material into ⅛-inch-thick cold-applied mastic. Then put on a top coat of ⅛-inch mastic. Make the patch 6 inches wider than the area you're repairing.

Flashing failure can be caused by equipment vibration and/or thermal expansion and contraction of dissimilar flashing materials such as membrane and metal. To repair flashing, embed and cover the loose material with roofing cement.

A primary location for leaks in a built-up roof is at the roof edge. As with other flashing repairs, cover the problem area with roofing cement.

Insulation can get wet if the roof covering or flashings aren't waterproof. Also, the insulation won't dry if it's not properly vented.

If you catch these things early on, you can avoid replacing the whole roof. A roof that has a lot of splitting, brittle surface material and saturated insulation will have to be replaced.

If there's no insulation under the existing roof covering, remove all of the roof covering before you install any new roof covering. You'll also have to completely remove the existing roof system if the extra weight of the additional roof system exceeds the safe design load.

If there is insulation under the roof, remove all aggregate, blisters and loose felts and sweep the roof clean. Replace all wet or damaged insulation. Some manufacturers recommend that you remove all insulation to avoid

Note: Celotex will not issue its roofing warrenty for reroofing jobs over existing built-up roofing membrane without prior written approval from Celotex. Consult the nearest Celotex Regional Office.

Courtesy of Celotex

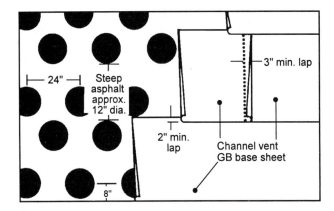

Figure 10-30 Vented base sheet installed over an old roof

Courtesy of Manville Corporation

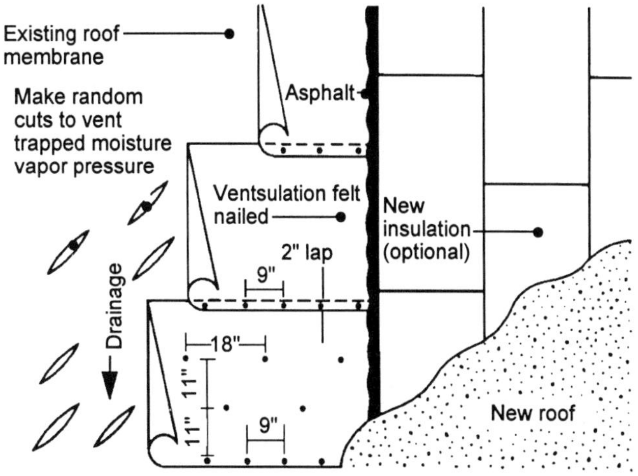

Figure 10-31 Venting an old roof

problems with trapped moisture. If you damage good insulation when you remove the original roof, sandwich a recovery board between the insulation and the new BUR system.

Over non-nailable decks, prime the old roof membrane and install a vented base sheet embedded in spot moppings of hot steep asphalt. Install the mopped spots on 18-inch centers as shown in Figure 10-30. Over nailable decks, nail the base sheets at laps on 9-inch centers and stagger-nail the sheets in two rows along the center on 18-inch centers.

When you're re-roofing adjacent to a roof requiring no work, protect the existing roof from damage due to traffic and impact. Never solidly mop a new roof over an old roof. Also, when you're in doubt about trapped moisture in the old insulation, cut holes in the old roof at 3-foot centers in both directions so trapped moisture in the old roof can escape before you install a vent sheet. That's shown in Figure 10-31.

You can test questionable substrates for dryness and adhesion by pouring one pint of bitumen, heated to application temperature, over a representative area of the deck. If the bitumen foams or bubbles, there's too much moisture. If you can strip the cooled sample cleanly from the deck, it didn't stick properly. Apply a primer when a deck can't be cleaned to permit proper adhesion.

Now that we've covered built-up membrane roofing, we're ready to move on the other membrane system, elastomeric roofing. That's the subject of the next chapter.

11 Elastomeric Roofing

▶ An elastomer is any material that has the elasticity to return to its original shape after being repeatedly stretched to twice its size at room temperature. Some elastomeric roofing materials will stretch up to 450 percent, compared to a felt built-up roof system which stretches only ½ to 1½ percent.

Elastomeric roofing is compatible with clean concrete and exterior-grade plywood decks, smooth metal, glass, flagstone, wood, and in some cases, asphalt or tar. Some types of elastomeric roofing aren't compatible with lightweight concrete or gypsum decks due to blistering that results from the residual moisture within the deck. However, some varieties are vapor permeable and allow moisture to escape.

Many elastomeric materials require a primer brushed, rolled, squeegeed or sprayed over the substrate. The rate of application varies from ½ to ¾ gallon per square. I strongly recommend a primer when the roof deck is especially dry or if felts show excessive alligatoring.

Always follow manufacturer's instructions with regard to substrate preparation. For example, petroleum-based materials such as bitumen or coal-tar pitch adversely affect PVC membranes. You have to install a slip sheet (also called a separation sheet) between these materials. Also, lumber in contact with PVC material must be wolmanized (pressure-treated).

If you're installing roof walkways, elastomeric roofing manufacturers recommend that you install an additional layer of membrane beneath the walkways. Install walkway panels over an elastomeric roof with the adhesive recommended for the particular elastomeric roofing system.

The Advantages of Elastomeric Systems

Elastomeric roofing material is more expensive than a BUR, but elastomeric roofing also has many advantages. It:

- expands and contracts without tearing
- remains flexible at low temperatures
- bonds well to the substrate
- weighs less (some systems)
- conforms to any roof shape or slope
- withstands water ponding (many systems but not all)
- is self-healing and self-flashing (many systems but not all)

Elastomeric roofing also keeps your labor costs down because it's easier to install than a built-up roof. Since there are fewer operations, there are fewer mistakes and your workers turn out a better job. It's also safer. Hot roofing is responsible for at least 23 percent of all roofing injuries, but elastomeric roofing doesn't require hot kettles. There's no risk of asphalt stains on the building face, and you can install polystyrene foam insulation — the most efficient insulation available.

Another advantage is that you can prefabricate elastomeric sheets off-site. Just cut the rolls to the proper lengths, re-roll them and take them to the site. Then unroll and install them in the proper order. That reduces time on the job, which can help you meet a tight project deadline.

On re-roofs over BUR systems, you don't have to completely remove the old roof before you install many elastomeric roof systems. You only have to remove loose aggregate and install a layer of rigid insulation before you apply the membrane.

Most elastomeric roofs are virtually maintenance-free and easy to repair. You can install many loosely-laid systems in any type of weather. You can put them over an icy, snow-covered, wet or humid surface, so bad weather won't delay the job.

Elastomeric roofing comes in a wide variety of colors, including some with a reflective surface which deflects solar heat. It's highly resistant to ultraviolet light and industrial environments containing dirt and mild acid fumes. Builders also use elastomeric materials to waterproof basements, and for flashing.

Other than its higher initial cost, the only disadvantage of elastomeric systems in the limited number of roofers who know how to install them. If you're one of the ones who has experience, it can open a new world of opportunities for you.

Liquid-applied Elastomers

You can get liquid-applied elastomers in four consistencies: caulk, trowel, brush and spray. You can apply many liquid elastomers to virtually any substrate, including aged tar or asphalt. However, alligatored areas sometimes require a primer coat. Most liquid elastomers cure within 24 hours. Then they're hard and dry enough to walk on without feeling tacky.

Apply the troweled variety over reinforced mats at a rate of about 4 gallons per square. Next install a reinforced polyester mat, then cover that with an additional coat of troweled material, also at a rate of 4 gallons per square. Reinforced mats come in 5-, 10- and 30-square rolls 36 inches wide.

Use the brushed variety to rejuvenate dried-out felts. Apply both the troweled and brushed materials $1/8$ to $3/16$ inch thick.

Silicone, Urethane and Vinyl

Some of the most commonly-used liquid elastomers include silicone (polysiloxane), silicone rubber, urethane or vinyl liquid applied directly to the substrate. Normally, two layers of silicone rubber coatings are sprayed on over polyurethane insulation. This process provides a UL Class A fire rating. Apply a total minimum coating thickness of 15 dry mils. Use a micrometer to make sure.

You can embed a layer of ceramic-coated granules in the upper coating to make the roof more attractive, minimize discoloration, and provide a non-skid surface that's resistant to foot traffic, UV radiation and hail. Silicone rubber repels water from the outside, but allows vapor to escape from the substrate below. You can also repair damaged surfaces with silicone building sealant.

Urethane is also a suitable protective coating over polyurethane foam insulation. Apply urethane in two coats at the rate of $1 1/2$ gallons per coat per square. Apply a total minimum coating thickness of 25 dry mils. Urethane is a "breather" type of coating which allows water vapor to escape. Urethane coatings stretch up to 400 percent and can span structural cracks up to $1/8$ inch wide. Standard colors are black, white or aluminum.

Neoprene

Another popular elastomeric system includes neoprene (chloroprene) applied as a liquid or in sheet form fastened to the deck with adhesives. Polyester-reinforced fabrics are also popular. Cover the liquid or sheet with liquid coats of hypalon (chlorosulfonated polyethylene) or vinyl acrylic to get the minimum required dry film thickness. Fiberglass embedded in the elastomeric material is an excellent reinforcement for roof joints under liquid roofing.

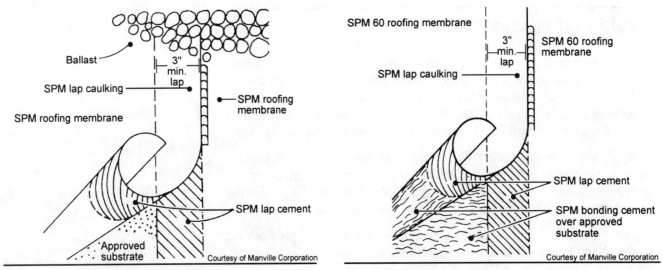

Figure 11-1 Loosely-laid and ballasted system

Figure 11-2 Fully-adhered system

Single-Ply Roofing Systems

The most common elastomeric sheet roofing materials are butyl rubber, neoprene, neoprene-hypalon, chlorinated polyethylene (CPE), polyvinyl chloride (PVC), ethylene propylene diene monomer (EPDM), polyisobutylene (PIB) and chlorosulfonated polyethylene (CSPE). A coal-tar elastomeric membrane reinforced with polyester fiber is also manufactured.

You can install preformed elastomeric sheets in any of three ways:

1) Loosely laid and ballasted with aggregate or pavers (Figure 11-1)

2) Fully adhered (Figure 11-2)

3) Mechanically fastened (Figure 11-3)

In any system, you cement or tape the joints between sheets or "weld" them with hot air or solvent. You must also caulk the edge of the joint.

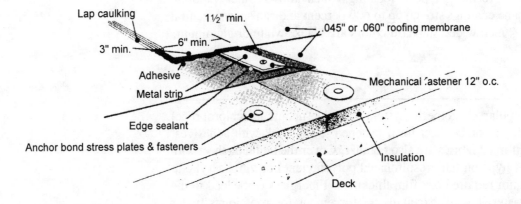

Figure 11-3 Mechanically-fastened system

Ballasted Roof Systems

If the roofing system is loosely-laid and ballasted with aggregate or pavers, the ballast adds about 10 pounds per square foot of dead load over the structure. This is a very important design consideration, especially when you're re-roofing over a structure originally designed for a specific roof dead load. When in doubt, consult a structural engineer. It's OK to install an unballasted elastomeric roofing system for retrofit on roof decks that can't support the weight of the ballast.

One disadvantage of a aggregate-ballasted roof system is that the membrane is vulnerable to physical damage from the ballast. Also, there are height limitations (80 feet), slope limitations (2 in 12), and the system can't withstand high winds. Furthermore, the aggregate ballast collects dirt, oil, chemicals and pollutants which eventually damage the membrane.

Ballast systems have advantages, too. First, you can remove the system and reuse it when more stories are added to an existing building. Second, the membrane moves independently of the deck and allows moisture venting. Since many elastomeric sheets are virtually impervious, venting is sometimes necessary.

Fully-Adhered Roof Systems

You can install a cemented system two ways. Either bond the entire membrane to the deck, or spot-bond only 40 percent to allow for expansion and contraction in all directions. Check the specifications of the bonding adhesive you plan to use. Some can't be applied when temperatures are extremely high or low, or in high humidity conditions. Many manufacturers recommend that you install a slip sheet (or separation sheet), typically made of unsaturated felt, between the substrate and the elastomeric sheet.

When you install elastomeric roofing directly to a deck, you must provide adequate air circulation between the deck and roofing material to prevent moisture from getting trapped. When you install elastomeric roofing over insulation, use a cold-process adhesive and put a vapor barrier between the deck and the insulation. You must install rigid insulation between elastomeric membranes and steel roof decks or aggregate-surfaced roofs.

EPDM Elastomeric Roofing

EPDM (ethylene propylene diene monomer) consists of vulcanized rubber-based membranes that are compatible with almost any roofing surface, including foam or asphalt-faced insulation products. But EPDM membranes are not compatible with and should never be installed with either plastic cement or dead-level asphalt. Never apply EPDM roofing to any roof that's subject to chemical discharge or covering a cold storage or freezer facility. The chemicals may react with the EPDM, and cold spaces vent too much moisture.

EPDM is very durable and has a service temperature range of -60° F to 200° F. Standard thicknesses are 0.045 inch, stretchable by as much as 425 percent, and 0.06 inch stretchable up to 450 percent.

EPDM sheets come in a variety of sizes. Standard widths are 4'4", 5'6", 7, 10, 20, 40 or 50 feet; standard lengths are 50, 100, 125, 150 or 300 feet. You can also special-order membranes that measure as much as 50' x 200'.

Your color choice for EPDM is limited; it's only available in black and in white. If the job requires a light-reflective surface, you can apply a hypalon coating over the face of EPDM membranes on fully-adhered or mechanically-fastened systems. However, before you apply the coating, allow the membrane to weather in place for at least two weeks. Then you apply the color coating to the EPDM membrane in two coats. You can put on the second coat as soon as the first one is dry.

Some systems come with a 15-year limited service warranty, although some membranes last up to 20 years or more.

You can install EPDM systems loosely laid and ballasted, fully adhered, or mechanically fastened. But no matter which method you use, install a butted insulation protection course over any substrate that has cracks or joints ¼ inch wide or wider. On re-roofs, remove all gravel before you install insulation. Overlap adjacent EPDM sheets at least 3 inches (6 inches if mechanically fastened), and offset joints at least 12 inches. Position all sheets with a 2- to 3-inch overhang at the roof's edge and a 2- to 3-inch run-up at all vertical surfaces.

Loosely-Laid EPDM Systems

Install a loosely-laid system in new construction or as re-roofing over plywood, wood plank, approved insulation (confirm with manufacturer) and smooth-surfaced roofs. The system is also compatible with lightweight concrete decks, provided:

- The deck is at least 3 inches thick.

- It has a dry density of at least 22 pounds per cubic foot. A testing lab can determine this.

- It has a minimum compressive strength of 125 pounds per square inch.

- It's installed over a substrate that allows the deck to dry from underneath, for example, form boards or vented metal panels.

Insulation materials such as urethane, fiberglass, polyisocyanurate, phenol formaldehyde, perlite and polystyrene are all compatible with EPDM roofing membranes. Install a slip sheet between the membrane and incompatible substrates, like insulation covered with dead-level asphalt.

Install a ballasted EPDM system over loosely-laid insulation boards or mechanically fasten the insulation board to the deck or install it in hot asphalt. The method depends on the deck type and roof slope.

Membrane thickness for a ballasted system is normally 0.045 inch. Some owners may feel the extra protection the 0.06-inch thickness provides is worth the cost. Seal and caulk all membrane laps as shown in Figure 11-1. Use ¾-inch river-washed gravel or paver blocks weighing up to 60 pounds each for ballast. Using crushed stone for aggregate ballast will damage the membrane. Use smooth aggregate designated as ½ to 1½ inch, where no more than 10 percent is retained in a sieve with 2-inch openings, and no more than 10 percent passes through a sieve with ½-inch openings.

On roofs 80 feet or higher above the ground, you must ballast the roof with 2' x 2' x 2" precast concrete pavers installed to provide a minimum of 10 pounds per square foot. Install aggregate or pavers at the end of each day, or at times necessary to prevent wind damage.

The ballasted system is given a UL Class A fire rating.

Fully-Adhered EPDM Systems

Here are some installation tips for a fully-adhered EPDM system:

- Install the system in new construction or as re-roofing on inclines through 6 in 12. You can install this system over wood, plywood, high-density fiberboard, concrete or approved insulation (confirm with manufacturer).

- If you install the system over insulation, mechanically attach the insulation to the deck with one fastener for each 2 square feet to make sure the insulation won't move.

- Over non-nailable decks, install the insulation in a solid mopping of hot steep asphalt.

- Mechanically fasten a fully-adhered system at the roof perimeter and around openings of 16 square feet or more.

- Apply membrane 0.06 inch thick. Apply the adhesive used for bonding the membrane to the deck or insulation (bonding adhesive) to both the bottom of the membrane and the top of the substrate at the rate of 1⅔ gallons per square. That's about 1 gallon for every 60 square feet of membrane. Note: A fully-adhered system weighs about ⅓ pound per square foot.

- Apply the adhesive using a 9- to 12-inch-wide paint roller with a solvent-resistant core.

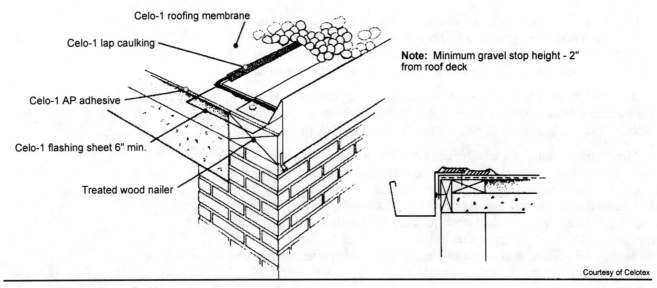

Figure 11-4 Gravel stop flashing

Mechanically-Fastened EPDM Systems

Install a mechanically-fastened (bar-anchored) EPDM system in new construction or as re-roofing on inclines up through 6 in 12 as follows:

- Install this system over wood plank, plywood, approved insulation (confirm with manufacturer) or smooth-surfaced roofs over any deck that will accept and hold mechanical fasteners.

- Install membrane 0.045 inch thick (0.06 inch in high-wind areas). Note: This system weighs approximately ⅓ pound per square foot.

- Secure the underlying lapped portion of each sheet under an aluminum or galvanized cap strip (termination bar or anchor bar) set in sealant and mechanically fastened to the deck on 12-inch centers.

- Lay the lapped-over portion of the successive sheet over the metal strip, adhere it with an adhesive and seal the lap with caulk.

- Anchor the membranes under metal strips at the roof perimeter and around openings.

- Use mechanical fasteners long enough to penetrate at least 1 inch into concrete decks and ¾ inch into wood or plywood decks.

- Use narrow EPDM sheets (5'6" wide) in a mechanically-fastened system.

Mechanically fasten the membrane on 8-inch centers over treated wood nailers located at the roof perimeter (Figure 11-4), at roof penetrations (Figure 11-5), and vertical abutments (Figure 11-6). Use nailers (No. 2 or better lumber) or plywood. Either must be pressure-treated with salt

Elastomeric Roofing

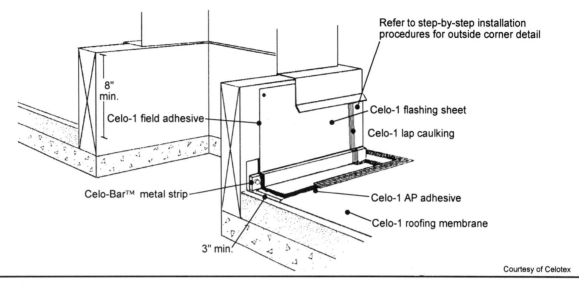

Figure 11-5 Curb flashing (vertical nailer)

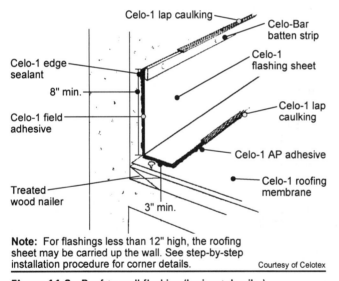

Figure 11-6 Roof-to-wall flashing (horizontal nailer)

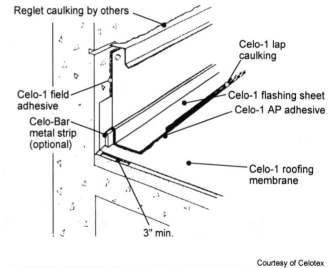

Figure 11-7 Roof-to-wall flashing with counterflashing

preservatives (wolmanized) and at least ½ inch thick. Creosote- or asphalt-treated lumber is incompatible with the membrane. Use 1-inch galvanized nails at wood nailers with 1-inch-diameter (minimum) heads.

You can also install metal strips instead of wood nailers as shown in Figures 11-7 and 11-8.

Adhere all laps with an approved lap cement or joint tape and caulk the joint with a ¼-inch-diameter bead of seam caulk. Brush-apply lap cement at the rate of 1 gallon per 150 to 200 linear feet for a 3-inch lap, or 1 gallon

341

per 75 to 100 linear feet for a 6-inch lap. Apply seam caulk at the rate of 1/10-gallon tube per 20 to 25 linear feet. Be sure that lap cement (at seams) and bonding adhesives (in the field) are compatible with the EPDM material. Don't use coal tar, asphalt- or oil-based roofing cements. Use lap cement (seaming cement) only for field seaming membranes and flashing; never use it to adhere membranes and flashing to any other surface.

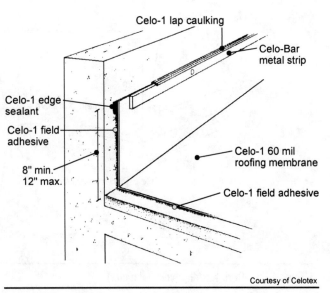

Figure 11-8 Roof-to-wall flashing without counterflashing

CPE Elastomeric Roofing

Chlorinate polyethylene (CPE) elastomeric roofing is usually mechanically attached to the roofing system with plates and screws, following the same rules as for EPDM. Seams are heat-welded to form a one-piece roofing membrane.

CPE membranes are resistant to ultraviolet and infrared radiation, ozone, oil and many chemicals and microorganisms. CPE is also fire-resistant. CPE retains flexibility at below-freezing temperatures and can withstand heat up to 150° F. Some manufacturers offer a 10-year warranty.

The inclusion of titanium dioxide during manufacture gives a highly reflective white color to the membrane which saves air conditioning costs. CPE roofing is normally manufactured in rolls 5'2" wide and 103'6" long and weighs about 5 ounces per square foot. Manufacturers produce a 40-mil-thick roll for a mechanically-fastened roof system and a 32-mil extrusion-coated polyester reinforced membrane for a ballasted system.

CPE membranes are compatible with many substrates including asphalt and coal-tar pitch. And you can install CPE membranes over insulation without applying a separation sheet.

CSPE Elastomeric Roofing

Chlorosulfonated polyethylene (CSPE) is basically a combination of plastic and synthetic rubber. You install it the same way as EPDM roofing. CSPE is compatible with bituminous materials and is resistant to water, ozone and ultraviolet light. CSPE membranes are most commonly available in black or white, but you can also get custom colors. White is recommended in the summer heat of the southern states. CSPE comes in a nominal thickness of 0.045 inch and weighs 0.29 pounds per square foot.

Hypalon Roofing

You can install polyester-reinforced hypalon membranes mechanically attached or fully adhered to a substrate. This material is unaffected by asphalt or coal-tar pitch and can be bonded using these materials. Heat- or solvent-weld lap joints. Some systems come with a 10-year warranty. Sheets usually come with a white surface, but you can spray coat the surface with hypalon to produce another color.

PVC Elastomeric Roofing

Polyvinyl chloride (PVC) elastomeric roofing is normally manufactured in rolls 5'10" wide and 72 feet long. Rolls are 0.045 or 0.06 inch thick and weigh 0.3 pounds per square foot. Stabilized PVC is resistant to ultraviolet light and ozone and is virtually impervious to water vapor. But PVC frequently loses its plasticizer with heat aging and becomes more brittle over time. Before heat aging, some PVC membranes stretch up to 3 times their original size in any direction. You can also use reinforced PVC material impregnated with woven glass or polyester fabric.

PVC elastomeric roofing is appropriate for ballasted, mechanically-fastened or fully-adhered systems. You can fully adhere PVC roofing to concrete, plywood or approved insulations. You can install it loosely laid or mechanically attached over concrete, wood, metal or approved insulation. Compatible insulations include perlite and polyurethane boards.

PVC is incompatible with bituminous material (including bituminous-coated insulation), foamed glass or polystyrene insulation, unless you separate them with a slip sheet before you install the PVC membrane. But never fully adhere PVC to incompatible substrates even when a slip sheet is installed.

Lap and seal membrane seams with a seam solvent, or heat-weld and caulk the seams. Use 3-inch laps for the ballasted system and 6-inch laps for the bar-anchored system.

Composite Roofing Systems

Elastomeric roofing composites are plastic films laminated to fabrics, paper, metal, felts or rubberized asphalt of various types. Metals incorporated in composite elastomeric membranes include aluminum, copper and stainless steel. Some composite membranes are coated with reflective white vinyl acrylic after installation. It reflects heat better — and looks better. Many manufacturers recommended that you reapply the coating over flat roofs at 5-year intervals.

Composite membranes, like many other elastic membranes, are installed ballasted, mechanically-fastened or fully-adhered. Some materials come with release paper covering the bottom surface. You just remove the paper and adhere the membrane to the substrate without additional adhesives. Bond the other types of composite materials to the substrate with cold-applied adhesives.

Rubber-vinyl composite membranes come in 50-mil-thick rolls 74'6" long and 3 feet wide. Each roll weighs 83 pounds which equates to a system weighing about 37 pounds per square. Rubberized composite membranes are self-healing and remain watertight even under ponded water.

Plastic-bitumen composites come in rolls 3'7" wide and 33 feet long. The material is 160 mils thick and each roll weighs 99 pounds. Plastic-bitumen-aluminum composites come in rolls 3'7" wide and 33 feet long. These are 120 mils thick and each roll weighs 76 pounds. You can install a plastic-bitumen composite loosely-laid and ballasted. Use asphalt cement or steep asphalt to adhere plastic-bitumen-aluminum membrane. Never apply the aluminum sheet directly to a wood deck because wood resins react with it.

Flashings for Elastomeric Roofs

Flashings installed on elastomeric roofs vary with the particular roof material. In most cases, you'll install elastomeric sheet base flashing at vertical surfaces by continuing the membrane up the wall as in Figure 11-7. Extend the base flashing up the vertical surface at least 8 inches and 3 inches onto the field of the roof (Figure 11-6). I only recommend separate base flashing for areas that are hard to flash, like corners and tight areas.

Install flashing material in adhesive applied at the rate of 1 gallon per 120 to 140 square feet. Lap seams at least 4 inches into seaming cement and caulk the joint and all edges of flashing material. Use flashing cement compatible with the membrane material. Don't use coal-tar, asphalt- or oil-based roofing cements.

Use a prefabricated vent flashing boot to waterproof vent stacks as in Figure 11-9 and flash roof drains as shown in Figure 11-10.

Flashing material used on EPDM roofs consists of uncured (or unvulcanized) neoprene or EPDM rubber. EPDM flashings usually have a nominal thickness of 0.06 inch. They come 6, 12, 18, 24, 36 and 48 inches wide, in 50-foot lengths. Install these uncured materials so they'll adapt to irregular shapes and surfaces. With aging, they vulcanize and become tough rubber membranes.

Elastomeric Roofing

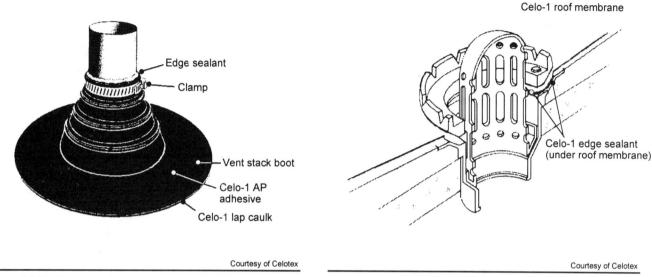

Figure 11-9 Vent stack boot

Figure 11-10 Roof drain flashing

Estimating Elastomeric Roofing

Basically, you can use the same rules to estimate elastomeric roofing that you use for BUR systems. But since elastomeric roofs aren't as common as BUR systems, you've got to consider whether or not you'll be able to use left-over membrane for future projects. If not, figure the most cost-efficient roll width to use on the current project.

For example, let's assume that you're going to install 100-foot-long rolls of fully-adhered EDPM elastomeric roofing (using a 3-inch lap) over the roof shown in Chapter 10, Figure 10-28. Using the standard roll widths available, you can calculate the waste.

For 4'4" rolls:

$$\text{No. Rolls} = \frac{50 \text{ feet}}{4.08 \text{ feet}} = 12.25 \text{ rolls}$$

The waste factor is: $\frac{13 \text{ rolls}}{12.25 \text{ rolls}} = 1.06$ or 6% waste

For 5'6" rolls:

$$\text{No. Rolls} = \frac{50 \text{ feet}}{5.25 \text{ feet}} = 9.52 \text{ rolls}$$

The waste factor is 5%

The waste for other roll widths would be:

Roll Width	No. Rolls	Waste
7'	8	8%
10'	6	17%
20'	3	19%

Using 5'6" rolls, you'd have 10 seams to splice, but the high cost of the membrane more than makes up for the added labor.

▼ **Example 11-1:** Figure the cost for installing 5'6" rolls of fully-adhered EPDM elastomeric roofing over the roof shown in Figure 10-28 (in Chapter 10). Use the surface area and flashing quantities from Example 10-3.

The insulation material will cost $3,696.00.

Other materials will cost:

EPDM membrane	= 49.3 SQ x $92.50/SQ	=$4,560.00
Lap cement	= 10 seams x 100 LF x $0.22/LF =	$220.00
Sheet flashing	= 483 LF x 1 SF/LF x $2.00/SF =	$966.00
Total material cost		$9,442.00

According to the NCE, an R3 crew consisting of two roofers and one laborer can install roofing materials at the following rate:

EPDM membrane: 1 SQ in 1.05 hours
Lap cementing: 1 LF in 0.00223 hours
Sheet flashing: 1 SF in 0.021 hours

From Example 10-3, the average cost per manhour is $18.33. Then the labor costs would be:

2" insulation (from Example 10-3) = $663.00

EPDM membrane = 49.3 SQ x 1.05 manhours/SQ x $18.33/manhour
 = $949.00
Lap cementing = 1000 LF x 0.00223 manhours/LF x $18.33/manhour
 = $41.00
Sheet flashing = 460 LF x 0.021 manhours/LF x $18.33/manhour
 = $177.00

The total labor cost is: $663 + $949 + $41 + $177 = $1,830.

The total cost is: $9,442 + $1,830 = $11,272.

In the next chapter, we'll take an in-depth look at other materials important to the roofing contractors — insulation and vapor retarders.

12 Insulation, Vapor Retarders and Waterproofing

▶ As a roofing contractor, you're usually responsible only for installing roofing insulation when it's required. But in this chapter I'm including a general discussion of building insulation and vapor barriers. In view of rising energy costs and the need for universal conservation, I think all technical construction authors should use every chance to stress the importance of a properly insulated and moisture-resistant building.

I'll also discuss waterproofing and dampproofing. While those operations most often apply to slabs, foundations, basements and exterior walls, the materials and techniques are familiar to the roofing contractor. These jobs can increase your business in slack times, or open the door to work in your specialty. Let's begin with insulation.

The Benefits of Insulation

Figure 12-1 shows how to calculate cost-in-use benefits of a properly insulated building. Note that these numbers are only examples. They don't reflect actual costs at today's rates. The chart assumes the following:

- Building size: 15,000 square feet

- Fuel source: gas heat (winter design U-value); electric cooling (summer design U-value)

City, State	Heating only			Heating + cooling		
	Potential HVAC equipment savings	Estimated first year fuel cost savings	Total estimated first year savings	Potential HVAC equipment savings	Estimated first year fuel cost savings	Total estimated first year savings
Minneapolis, MN	$812	$863	$1,675	$2,790	$1,071	$3,861
Kansas City, MO	$941	$610	$1,551	$3,348	$1,047	$4,395
Houston, TX	$385	$198	$689	$3,339	$583	$4,028
Boston, MA	$598	$1,335	$1,933	$2,492	$1,620	$4,112
Fresno, CA	$342	$341	$683	$3,162	$779	$3,941
San Francisco, CA	$812	$497	$1,309	$2,325	$554	$2,879
Raleigh, NC	$499	$566	$1,065	$3,720	$908	$4,628
Atlanta, GA	$484	$485	$969	$3,068	$851	$3,919
Cleveland, OH	$655	$732	$1,387	$2,790	$965	$3,755
Seattle, WA	$456	$998	$1,454	$2,232	$1,107	$3,339

Courtesy of Owens Corning

Figure 12-1 Cost-in-use benefits of a properly insulated building

- Fuel cost: estimated regional rates
- Heating equipment cost: $1,425.00 per 100,000 BtuH
- Air conditioning equipment cost: $930.00 per ton

Insulation Materials

You can buy building insulation as batts or blankets, loose fill, foam and rigid board. Both the batts and rigid insulation may have reflective surfaces. Insulation materials prevent heat loss and resist the transmission of water vapor, dust and sound. The different types of insulation are shown in Figure 12-2.

Batts and Blankets

Builders install batt and blanket insulation between the structural members of the floors, walls and ceilings in new construction. Figure 12-3 shows batt insulation installed between studs.

The two types of batts or blankets are glass fiber and rock wool. Rock wool (mineral wool) is made from the fibers of various types of volcanic rock. Rock wool is

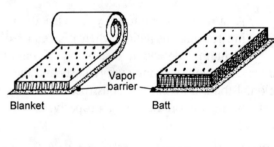
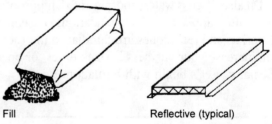
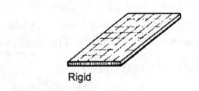

Figure 12-2 Types of insulation

Insulation, Vapor Retarders and Waterproofing

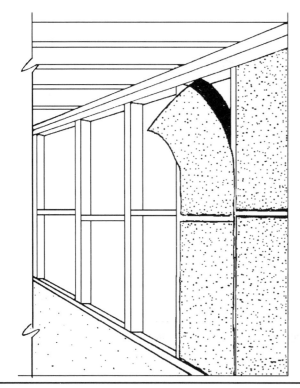

Figure 12-3 Fiberglass batts between studs

Thickness	Kraft paper faced	Foil faced
2	R-7	R-7
3	R-11	R-11
3⅝	R-13	--
5¼	R-19	R-19
6	R-22	R-22

Figure 12-4 R-values for rock wool insulation

| **A Unfaced or Kraft-faced fiberglass batt insulation** ||
Insulation thickness (inches)	R-values
3½ or 3⅝	R-11 and R-13
6¼ or 6½	R-19
7½	R-22
9¼ or 10½	R-30
12 or 13	R-38
B Mobile home fiberglass batt insulation	
Insulation thickness (inches)	R-values
1⅝	R-5
2½	R-7
3½ or 4	R-11
4½	R-14
C Foil-faced fiberglass batt insulation	
Insulation thickness (inches)	R-values
3½ or 4	R-11
6¼ or 6½	R-19

Figure 12-5 R-values of fiberglass batt insulation

denser and has a higher R-value per unit of thickness than glass fiber. Figure 12-4 shows R-values for rock wool insulation. I'll discuss R-values later in this chapter.

Fiberglass batt insulation products come in many forms, including unfaced, paper-faced and foil-faced insulation. Figure 12-5 shows R-value ratings for the common types of fiberglass batt insulation.

■ **Unfaced Batt Insulation** Use unfaced (rigid-fit) insulation alone where no vapor barrier is required, or install a separate vapor barrier. Unfaced batt insulation comes in widths which permit pressure-fit installation in cavity walls: 15, 15¼ and 16 inches for 16-inch stud spacing, or 23, 23¼ and 24 inches for 24-inch stud spacing. It comes 47, 48 and 93 inches long or in rolls approximately 39 feet long.

Unfaced fiberglass batts 2½ to 2¾ inches thick (R-7 to R-8), 3½ to 4 inches thick (R-11), or 6¼ inches thick (R-19) are often used to reduce sound transmission between adjacent rooms when thermal insulation isn't a factor. These batts are also called *sound attenuation blankets* or *sound control batts*. Blankets come 16 and 24 inches wide, and 96 inches long. Sound batts also come with a Kraft paper vapor-barrier facing.

349

You can also get unfaced batts for use between furring strips over masonry walls as shown in Figure 12-6. They're 15 or 23 inches wide, and ¾ to 1⅛ inches thick with an R-value of 3.4. You can install sill sealers between a foundation wall and sill plate in pier-and-beam construction. The standard size is 1 inch thick, 3½ or 6 inches wide, in 100-foot rolls.

Another unfaced variety of fiberglass insulation is available for mobile homes.

■ **Paper- or Foil-faced Batt Insulation** Kraft-faced batt insulation has the paper facing coated with asphalt to provide a vapor barrier. The facing paper has edge flanges to staple the batts to the studs. These batts come in widths of 15, 15¼ and 16 inches, or 23 and 24 inches. Lengths are 47, 48 and 93 inches, or rolls about 39 or 70 feet long.

Special Kraft-faced fiberglass batts are available for installation over suspended acoustical ceilings. Standard thickness for these batts is 3½ (R-11) or 6¼ (R-19) inches. They come 24 inches wide and 48 inches long.

Foil-faced fiberglass batt insulation has an aluminum foil facing that provides a vapor barrier and a reflective face to inhibit infrared heat passing across an air space. The aluminum facing has edge flanges for stapling to studs. Foil-faced batts come 15 and 23 inches wide, in rolls about 39 or 70 feet long. Flame-resistant reinforced foil-faced fiberglass batt insulation is also available. It comes in 4-foot lengths.

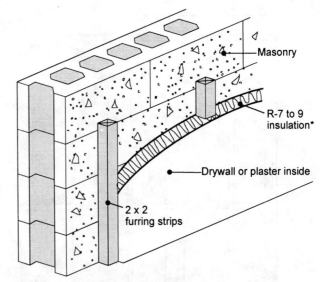

* **Note:** Use only insulating products whose thermal efficiency is not affected by moisture.

Figure 12-6 Masonry wall insulation

Loose Fill Insulation

Loose fill insulation is made of glass fibers, rock wool, cellulose fiber, vermiculite or perlite. The R-values per inch thickness of loose fill insulation are shown in Figure 12-7.

	R-value per inch	
Material type	**Poured**	**Blown**
Glass fiber	2.2	2.2
Rock wool	2.2 - 2.8	2.8
Cellulose fiber	3.7	3.7
Vermiculite	3.0	--

Figure 12-7 R-values of loose fill insulation

Insulating Frame Walls You can either pour or blow loose fill insulation. It's usually poured into accessible horizontal spaces in floors and ceilings as shown in Figure 12-8. All the insulating materials I've mentioned, except for vermiculite, can be blown. Blow insulation through holes into inaccessible spaces such as between studs in finished walls.

Figure 12-9 shows material and manhours required for poured insulation. You'll find the material and installation requirements to achieve required R-values with blown fiberglass or rock wool insulation in an attic in Figure 12-10.

Insulating Masonry Walls You can get concrete block with polystyrene core inserts, or sometimes polystyrene scrap is poured into cores. But inserts don't completely fill a core, and polystyrene beads don't completely fill the wall because of electrically-repelling static within the polystyrene material.

Water-resistant inorganic vermiculite or silicone-treated perlite insulation is a better choice to insulate masonry walls. You can pour it into the cores of concrete block or into brick and block wall cavities as shown in Figure 12-11. Vermiculite pours freely into wall heights up to 20 feet without the need for tamping or rodding and subsequent settlement is less than 1 percent. Another advantage of vermiculite is that it won't burn. The coverage capacity of vermiculite masonry insulation is shown in Figure 12-12.

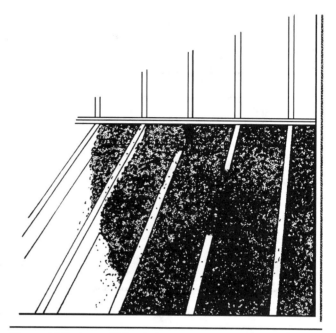

Figure 12-8 Loose fill insulation

Fill thickness (inches)	Number of SF covered by CF @ density rating					Manhours per 100 SF of ceiling
	6 lb	7 lb	8 lb	9 lb	10 lb	
1	21.1	18.0	15.9	14.1	13.0	0.6
2	10.6	9.1	8.0	7.1	6.4	0.6
3	7.1	6.1	5.3	4.7	4.2	0.7
4	5.3	4.6	4.0	3.5	3.2	0.8
5	4.2	3.6	3.2	2.8	2.6	0.9
6	3.6	3.0	2.7	2.4	2.2	1.0
7	3.1	2.6	2.3	2.0	1.9	1.1
8	2.6	2.3	2.0	1.8	1.6	1.2

Figure 12-9 Labor and materials for poured ceiling insulation

R-value	Bags per 1000 SF of net area	Maximum SF per bag	Minimum wt (pounds per SF)	Minimum thickness (inches)
Fiberglass insulation				
R-49	53	19	1.30	17
R-44	45	22	1.15	15
R-38	40	25	1.00	13
R-30	30	33	0.75	10½
R-22	22	45	0.55	7½
R-19	20	50	0.50	6½
R-13	14	73	0.34	4½
R-11	11	90	0.28	4
Rock wool insulation				
R-38	40	25	1.000	17½
R-30	30	33	0.758	13¾
R-22	22	45	0.556	10
R-19	20	51	0.490	8¼
R-13	13	75	0.333	6
R-11	11	90	0.278	5

The above thermal performances are achieved at weights and coverages specified when insulation is installed with pneumatic equipment with a horizontal open blow.

Figure 12-10   R-values and material requirements for blown insulation

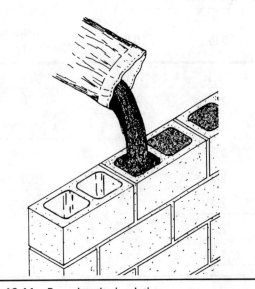

Figure 12-11   Poured cavity insulation

Wall type	Bags per 1000 SF of wall area
8" block	69
12" block	125
1" cavity	21
2" cavity	42
2½" cavity	50
4" cavity	95

Figure 12-12   Coverage capacity of vermiculite insulation (number of 4 CF bags required)

Sprayed Foam Insulation in Walls

Foam insulation is practical for insulating existing masonry cavity walls, curtain walls and stud walls. The best way to put foam insulation into existing walls is to blow it with compressed air into minimum 1-inch-diameter holes drilled between the studs.

You can also use foam insulation in pipe chases to eliminate condensation problems and to reduce noise generated by rushing water. Spray it in the cores of cement block during new construction. You can install foam insulation over a vapor barrier in ceilings, but don't let it cover a recessed light or electrical box. Excessive heat buildup could cause a fire.

Most foam insulation contains urea formaldehyde, although you can get some products without it. You can get a cementitious foam-in-place insulation that's fire-resistant and doesn't give off hazardous gases when burned.

Insulation type	R-value per inch of thickness
Urethane	7.14
Styrofoam	5.41
Glass fiber	4.30
Polystyrene	4.17
Foam glass	2.86
Fiberboard	2.77
Expanded perlite	2.56

Figure 12-13 R-values of various types of rigid insulation

Rigid Insulation

The most common types of rigid board insulation include mineral fiber, glass fiber, polystyrene, urethane, foamed glass, wood fiber boards, compressed particleboard, cork board and Styrofoam. The R-values per inch thickness of various rigid insulation materials are shown in Figure 12-13.

Fiberboard insulation comes in thicknesses through 2¼ inches, but the most common sizes are ½ or 1 inch thick. You can get greater thicknesses by building up two or more layers of insulation board. Rigid insulation boards made from organic fiber are often used for outside wall sheathing. Some boards come with aluminum foil faces for vapor protection. Standard sizes are 4' x 8' with thicknesses ranging from ⅜ through 1⅞ inches.

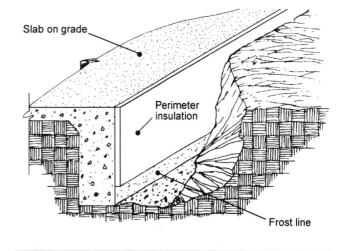

Figure 12-14 Rigid perimeter insulation

Use rigid insulation made of polystyrene, polyisocyanurate or urethane for perimeter insulation (Figure 12-14). Asphalt-impregnated cork boards are also popular. Apply perimeter insulation with spot-applied adhesives. Rigid perimeter insulation boards are normally 1 to 2 feet wide and 1 to 4 inches thick.

Rigid Styrofoam boards come 16, 24 and 48 inches wide, in lengths of 48 and 96 inches. They're ⅜ through 4 inches thick. You can also use rigid Styrofoam to insulate slabs, foundation walls, cavity walls and as an exterior sheathing component as shown in Figure 12-15.

■ **Tapered Rigid Insulation** Tapered rigid insulation usually comes in 2-foot widths with a taper across the 2-foot dimension. The minimum taper required is 1/8 inch per foot. I recommend a 1/4-inch taper to provide good drainage. Standard manufactured tapers are listed in Figure 12-16. To build up the insulation to the required thickness, install insulation boards in 1-inch-thick increments beneath the tapered insulation boards.

Reflective Insulation

Batt and rigid board insulation are available with reflective foil laminated to one or both sides. The reflective material increases the R-value of the insulation two ways. First, it reduces radiant heat loss from inside the building during winter. Second, it reduces radiant heat flow into the building during summer. Reflective faces also serve as a vapor barrier and increase fire resistance.

You can get a reflective insulation with a white-tinted or embossed foil face on one side and a reflective foil face on the other. It's installed in buildings such as warehouses where one face of the insulation is left exposed within the building. A heavy-duty foil-faced insulation is also available for exposed insulation that may be subject to abrasion or impact abuse.

Sprayed-on Roof Insulation

The most widely used sprayed-in-place insulation includes plastic foams such as expanded urethane or polystyrene, isocyanurate foam or fibered masses such as gypsum or perlite. Sprayed-on foams often require a primer coat appropriate for the substrate. Be sure the deck is clean and free of aggregate, sand, curing compounds, oil, grease or any other contaminant that might keep the foam from adhering. Don't spray foam during rain or fog, or in winds of more than 15 miles per hour. Be sure the deck is 50° F or warmer, and follow the manufacturer's instructions.

Sprayed-on plastic foams have several advantages. They're lightweight (less than 1 pound per square foot) and will conform to any roof surface contour. You can spray polyurethane foam into surface irregularities and in varying thicknesses to provide pitch to drain. The insulation provides a monolithic, self-flashing system. There are no joints to seal, and it's easy to patch broken or damaged areas. Sprayed insulation helps keep the deck from changing temperature, so there's less movement, and resists severe wind uplift.

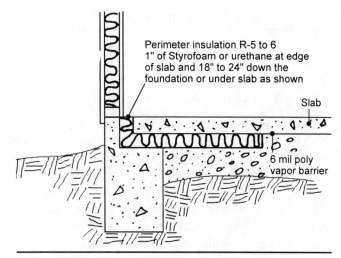

Figure 12-15 Perimeter insulation under slabs

Incline (inches per foot)	Thickness (high end)	Thickness (low end)
1/8	3/4"	1/2"
	1"	3/4"
	1 1/4"	1"
	1 1/2"	1 1/4"
1/4	1"	1/2"
	1 1/2"	1"

Figure 12-16 Tapered insulation thicknesses

One significant disadvantage of sprayed-on plastic foam is that it burns and gives off dense, black smoke. Other drawbacks are that cracks often occur if the material shrinks, and urethane turns brown after sun exposure, so you have to cover it with a liquid roof covering.

Polystyrene insulation is incompatible with asphalt and pitch. Use a minimum 6-mil-thick polyethylene separation sheet between the insulation and a BUR system.

Lightweight Concrete Fill Insulation

The most widely used poured-in-place insulation is lightweight concrete fill composed of lightweight aggregates and Portland cement. The lightweight aggregate used is either perlite (volcanic glass) or vermiculite (an expanded micaceous silicate mineral). The more aggregate, the greater the insulating value of the fill. But too high a ratio of aggregate to cement will weaken the fill to a point where it isn't considered nailable. Other types of cementitious insulation include cellular Portland cement, poured or sprayed-on gypsum concrete, or gypsum concrete mixed with a wood fiber binder or glass bead binder.

One advantage of lightweight concrete is that it retains heat, and it stores and releases heat slowly.

If a lightweight concrete deck is to receive a built-up roof, mix the cement-to-aggregate ratio 1:4 or 1:6 with a minimum dry density of 22 pounds per cubic foot. The deck must accept and retain nails.

Reducing Heat Loss

A properly insulated and weatherproofed building reduces energy costs and keeps the occupants more comfortable. Insulation also permits the use of smaller heating and cooling units and smaller ductwork. And since those heating and cooling units won't be used as much, they'll last longer. Figure 12-17 shows where to install insulation throughout a house for efficient energy conservation. Insulation R-values shown are relative to the location of the insulation. The actual R-values required depend on the local climate. Look at Figure 12-18.

Floors

At slabs on grade, install 1-inch-thick rigid Styrofoam or urethane perimeter insulation (R-5 or R-6) at the edge of the slab and 18 to 24 inches down the inside of the foundation wall, or under the slab as in Figure 12-15.

Floors above heated rooms or heated basements require no insulation, but insulate floors over unheated basements. The R-value of the insulation depends on the climate. Refer to Figure 12-18.

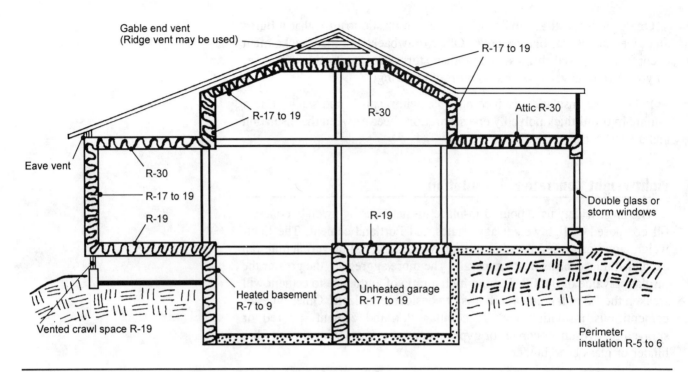

Figure 12-17 Recomended insulation locations

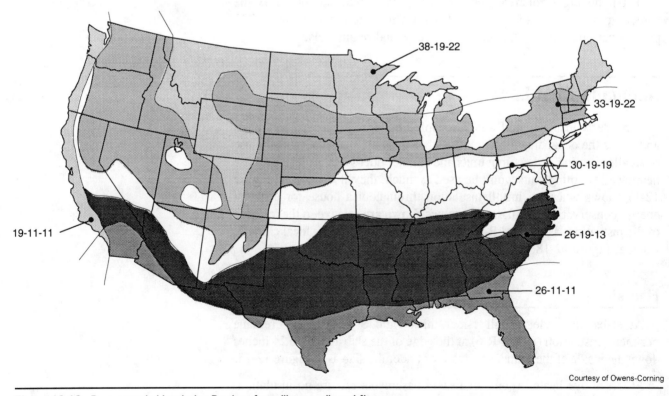

Courtesy of Owens-Corning

Figure 12-18 Recommended insulation R-values for ceilings, walls and floors

Walls

Outside walls (excluding windows and doors) account for about 13 to 14 percent of the total heat loss in a house. A brick veneer wall with R-11 insulation installed between studs has 69 percent less heat loss than the same type of wall without insulation. You can reduce the heat loss through a masonry wall by as much as 83 percent by installing moisture-resistant insulation (R-7 to R-9) between 2 x 2 furring strips along the inside face of the wall. (Look again at Figure 12-6.)

Windows

Windows and sliding glass doors account for about 40 percent of the total heat loss in a house. You can reduce heat loss by filling all voids around the opening with insulation. Also, cover the voids with a vapor barrier as shown in Figure 12-19 and caulk all cracks around windows. Later in this chapter there's a chart (Figure 12-24) that shows how much caulk you'll need.

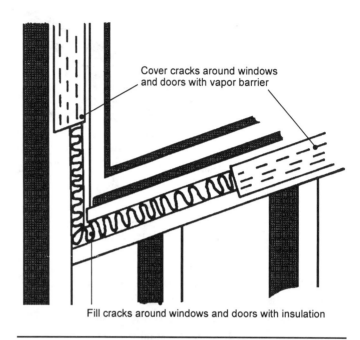

Figure 12-19 Insulating around doors and windows

You can reduce heat loss even more by installing a storm sash or insulating glass. A window with a single pane and storm sash has about 50 percent less heat loss than the same window without a storm sash. A window with insulating glass has about 46 percent less heat loss than a single-pane window without a storm sash.

Doors

Outside doors (excluding sliding glass doors) account for about 2 to 3 percent of the total heat loss in a house. As with windows, you can reduce heat loss by adding insulation and vapor barriers, and by caulking.

Also consider installing storm doors or insulated doors. A 1¾-inch-thick solid core door and storm door have about 35 percent less heat loss than the same solid core door with no storm door. A urethane-insulated steel door (R-13.8) with no storm door has about 85 percent less heat loss than a 1¾-inch-thick solid core door with no storm door. The same urethane-insulated steel door with no storm door has about 76 percent less heat loss than a 1¾-inch-thick solid core door *with* a storm door.

Ceilings

You can insulate a ceiling with batts, blankets or loose fill, or with a combination of these materials. The required R-value of the insulation depends on the local climate. For the sake of comparison, 8-inch-thick

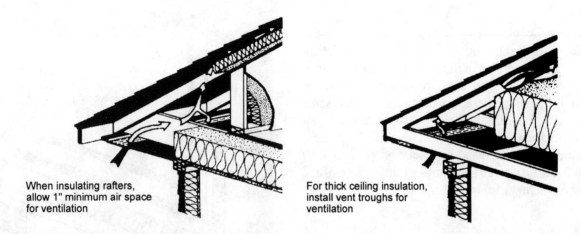

Figure 12-20 Provide for ventilation at eaves

insulation (R-25.3) has about 21 percent less heat loss than 6-inch-thick insulation (R-19); 10-inch insulation (R-31.7) has about 35 percent less heat loss than 6-inch insulation; 12-inch insulation (R-38) has about 46 percent less heat loss than 6-inch insulation.

When you install ceiling insulation, don't put insulation over the soffit, especially where soffit vents have been installed. Insulation there can lead to the formation of ice dams in cold climates (see Figure 3-38 back in Chapter 3). Also, if you install batt insulation between rafters, allow at least 1 inch of air space between the insulation and roof deck to provide adequate attic ventilation. That's shown in Figure 12-20.

Roof Insulation

Roof insulation reduces heat loss through the roof in cold weather and heat penetration when it's hot. Roof insulation also prevents condensation from forming on the inside surface of the roof deck. Roof insulation provides a good substrate for roofing felts and helps isolate the roof membrane from stresses caused by deck or structural movement. Insulation reduces deck expansion and contraction. It also keeps heat from passing through built-up roof membranes so the felts don't dry out so fast.

The biggest disadvantage to installing roof insulation beneath a built-up roof is that it causes the BUR system to age faster than it would if installed directly over the deck. That's because, during the heat of the day, the base ply becomes extremely hot since the insulation doesn't let the heat pass through to the inside of the building. At night, the base ply cools quickly, again because the insulation prevents heat from passing through to the BUR system. That extreme temperature range increases expansion and contraction of the system, making it age faster.

Most roof insulation is rigid or sprayed-on insulation. Common roof insulation products include polyurethane, glass fiber, expanded polystyrene bead board, foam glass, fiberboard and expanded perlite. Of all the insulation products available, polyurethane is the most efficient. But ultraviolet rays rapidly degrade exposed polyurethane, so it's necessary to cover it within 48 hours after application.

■ **Installing Rigid Roof Insulation** Insulation boards are either applied directly to the deck or over a vapor retarder sheet, depending on the amount of humidity expected inside the building. Insulation boards must be strong enough to withstand foot traffic, snow loads and hail impact. The boards should be attached to the deck well enough to resist wind uplift.

Some manufacturers recommend that you install insulation with the long dimension of the board parallel with the long roof dimension and stagger the end joints as in Figure 12-21. Others recommend installing the long dimension of the insulation board perpendicular to the direction of the membrane.

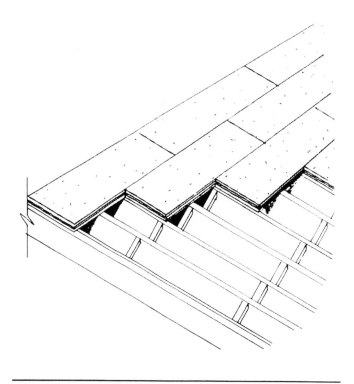

Figure 12-21 Application of rigid insulation

On metal decks, run the long dimension of the board parallel to the ribs and rest the edges firmly on top of a flute. If the insulation needs to be more than $1^1/2$ inches thick, install multiple layers. When two or more layers of insulation are used, install the upper layer with end and side joints staggered at least 6 inches from the bottom layer to prevent bitumen from dripping into the building.

To allow for expansion, install all insulation board with a 1-inch gap between the board and any vertical surface such as a wall. Fill all gaps wider than $1/8$ inch between insulation boards with roofing cement and strike the joints smooth. Install 4' x 8' insulation boards wherever possible to reduce the number of joints.

You'll usually anchor insulation board with a solid mopping of hot bitumen. As a rule, don't install rigid insulation with cold-applied adhesives. Note that some materials such as Styrofoam can't withstand hot bitumen so you have to use cold-applied adhesives or mechanical fasteners. Urethane boards can endure hot asphalt (425° F) only very briefly. Use mechanical fasteners over metal decks.

According to some membrane manufacturers, polystyrene insulation will damage bitumen roofing membranes. But you can install fiberboard, perlite board or glass fiber roof insulation with mechanical fasteners over the insulation followed by hot moppings of bitumen and roof membranes. Install fiberboard with mechanical fasteners over Styrofoam, urethane or isocyanurate insulation board followed by hot moppings of bitumen and roof

Roofing Construction & Estimating

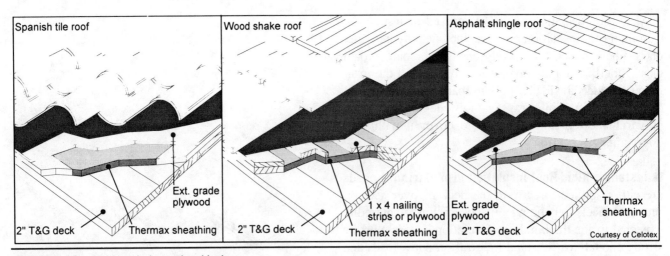

Figure 12-22 Rigid insulation under shingles

membranes. Some decks aren't suitable for hot bitumen because the bitumen will leak through joints. Then you'll have to apply rosin paper to the deck before you mop to keep bitumen from leaking through.

Don't use foam, sprayed-on insulation or polystyrene boards under fiberglass roofing membranes because they react with asphalt. Use fiberglass, perlite board, wood fiber board or cellular glass instead.

When you use mechanical fasteners to install insulation over a nailable deck, use at least 5 fasteners per 2' x 4' panel, 6 fasteners per 3' x 4' panel and 14 fasteners per 4' x 8' panel. Place one in each corner and a random pattern in the field. Use fasteners long enough to penetrate through all the layers of insulation and ¾ inch into the deck, but not so long that they poke through the underside of the deck. Drive all nails through tin caps.

When you install insulation over a non-nailable deck, embed the insulation into a solid mopping of hot steep asphalt applied at the rate of 23 pounds per square. Before you mop the asphalt, prime the deck at the rate of 1 gallon per square.

To prevent felts from slipping on decks steeper than 1 in 12 (asphalt bitumen BUR), or ½ in 12 (coal-tar pitch BUR), nail the rigid board insulation to the deck (if nailable) or install the insulation between treated wood nailing strips (if non-nailable). Refer back to Figure 10-3 in Chapter 10. Over wood decks, install the insulation with large-headed nails long enough to penetrate ½ inch into the deck. Use serrated nails on plywood decks.

You can install rigid insulation under all kinds of shingle roofs, but it's most common over vaulted or exposed-frame ceilings where you can't install insulation in the attic. You must install the insulation over a solid deck. Then install nailing strips or a solid plywood deck before you install the shingles. That's shown in Figure 12-22.

Insulation Values

The *insulation value* of a material describes how well the material conducts or resists heat transfer. Insulation value is expressed in British thermal units (Btu). A Btu is the energy required to raise the temperature of 1 pound of water 1 degree Fahrenheit.

Thermal conductivity (K) measures a material's ability to conduct heat and is the amount of heat (in Btu per hour) transmitted per square foot through each inch of material, per degree difference in temperature (F) on opposite sides of the material. The lower the value of K, the less the heat transmission and therefore, the better the insulation.

Most insulating products are rated and stamped with an R-value. The R-value is the total resistance of a material to conduct heat and is defined as:

$$R = \frac{1}{K} \times \text{Material Thickness (inches)}$$

Equation 12-1

A higher value for R means less heat transmission, or better insulation. You can't necessarily relate the thickness of a material to its insulating quality. For example, the R-value for an 18-inch-thick concrete wall is almost identical to that of 1 inch of rock wool insulation.

The coefficient of transmission (U) of a material is defined as:

$$U = \frac{1}{R}$$

Equation 12-2

The lower the U-value, the better the insulation. Both the thermal conductivity (K) and the coefficient of transmission (U) of a material indicate the material's ability to conduct heat. The difference between them is that the thermal conductivity is a measure of the heat flow per inch of material thickness, whereas the coefficient of transmission is a measure of heat flow through the entire thickness of the material.

▼ **Example 12-1:** The thermal conductivity of common brick is 5 Btu/hour/SF/inch/° F. Find the total resistance (R) and coefficient of transmission (U) if the brick is 4 inches thick.

From Equation 12-1, the total resistance is:

$$R = \frac{1}{5} \times 4$$
$$= 0.80$$

From Equation 12-2, the coefficient of transmission is:

$$U = \frac{1}{0.80}$$
$$= 1.25$$

▼ **Example 12-2:** Find the coefficient of transmission (U) for a material whose total resistance (R) is 30 Btu/hour/SF/° F.

From Equation 12-2, the coefficient of transmission is:

$$U = \frac{1}{30}$$
$$= 0.033$$

The amount of insulation you need depends on the local climate. Use the map back in Figure 12-18 to find a locality's temperature zone and the R-values required for ceiling, wall and floor insulation of buildings within that zone.

When two or more insulating materials are combined, the total R-value is the sum of the R-values of each individual material. For example, combining a foam board with an insulating value of R-15 and a batt with an insulating value of R-5 yields a total insulating value of R-20. Also, if a material is doubled, the R-value will be doubled. For instance, a 2-inch-thick material with an insulating value of R-5 has an R-value of R-10 if two layers are installed. (This isn't true for K-values, which aren't cumulative.)

Vapor Barriers

The first and most important reason to install a vapor barrier is to prevent warm humid air from contacting a cold surface, which results in condensation. The second reason is to reduce the flow of air between the inside of a building and the outdoors.

The main disadvantage of vapor barriers is that they can trap moist air within the building. To avoid that, install or recommend ceiling fans, kitchen and bath exhaust fans, and dryer vents to provide adequate air circulation and ventilation. Excess moisture within a building causes such things as lingering odors, dampness, mold, mildew, finish discoloration, peeling paint and decay of framing lumber.

Vapor barriers come in a variety of materials, including asphalt-coated felts, asphalt-laminated papers, aluminum foil and plastic films. Polyethylene film is the most commonly used vapor barrier, although it's hard to work with and tears easily. Polyethylene film comes in 100-foot rolls and in widths of 2, 3, 4, 6, 8, 10, 12, 14, 16 and 20 feet. Film used for vapor barriers is 0.004 or 0.006 inch thick. Tape for sealing joints in the film comes 2 inches wide, 100 feet long and 0.004 inch thick.

There's little room for error when you install vapor barriers. A hole in a vapor barrier is an escape route for moisture, which makes the vapor barrier virtually worthless. Putting a vapor barrier at the wrong location is even worse. Always install a vapor barrier in an exterior wall to the inside (warm side) of the wall to prevent moisture from getting into the wall from inside the building. Apply a porous material which breathes, such as asphalt-saturated felt, to the outside (cold side) of the wall so that moisture can escape. Install roofing felt to the outside that's at least five times more permeable than the vapor barrier. This configuration prevents moist air from condensing and becoming trapped within the wall. I also recommend this 1:5 ratio for sheathing materials used as the outer "skin" of a house.

Never install a vapor barrier on both sides of an exterior wall. No vapor barrier is required within interior partitions.

Measuring Vapor Barrier Effectiveness

The ability of a vapor barrier to resist vapor penetration is measured in *perms*. One perm equals the transmission of 1 grain (approximately 0.07 grams) of water vapor per square foot per hour per each inch of mercury vapor pressure difference between opposite sides of the vapor barrier. The lower the perm rating, the less the vapor transmission and the better the vapor barrier. Vapor barriers come with a rating of 0.1 perm or less. Polyethylene film has the lowest perm ratings, ranging from 0.02 to 0.08.

Vapor barriers aren't 100 percent efficient in stopping the transmission of water vapor; they only reduce the rate of transmission. Therefore, *vapor retarders* is perhaps a more appropriate term. Since it's impossible to totally prevent the passage of vapor, some moisture will finally penetrate an exterior wall. It's important that this moisture have an escape route. That's why the outer surface (cold side) of an exterior wall must be allowed to breathe.

Vapor Barriers and BUR

Install vapor barriers (only on insulated roofs) to prevent condensation from forming within a BUR system. Condensation causes insulation to swell and then shrink as it dries. Most types of cellular insulation are quickly destroyed by moisture. If this type of insulation gets wet, the roof system is soon completely undermined. That causes the roofing felts to split and then a leaky roof.

Install vapor barriers only on roofs with insulation sandwiched between the deck and the BUR membranes. Install vapor barriers between the deck and the insulation.

Always put vapor barriers over the deck of an insulated roof for any building with a high interior humidity and little or no ventilation. Without a vapor barrier, the insulation will shift the dew point from under the roofing system to within the roofing system. (The dew point is the temperature at which water vapor will condense.) This is one disadvantage of roof insulation. If insulation absorbs any moisture, the vapor barrier and BUR membranes prevent the vapor from escaping. In order to lessen this effect, install insulation board with beveled edges to provide a horizontal passage for trapped water vapor. Also leave a ¼-inch gap between insulation boards. Moisture in the insulation is less of a problem when a vented base sheet is installed.

Weatherproofing Existing Homes

Many older homes have had insulation blown into exterior wall cavities, but no vapor barriers were installed. We used to think that this blown-in insulation allowed condensation within the walls, creating the potential for rot and decay. But recent studies have shown that condensation isn't serious and that the problem tends to correct itself. When insulation gets wet, it loses some effectiveness and increases heat loss through the wall. This increases the temperature within the wall, resulting in reduced condensation and increased evaporation.

Because it's not practical to install standard vapor barrier materials in an existing wall, you can use vapor-retarder paint over interior drywall to help block the transmission of interior moisture into the wall.

In homes framed on a pier-and-beam foundation, you can install insulation and vapor barriers in crawl spaces and under the floors. Place puncture-resistant (6-mil) polyethylene sheets over the soil. Hold the sheets in place with bricks and cover the sheets with sand for protection in traffic areas. This reduces the amount of under-floor ventilation required. Figure 12-23 shows how to install under-floor vapor barriers.

When a vapor barrier is installed under the floor in cold climates, install it on the warm side of the floor above the insulation. In climates that require air conditioning during the summer, install the vapor barrier beneath the insulation.

I also recommend a vapor barrier over poorly-vented ceilings in homes with flat roofs or cathedral ceilings.

Caulking and Sealants

In addition to insulation and vapor barriers, you can make any home more energy-efficient by keeping air leakage to a minimum. That's where caulking comes in.

The main difference between a caulk and a sealant is the grade of material. A sealant is a lower grade and I recommend it only for interior work. Most sealants are oil-based and will bleed through latex paint.

Insulation, Vapor Retarders and Waterproofing

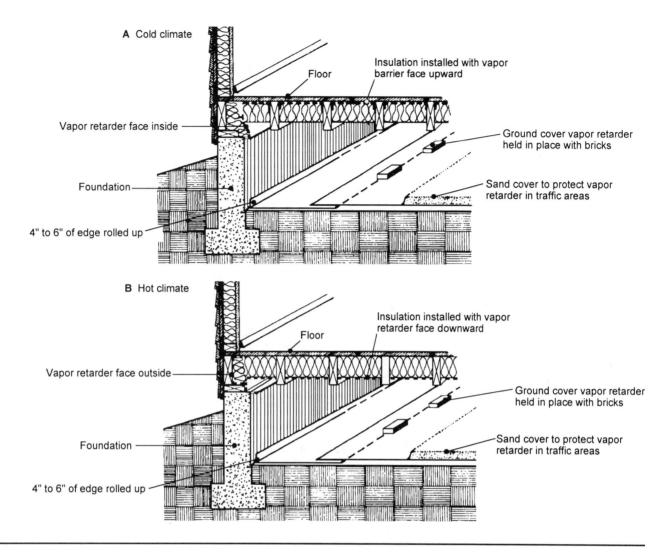

Figure 12-23 Vapor retarder locations for crawl spaces and floors

There's a wide variety of caulking and sealing compounds on the market, including butyl, latex, polysulfide, polyurethane, solvent acrylic, vinyl acrylic, epoxy and silicone. We'll look at them one at a time.

Butyl caulk and sealant doesn't have good expansion-contraction qualities (10 to 12 percent) and eventually becomes brittle. It does have good resistance to ultraviolet light and weathering. Manufacturers claim it has a life expectancy of 10 to 20 years, depending on quality. It comes in a wide variety of colors and no priming is required.

Acrylic latex caulk is a water-based acrylic compound. Latex caulk shows good resistance to ultraviolet light and weathering. Since it takes latex or oil paint very well, it's often used as a cap bead over other caulking compounds to make them paintable, but wood surfaces require priming to increase adherence. This caulk gives off very little odor and cleanup only requires water. Latex caulk comes in several colors.

Polysulfide caulk and sealant comes either as a one-part product requiring no mixing, or a two-part product. The one-part type requires 6 to 12 months cure time and the two-part variety requires 2 to 4 months. For all practical purposes, it's weatherproof in a few hours. But until it cures, its expansion capabilities aren't at the maximum. Both types allow up to 25 percent expansion and have a life expectancy of 20 years. They require a surface primer on most surfaces, but both have good adhesion. Both have poor to fair resistance to ultraviolet light, and only fair to good resistance to weather, so they'll crack eventually. They come in a variety of colors, and polysulfide caulks take paint well. A two-component immersion polysulfide sealant is also available to seal swimming pools, reservoirs, dams or other water-submerged environments.

Polyurethane caulk and sealant also comes as a one- or two-part product. Both varieties allow up to 50 percent expansion and have a life expectancy of 15 years. The one-part type requires 2 to 6 months cure time and the two-part variety requires 3 to 8 months. Both require a surface primer on some surfaces. The one-part product (urethane) has good adhesion and the two-part (polyurethane) variety has fair to good adhesion. Both types show good resistance to ultraviolet light and to weathering. Both products come in a wide range of colors. A two-component urethane-asphalt sealant is available for sealing horizontal joints in salt water or high-abrasion industrial environments.

Solvent acrylic caulk allows up to 10 percent expansion and has a life expectancy of 10 years. No surface primer is required and adhesion is good. This product exhibits good resistance to ultraviolet light and fair resistance to weathering. Many colors are available and solvent acrylic caulk doesn't require curing.

Silicone caulk and sealant comes with a wide variety of physical properties. Silicone caulk allows expansion ranging from 25 to 100 percent and has a life expectancy of 30 years or more. Most products don't require a primer on most surfaces (primer is recommended over concrete, and some plastics) and all exhibit good adhesion. All products show good resistance to ultraviolet light and to weathering. Most silicone caulk comes in a wide range of colors, but the mildew-resistant kind comes only in white. Due to its expansion-contraction properties, paint does not adhere well to silicone sealants. Most silicone caulks cure in only 7 to 14 days. Silicone caulks are used for a variety of purposes including interior environments of extreme humidity and temperatures, to seal construction joints where movement is extreme, to seal glass, metal and plastic surfaces, and for sealing insulating glass.

Vinyl acrylic caulk can also be used as an adhesive to bond virtually any two materials and no surface primer is required. This product exhibits good resistance to ultraviolet light and to weathering. Vinyl caulk is available in many colors and cures in 48 hours. Don't use this product under water.

Joint dimensions (inches)									
	1/8	3/16	1/4	3/8	1/2	5/8	3/4	7/8	1
1/8	1232	821	616	411	307	246	205	176	154
3/16	821	547	411	275	205	164	137	117	103
1/4	616	411	307	205	154	123	103	88	77
3/8	411	275	205	137	103	82	68	58	51
1/2	307	205	154	103	77	62	51	44	39
5/8	246	164	123	82	62	49	41	35	30
3/4	205	137	103	68	51	41	34	29	25
7/8	176	117	88	58	44	35	29	25	22
1	154	103	77	51	39	30	25	22	19

Figure 12-24 Caulking material requirements (linear feet per gallon [231 cubic inches] of caulk)

Epoxy sealant comes as a two-part product. It sticks to metal, glass, marble, concrete, wood and many other surfaces. Epoxy sealant is normally used to seal joints in high-moisture areas and facilities subject to submersion such as reservoirs, sewer treatment facilities and swimming pools. It's not affected by jet fuel, oil, gasoline, caustics, salts and most acids. The life expectancy varies from 10 to 20 years, depending on service conditions. This product comes in white or gray, with custom colors available.

Caulking compounds come in 11-ounce cartridges, 1-gallon containers, 5-gallon pails and 55-gallon drums. The volume of one tube is 315 cubic centimeters (about 19.22 cubic inches). This means that 1 gallon is about equal to 12 tubes (231 cubic inches). Use Figure 12-24 to find how much caulking material you need. Volumes vary depending on joint design, tooling, backer rod placement and waste.

▼ **Example 12-3:** Assuming a 1/4-inch-wide by 3/8-inch-deep bead of caulk applied to both sides of all exterior openings, how much caulk is required for the building diagrammed in Figure 12-25?

From Figure 12-25, the total linear feet of caulk required is:

```
3070 Door:    2 Ea x 17 LF/Side x 2 Sides =    36 LF
2630 Window:  1 Ea x 11 LF/Side x 2 Sides =    22 LF
3040 Window:  4 Ea x 14 LF/Side x 2 Sides =   112 LF
4040 Window:  4 Ea x 16 LF/Side x 2 Sides =   128 LF
                                              ------
Total                                         298 LF
```

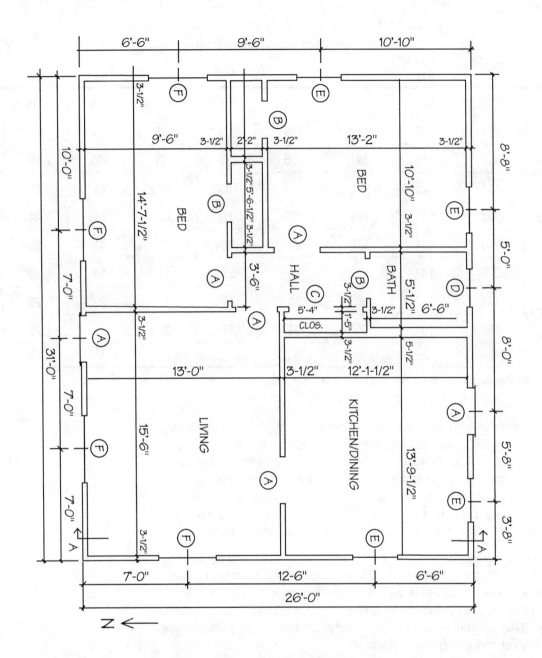

Figure 12-25 Caulking example

Insulation, Vapor Retarders and Waterproofing

Pipe size (Schedule 40 pipe)	Volume (cu. in.) required per 1" depth	Tubes of caulk required	Gallons of putty required
1"	1.04	0.06	0.005
2"	1.49	0.08	0.007
3"	2.94	0.16	0.010
4"	3.73	0.21	0.020
5"	3.92	0.22	0.020
6"	6.80	0.38	0.030
7"	11.00	0.61	0.050
8"	12.40	0.69	0.060

Figure 12-26 Fireproof caulk or putty required for various pipe penetrations

To find how much caulk to buy:

Caulk required = LF ÷ LF/Gal (linear feet per gallon of caulk)

298 LF ÷ 205 LF/Gal = 1.45 gallons

Since there are 12 tubes per gallon:

Caulk required = 1.45 gallons x 12 tubes/gallon
= 18 tubes

Backer Rods

Install a backer rod of oakum or polyethylene rope before you caulk joints wider than ¼ inch. Then moisten your finger to smooth and feather the bead of caulk. Or you can use a small paint brush dipped in soapy water.

Concrete expansion joint sealers consist of a polyurethane-based material applied over a primed backer rod. Apply a sealant tape over the backer rod material before applying the caulking compound to create a barrier between the two materials. Some materials have additional components such as bitumen or rubber for use in joints exposed to fuel spillage.

Fireproof Caulk and Putty

You can get fireproof caulk and putty to seal pipe penetrations and stop the spread of fire, heat, smoke and gas. This synthetic elastomeric one-part material expands by 10,000 percent when heated, and bonds to concrete, metals, wood and plastic. It also seals against water in an unexpanded state. You can get caulk in tubes and putty in quart and 1- or 5-gallon cans. For a three-hour fire rating, a 1-inch depth of caulking is required. You can use Figure 12-26 to find how much fireproof caulking or putty you need.

Foam Sealants

Foam sealant tape is a self-adhesive rolled tape used to seal joints in virtually any type of construction system. It offers protection against weather, vapor, sound and dust. Foam sealant tape is an open-cell polyester polyurethane foam impregnated with neoprene rubber. You can also get butyl-coated PVC foam sealants. The tape comes in widths ranging from ⅜ to 2 inches and in 15-, 20- and 25-foot lengths. Sealant tape will not shrink or dry out and expands to fit the joint in which it is installed.

Polyurethane foam sealants are applied from a container with a hose. This sealant bonds to wood, metal, concrete, brick, glass and most plastics. It withstands expansion and contraction and cures within 24 hours. Foam sealants seal cracks as well as electrical and plumbing penetrations.

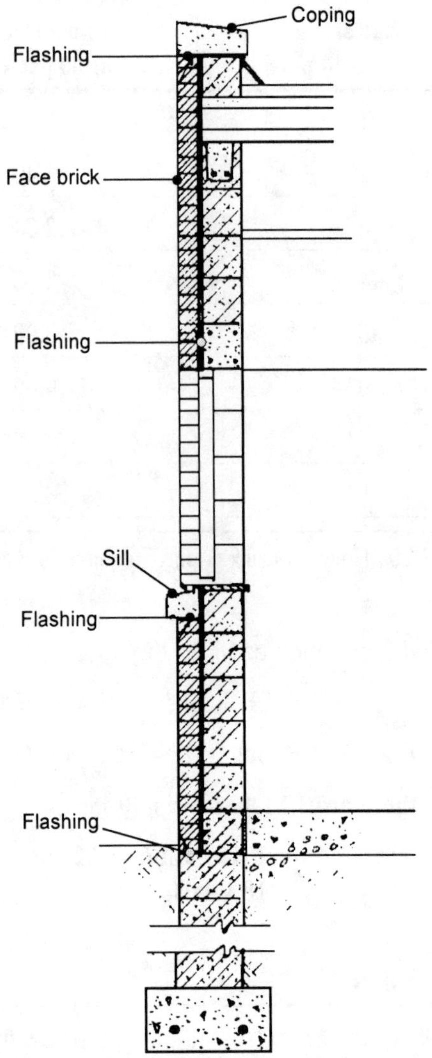

Wall Flashing

The roofing subcontractor normally provides wall flashing, but it's usually installed by the general contractor (in wood frame construction) or the masonry subcontractor (in masonry construction).

Wall flashing helps prevent moisture from penetrating outside walls. Moisture penetrates at parapets, sills, projections, recesses and through mortar joints. Without wall flashing, the trapped moisture causes problems, including alkali action from the mortar and dilute sulfuric acid from atmospheric sulfur dioxide. Sulfuric acid is especially harmful to the metal components of a building.

Figure 12-27 Various wall flashings in a masonry wall

Install wall flashing at each foundation, spandrel, sill, head, parapet, corner and door, plus through-wall flashing at various locations. Those flashings appear in Figures 12-27 and 12-28.

Wall flashing is made from a variety of materials including flexible plastic, asphalt-saturated cotton fabric, saturated or non-saturated woven fiberglass, asphalt felt and cotton fabric, copper, or copper plus lead bonded to asphalt-saturated cotton fabric, fiberglass, flexible rubbery bituminous compound, Kraft paper or polyethylene film. It can also be made of aluminum bonded to and between two layers of asphalt-saturated cotton fabric, polyvinyl chloride resin alloyed with other elastomeric substances, stainless steel, terne-coated stainless steel, and plain or lead-coated copper. Wall flashing comes in strips, interlocking pieces or in rolls.

Estimating Wall Flashing Material

Take off flashing material by the linear foot for widths up to 12 inches, and by the square foot for widths greater than 12 inches. Some items, like flashing over, under and around doors and windows, are estimated by the piece. When you prepare an estimate, note the flashing gauge and location.

Waterproofing

Roofing contractors routinely install waterproofing or dampproofing materials to seal vertical surfaces on concrete or masonry basement walls beneath finish grade. And because the materials and installation methods are so similar to many roofing procedures, I'll discuss other applications as well. Components which come under the heading of waterproofing and dampproofing include shower pans, waterstops, linings for planter boxes, and lining concealed under concrete, terrazzo, ceramic tile, clay pavers and walking surfaces of roofs.

Waterproofing seals out water that's under pressure, called *hydrostatic head:* the pressure produced by the weight of liquid water against the foundation walls and slabs of a structure.

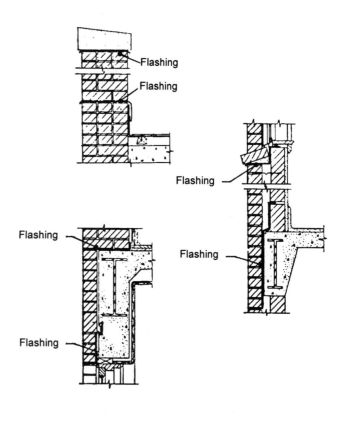

Figure 12-28 Typical flashing details

The most common waterproofing is the application of multiple layers of asphalt-saturated roofing felts sandwiched between moppings of hot asphalt or pitch. Other methods include troweled-on mastic, brush- or spray-applied liquid water repellents, single or multiple layers of plastic or rubberized sheet material, or asphalt-coated protective board embedded in cold-applied adhesive. You can buy special waterproof foundation-coating cements in 5-gallon pails or 55-gallon drums.

Cover waterproofing material as soon as possible after installation. If the backfill material contains stones or other sharp objects, install a layer of paper-pulp board over the waterproofing for protection during backfilling operations. If you're careful during compaction to prevent damage to the waterproofing, you can omit the protective board when backfill material is sand or clean loam.

It's often necessary to install a drainage system in addition to waterproofing to prevent standing water from penetrating a basement or foundation wall. A drainage system removes most of the water before it reaches the wall, or lowers the water table near the building to a level below the slab.

That's shown in Figure 12-29. The most commonly used drainage pipe is 4-inch-diameter farm tile, bituminous fiber pipe or perforated plastic pipe. The first comes in 12-inch lengths, the other two in pieces 10 feet long.

Lay at least 2 inches of crushed stone under the pipe and place at least 6 inches over the top of the pipe. Severe groundwater conditions could require the use of well-points and continuous pumping.

Integral Waterproofing

You can use a variety of liquid, paste and powder chemical admixtures to make concrete, mortar and stucco denser and therefore less permeable to water through increased hydration of the cement. This is called the integral method of waterproofing. Estimate quantities for this type of waterproofing according to the cubic yard of concrete poured.

Built-up Waterproofing

Fabrics

Fabrics (cotton, glass fiber or jute) are more elastic than felts and have greater tensile strength. It also takes fewer coats of waterproofing bitumens to reach the desired thickness because fabrics hold the bitumens in place.

These fabrics come in rolls that are 180 feet long (nominal length 150 feet) and 3, 4, 6, 9, 12, 18, 24 and 36 inches wide. Some glass fabric rolls are 48 inches wide.

You can get bitumen-saturated cotton fabric in standard or specification grade. Specification grade contains more cotton. Cotton fabric is the most commonly-used material for covering corrugated metal roofs.

Saturated jute fabric (treated burlap) is very coarse so it holds a lot of field-applied bitumen coating material. That results in a heavy build-up of bitumen coating with fewer applications, lowering your labor cost.

Saturated woven glass fabric is widely used with cold-applied bitumens. It has a higher tensile strength than cotton or jute fabric. Standard glass fabric consists of randomly arranged glass fibers bonded with a resinous binder. Heavy-duty glass fabric consists of heavy glass yarn bonded and coated with bitumen.

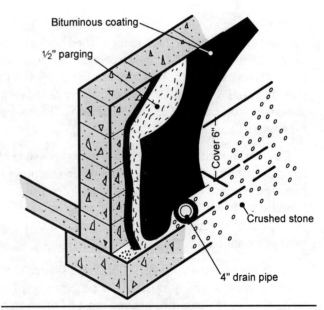

Figure 12-29 Foundation waterproofing

Membranes

Waterproof membranes are made from a variety of materials, including liquid-applied or sheet elastomeric plastic films, built-up perforated asphalt- or tar-saturated felts (organic or inorganic) or fabrics. Asphalt- or tar-saturated felts are suitable for all subgrade conditions. Apply glass fiber felts to horizontal surfaces only. Other materials include spray-applied glass fiber reinforced bituminous material and composites.

Installing Built-up Membranes

Prime the wall or slab before you install a built-up felt membrane waterproofing system. Use creosote to prime for pitch, and cutback asphalt to prime for asphalt. Apply the primer coat by spray, roller or brush. Apply asphalt primer at the rate of approximately 1 to $1\frac{1}{3}$ gallons per 100 square feet.

Following the primer coat, install the first and subsequent layers of felt into hot interply moppings of bitumen and coat the last layer of felt with hot bitumen. Then cover the entire membrane system with at least $\frac{1}{8}$-inch protection board embedded in the outer mopping of bitumen. That's shown in Figures 12-30 and 12-31. You can also install the protection board into a cold-applied adhesive.

Use dead level asphalt (Type I) or coal tar pitch on horizontal or vertical surfaces below grade. Use flat asphalt (Type II) below grade or above grade when temperatures don't exceed 125° F. Use steep asphalt (Type III) above grade on vertical surfaces exposed to direct sunlight. Due to the low softening point of dead-level and flat asphalt or pitch, shelter the finished membrane from the sun during hot weather.

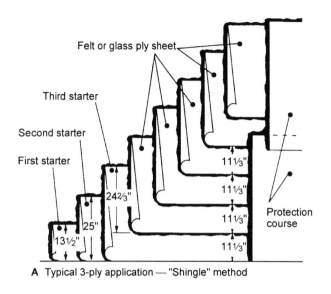

A Typical 3-ply application — "Shingle" method

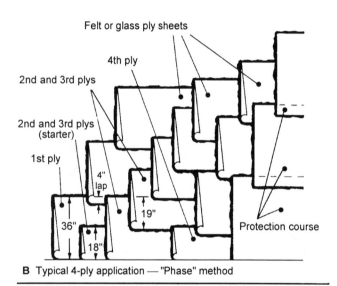

B Typical 4-ply application — "Phase" method

Figure 12-30 Installing built-up membranes

Fibered asphalt mastic contains cutback or emulsified asphalts and mineral fibers that are mixed either at the factory or within a spray nozzle during application. Use a trowel to apply factory-mixed fibered asphalt over a thinned asphaltic primer coat. To reinforce a fibered asphalt membrane coat, embed glass fabric in an asphalt undercoat before you install the membrane. Job-mixed fibered asphalt is especially practical when used over base felts on free-form roofs.

When you install organic felts, apply interply asphalt or pitch at the rate of 23 and 25 pounds per 100 square feet, respectively. With glass fiber felts, use interply asphalt or pitch at the rate of 25 pounds per 100 square

Roofing Construction & Estimating

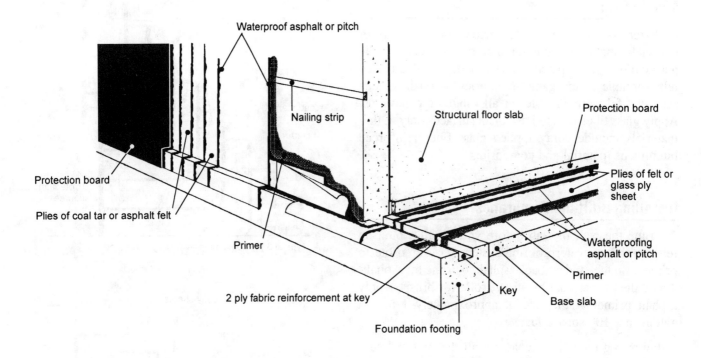

Figure 12-31 Typical slab and foundation waterproofing

feet. You can install cold-applied cutback or emulsified asphalt instead of hot bitumen in temperatures of 40° F and higher, and when dry weather is expected for 24 hours after application. Apply cutback or emulsified bitumens over a thinned primer coat.

In order to allow the waterproofing system to freely expand and contract and thus adapt to building settlement, install the first membrane over spots of cold-applied adhesive. I recommend you use waterproof fabric that's embedded in bitumen instead of felts, even though it's more expensive. However, on relatively flat surfaces (horizontal slope less than 1 in 12), saturated felts are more water-resistant than fabrics.

You can strike a compromise between expense and utility by installing a system composed of felts overlaid by an outer ply of fabric. Whether you use felts or fabrics, reinforce all inside and outside corners with at least two plies of membrane or fabric installed into hot moppings of bitumen. Install the reinforcement either as cushion strips extending 4 inches in each direction from the corner, or by lapping each membrane course (from both sides) around the corner 4 inches.

Built-up membrane waterproofing for slabs on grade is most often a 3-ply "shingle" method (Figure 12-30 A) or a 4-ply "phase" method (Figure 12-30 B). You can use these methods to waterproof foundation walls, also, but only with asphalt felts, not glass. Nail the membranes to the wall with a nailing strip near the top of each course to help hold them in place. That's shown in Figure 12-31.

When you install waterproofing to slabs above grade, prime the sub-slab and install a vapor barrier. Then install two plies of asphalt or glass felts overlaid by another vapor barrier and a protection board. Embed all membranes and protection board in moppings of hot bitumen. This is shown in Figure 12-32. Make sure the topping slab is poured within seven days after you finish the waterproofing system.

Estimating Built-up Waterproofing and Dampproofing

For a 1-ply waterproofing system, allow 12 percent waste for 4-inch side and end laps, and 18 percent for 6-inch laps. You don't have to lap the felts when you install two or more plies. With no laps, you don't have to allow for waste in your estimate because felts come in factory squares (108-square-foot rolls).

On basement walls, start the waterproofing at about 6 inches below the top of the lowest slab. At the top of the system, extend the bitumen moppings 2 inches above finish grade, and extend felts to 2 inches below the finish grade elevation. Use these upper and lower system boundaries when you estimate waterproofing or dampproofing materials.

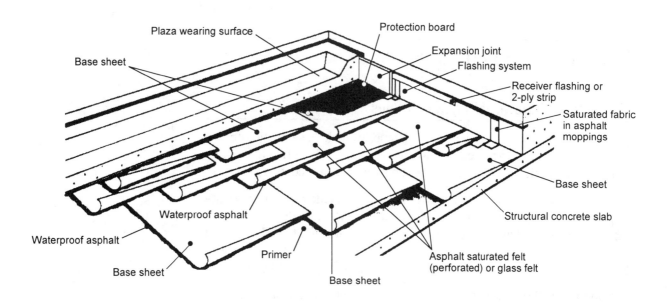

Figure 12-32 Waterproofing a slab above grade

	Head of water (feet)			
	1-3	4-10	11-25	26-50
No. plies felt or glass ply sheets	2	3	4	5
No. bitumen moppings*	3	4	5	6

*Asphalt approximately 23 pounds per mopping (25 pounds with glass ply sheets). Pitch approximately 25 pounds per mopping.

Figure 12-33 Plies of membrane required

Apply bitumen with a hand mop at the rate of about 63 pounds per 100 square feet. Use cold-applied asphalt only in areas where it would be hazardous to use hot material or where the coated area is too small to justify using a kettle. Apply cold asphalt by brush or spray at the rate of about 1½ gallons per 100 square feet.

The number of plies of membranes and bitumen moppings required for adequate waterproofing depends on the water pressure against the waterproofing system. Use Figure 12-33 as a guide to determine waterproofing requirements.

Other Types of Waterproofing

Elastomeric Waterproofing Membranes

The most common liquid-applied elastomerics include synthetic rubbers such as chlorophene (neoprene), polyurethane, and polysulfide (Thiokol). Liquid elastomerics must often be applied over a primer coat.

Cured preformed sheet elastomerics normally include butyl (isbutylene-isoprene), chlorophene, neoprene, ethylene propylene (EPDM) and urethane bitumen. Rubberized membrane sheets come in 40-, 30- and 22-gauge thicknesses. Sheets come in standard sizes up to 45' x 100' and larger. Plastic sheets come 0.004, 0.006 and 0.010 inch thick.

Butyl and EPDM elastomeric sheets are resistant to bacteria, fungi, soil acids, ozone and ultraviolet light. Neoprene elastomeric sheets have an additional resistance to oil and certain other chemicals.

You usually apply elastomeric sheets as a single-ply material, so installation goes faster than built-up applications.

Composites

Composites are metal foils or plastic films laminated to fabrics, paper, felts or rubberized asphalt of various types. Plated or coated metals are not considered composites although they produce similar results.

Liquid Plastic Waterproofing

Liquid plastic waterproofing consists of a powdered iron oxide compound which you trowel onto the surface of a concrete wall. Application thickness varies from 5/8 to 1 inch. Some surfaces, including spalling brick or concrete, require roughening with a bush hammer before you can apply the waterproofing.

Waterproofing Panels

Asphalt-coated protective board comes in 4' x 4' and 4' x 8' sheets in thicknesses of 1/4, 1/2, 3/4 and 1 inch.

Bentonite panels are very popular and cost-efficient waterproofing systems. Bentonite is a clay material that swells when wet, so panels are self-healing even when building settlement causes cracks. You can place the panels within the concrete forms. A biodegradable Kraft paper on the panels prevents them from sticking to the forms. This eliminates the need for the waterproof application after the forms are stripped. Install a protection board when backfilling with coarse material.

You can fasten panels over existing concrete walls with masonry nails. Lap panel edges at least 1½ inches and staple laps to prevent displacement during concrete placement. You can install this type of panel over vertical walls, under floor slabs and over below-grade roofs. Bentonite panels come in 4' x 4' sheets, 3/16, 3/8, 1/2 and 5/8 inch thick.

Dampproofing

Use dampproofing materials where there's no hydrostatic head, but where you want to prevent moisture from entering due to capillary action resulting from occasional exposure. Such conditions occur where there's good drainage or where the moisture content of the soil is low. The most common types of dampproofing include layers of bitumen, heavy asphaltic paint or other liquid water repellents. These coatings are sometimes called *hydrolithic coatings*.

In below-grade dampproofing, apply water-resistant coatings to exterior surfaces. The coating penetrates and seals pores to form a continuous protective film. Above grade, apply protective coatings over the inside surfaces of walls before plastering or tiling, or to the outside of back-up walls. Colorless sealers are available for coating exposed masonry surfaces.

Bituminous Dampproofing

The most common form of dampproofing consists of two or more coats of cold- or hot-applied asphalt or pitch installed over a primer coat. You can also use troweled-on hot fibrous asphalt. Install a layer of asphalt-saturated felt over the outer coat to protect the dampproofing from damage during backfilling. Some manufacturers recommend that you install a protection board over dampproofing materials, but don't count on dampproofing materials to act as an adhesive for the protection course. Install glass fabric to reinforce asphalt dampproofing applications when you expect ground movement or vibration.

You can use emulsified asphalts for dampproofing, but don't apply these products in cold weather (below 40° F) or when you expect rain within 24 hours after application. You can apply fibrated emulsion coatings over "green" or damp concrete, but apply cutback asphalts over dry surfaces only. Brush or spray emulsion coatings in two coats at an application rate of 1 to 3 gallons per 100 square feet, per coat, depending on the amount of dampproofing desired.

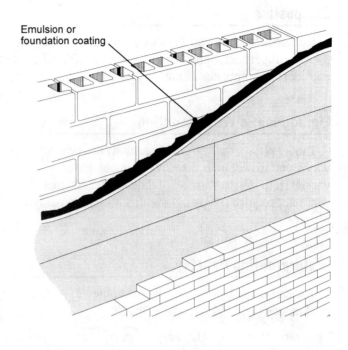

Figure 12-34 Dampproofing a cavity wall

You can also use emulsion coatings to dampproof masonry cavity walls as shown in Figure 12-34. You can use troweled bituminous mastic in cavity-wall construction or at foundation walls. Apply the mastic at the rate of 8 to 9 gallons per 100 square feet to provide an ⅛-inch coat (wet). When installed over foundation walls, cover the mastic with a protection board before backfilling.

Dampproofing and Waterproofing Existing Walls

Apply plaster bond paint to the inside or outside of an exterior basement wall, followed by plaster. You don't need lath when you use plaster bond paint. The number of coats and covering capacity of the paint will vary, depending on the porosity of the surface of the wall. Labor cost varies depending on the consistency of the paint, the condition of the wall and the application method.

You can apply waterproof wax into cut and repointed mortar joints in an existing masonry wall. The labor varies depending on the type of mortar originally used.

When you install certain types of waterproofing or dampproofing materials over a brick or block masonry wall, the wall is sometimes given a cement *parge coat*, ½-inch-thick cement plaster coat applied by plasterers. Follow the parge coat with a bituminous coating applied at the rate of 1 gallon per 100 square feet. Refer back to Figure 12-29. The parge coat takes about 1.9 bags of masonry cement and 0.23 tons of sand per 100 square feet of wall.

There are many specially formulated cement (hydrolithic) coatings now available that provide a waterproof envelope in interior applications. Hydrolithic coatings are formulated using various additives to improve different properties. These additives include:

- purified iron particles and various chemical agents that improve workability and oxidation.

- compounds that promote crystalline growth within concrete capillaries, where only the interior side of an existing building can be treated.

Some hydrolithic cement coatings are installed over layers of reinforced glass fiber fabric. Your budget, the architect, and the degree of protection you need determine the number of fabric layers required. Apply hydrolithic coatings to surfaces using a brush, roller or trowel.

Penetrating acrylic or quartz carbide sealers are available for sealing and protecting existing architectural concrete, masonry, block or stone. You can also use transparent silicone-based sealers over masonry surfaces to protect the surface and minimize efflorescence. They also help minimize damage due to freeze-thaw cycles.

There are sealers made of a mixture of organosilane and ethyl alcohol. This type of sealer is resistant to salt which contributes to concrete spalling and the corrosion of reinforcing steel.

Epoxy Waterproof Coatings

Epoxy coatings consist of a mixture of epoxy and emery. Use epoxy coatings on concrete floors to provide traction and to improve chemical and wear resistance. Likely candidates for this application include chemical, food, beverage and meat processing and industrial plants.

Coal tar epoxy coatings are available for application to floors and walls in sewer treatment facilities. You can also use epoxy materials which are made for patching, sealing and waterproofing cracks in existing concrete structures.

Now that we've covered all the information you need for roofing and waterproofing in new construction, let's see how to maintain and repair existing roofs. That's the subject of the next chapter.

13 Roofing Repair and Maintenance

▶ Leaks can occur almost anywhere on a roof, and for any number of reasons. Severe weather which brings rain, snow, wind, and hail is hard on roofs. Falling objects, foot traffic, and accumulated debris all take a toll. And some roofs leak because they're just plain worn out. Proper installation is the best prevention for many kinds of damage, and helps keep a roof sound over many years. Maintenance is the second best. Offer your customers periodic follow-up inspection and maintenance services, and you'll detect most roof damage before it becomes a major disaster.

People usually only notice a roof leak after water begins to drip through the ceiling. But the source of that leak is rarely directly above where you see the evidence. Often, water comes in under flashing or shingles, then runs over or below the sheathing and down a rafter. Sometimes it even runs the length of a rafter and into an outside wall. That kind of leak can go unnoticed until it does a lot of damage.

Finding the Source of Leaks

When you look for a leak on an asphalt shingle roof, start by looking for worn shingles with dark patches on them. This indicates the loss of surface granules. Also watch for curled or damaged shingles.

On wood shingle or shake roofs, look for missing or rotted shingles. Watch for cupped or curled shingles which could allow the entry of wind-driven rain under the shingle.

On built-up roofs, search for deterioration or delamination of felts, especially in low spots where ponds occur. Look for blisters, and inspect for loose or deteriorated flashing and gravel stops.

On tile roofs, watch for broken or missing tiles. Remove and replace broken tiles and replace missing ones.

On metal roofs, inspect for loose or damaged panels. Remove and replace panels which aren't sound.

On any roof with a valley, clean out all debris trapped in the valley. Search for worn or broken flashing material, or flashing material installed off-center.

Examine every roof for loose or missing ridge shingles. Those are more susceptible to loosening than field shingles. Around flashing and vent pipes, look for cracks and gaps in roofing cement. At chimneys, watch for cracked and loose mortar at the cap flashing.

Attic Inspection

When you're trying to locate a roof leak from the attic, start above the ceiling leak and follow water stains on the deck to the potential source. Darken the attic and watch for a ray of sunlight which could indicate a nail hole. Also, have a helper hose down the roof while you watch for a leak from inside the attic. If you find a nail hole, mark the location by pushing a nail or straightened coat hanger through the hole, where you'll be able to see it from atop the roof.

At Chimneys

Leaks often occur around chimneys when water runs down the flue or through the mortar on top of the chimney (the seal cap), detouring through an open joint and onto the sheathing. You can prevent this by installing a flue cap or by sealing the flue cap with roofing cement, followed by an application of felt painted with aluminum paint. Sometimes, a new layer of mortar is required at the top of the chimney. You can prevent water from entering mortar joints by tuck-pointing the joints and applying a brush coating of clear silicone sealer.

You can keep water from entering around flashing with a heavy application of plastic cement, followed by a strip of roofing felt painted with aluminum paint. Use roofing tape layered with asphalt cement to patch flashing and asphalt roofing materials. The tape is made from asphalt-saturated cotton, fiberglass, or some other porous fabric. Roofing tape comes in rolls up to 50 yards long and 4 to 36 inches wide.

Leaks occur around counterflashing during blowing rain because the counterflashing doesn't lap far enough over the base flashing. Loose mortar above or in flashing joints can also allow leaks at counterflashing. Scrape the joints clean to a depth of ½ inch and fill them with butyl rubber caulk or with fresh mortar.

At Eaves

Roof leaks occurring at the eaves of a roof are usually the result of a roofing contractor installing shingles with inadequate overhang. The end result is the rotting of fascia trim, roof deck and soffit. If the roof is framed with a close cornice, water can also seep in and damage the outside walls, floor covering and wood floors.

At Valleys

Figure 13-1 Roof penetration in a valley

Valleys are especially vulnerable to leakage. You can prevent most of these problems during installation. For instance, clip (*dub off*) sharp shingle corners in the valley. (Refer back to Chapters 4 and 6 for details about how to shingle a valley.) If you don't cut the corners, any water flowing along the edge of the shingle can hit a sharp corner and run along the top of the shingle until it finds a place to leak. The location of the leak is often far from the valley, so it's hard to find and repair. That's why it's sometimes a waste of time to try to find a leak by searching the attic for evidence of water entry.

Other valley problems that result from poor installation include:

- roof penetrations in a valley, as shown in Figure 13-1
- valley flashing installed off-center
- full-lace valley with incorrect installation of overlapping shingles

Leaves and other debris collecting in a valley or gutter can also cause a valley to leak. Water dams up and backs up under shingles adjacent to the valley. Also, debris retains moisture, which promotes the growth of fungus and mildew. Either will degrade shingles and metal. You can prevent this by periodically cleaning the valley and gutter, especially during autumn when leaves fall.

To repair a break in an asphalt shingle valley flashed with roll roofing, apply a coating of roofing cement. Repair an opening in a metal valley by embedding aluminum patching tape into roofing cement. To prevent potential leaks in valleys, embed the corners of all shingles within a valley in roofing cement, then caulk the edges of the shingles.

A valley must sometimes be replaced because of hail damage or sloppy installation. If the valley is 90-pound roll roofing, leave the old valley intact and just add to it. Apply a ¼-inch-thick layer of roofing cement, then slide a new valley strip over the old valley and under loosened shingles adjacent to the valley, starting at the bottom of the valley. Renail the loose shingles and replace any missing ones.

At Vent Flashings

Leaks around vent flashings are very common and are usually the result of improper shingling around the flashing. Install shingles *under* the lower end of the flashing instead of on top. Otherwise, water can be diverted under the shingles. Also, dirt and granules can collect under the tabs, leading to rusting and holes in the metal flashing.

Leaks at vent pipes can also result from roofing cement that has dried and cracked. Remove all loose material and reapply roofing cement.

On tile roofs, leaks at vent pipes are usually caused by mortar that's cracked around the base of the vent. As a temporary remedy, apply roofing cement over the problem area. However, a long-term solution is to seal the base of the vent with new mortar.

Miscellaneous Leaks

When a leak occurs at mid-roof, the cause is often an exposed nail hole, a joint occurring in a low spot or a joint in the shingles where the exposure is greater than 5 inches. One type of leak that's easy to find from inside the attic is one caused by nails driven through the shingles and deck where the homeowner installed Christmas decorations or antenna guy wires on the roof.

One big source of leaks and roof damage is low tree branches that sweep across the roof during high winds. The damage is even greater when branches sag under the weight of snow or rain.

Repairing Leaks

Asphalt Shingle Repair

You can spot repair minor cracks or worn areas on asphalt shingles by applying roofing cement and sprinkling on loose granules and rubbing them into the cement. Either buy tubes of granules from a roofing supplier, or make your own by rubbing two shingles together. If appearance isn't important, you can leave off the mineral granules.

If an asphalt shingle is curled, apply a coating of roofing cement beneath the shingle, and press the shingle into the cement. If you have to use nails to keep the shingle in place, cover the nail holes with roofing cement.

If only a single tab is damaged, raise the tab with a flat pry bar and remove the nails. You might also have to loosen the shingle above the damaged tab. Cut out the damaged tab and nail down a replacement. You can buy special asphalt plastic roofing cement for sealing and patching cracks and breaks in roofing. It comes in 1-quart or 1-gallon cans, or in 5-gallon pails. Apply this cement over nail holes and the cut joint.

You can use this same procedure to replace an entire 3-tab shingle. I recommend that you replace the entire shingle, never just a single tab. Either use a hacksaw to cut nails which hold shingles immediately above the damaged shingle, or notch the top of the new shingle to fit around the nails. Cement the new shingle in place. If you notch the shingle, cement the overlying shingle on top of it.

You can also repair a damaged shingle by installing galvanized sheet metal beneath the shingle and nailing the metal into a bed of roofing cement. Cover the nail heads with plastic cement.

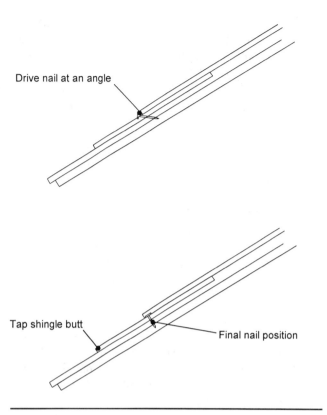

Figure 13-2 Replacing a wood shingle

Warm weather is the best time to repair asphalt shingles. When they're warm, asphalt shingles become pliable, so you're less likely to damage them. Never apply plastic cement to the bottom edge of a shingle to seal it, since this will allow water to be trapped and back up to cause a leak.

Repairing Wood Shingles and Shakes

You can patch small cracks in wood shingles with roofing cement. Small cracks in shakes are rarely a source for roof leaks since the underlayment (not the shake) provides the primary weather protection.

To replace a wood shingle or shake, use a hacksaw to cut the nails that hold it in place. Then you can usually pull it out by hand. Trim a new shingle to size, push it into place and nail it at an angle, just below the overlying shingle. Tap the butt of the new shingle to align it with the existing shingles and hide the nail. See Figure 13-2. Cover all exposed nail heads with roofing cement.

The most common leak on a wood shingle roof occurs where joints or splits line up in three adjacent courses. Splits in an individual wood shingle can also be a source of a leak. The easiest way to repair leaks caused by splits or joints is to install a flat piece of galvanized metal under the joints and splits. To keep the metal sheet in place, make a 90-degree bend in the bottom edge, just enough to give it friction to stay where you put it — about ½ inch. Don't nail the metal because the nails will gradually work loose and leave another opening for water to get in.

Wind-blown rain can enter under bowed or cupped wood shingles or shakes. To eliminate this problem, split the shingle with a chisel and remove a ¼-inch-wide sliver, then nail down the two resulting shingles with a nail on each side of the split. Cover the nail heads with roofing cement and drive the nails as close to the top edge of the shingle as possible. Don't nail the shingle near the butt because that causes the shingle to raise at the tapered end and dislodge the shingle above it.

Slate Repairs

To replace a broken slate, insert the new slate underneath the two overlying courses and nail it into place. See Figure 13-3. Use these nail location guidelines to place nails correctly:

- 5 inches down from the head of the *new* slate

- 2 inches down from the butt of the *second* overlying course

- through the vertical joint between slates in the *first* overlying course

In Figure 13-3, the nail is covered with a 3-inch by 8-inch copper strip. This metal strip is bent so that it acts like a flat spring to hold the new slate tightly in place beneath the overlying courses.

Be sure to use a slate that's compatible with the original roof. Match the color as closely as possible. Use a slate with the same weathering properties (weathering vs. unfading) and surface texture as the existing slates. If you're able to identify the quarry that made the original slates and they're still operating, order the new and replacement slates from the same place.

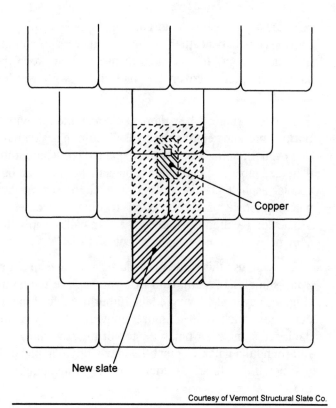

Courtesy of Vermont Structural Slate Co.

Figure 13-3 Proper method of inserting a new slate

Roof Maintenance

Encourage your customers to regularly inspect and maintain their roofs (or hire you to do it). Here are some tips to help make a roof look good, perform well and last for years:

- Never paint or coat a roof to change the color.

- Keep roof surfaces and gutters clean so water will drain quickly and freely.

- Never let water from a downspout pour directly onto a roof below.

- Keep trees trimmed so they don't rub against the roof covering. Also, keep climbing plants trimmed back from the roof.

- Remove snow and ice from the roof carefully so you don't damage the roof. Use a broom with an extension pole. Never climb onto a wet or snow-covered roof.

- Being walked on is never good for a roof. Keep it to a minimum. Remind service people to be careful and watch them to make sure they do so.

- Make sure that accessories such as antenna wire or anchors are made of noncorrosive material to prevent metal discoloration or "iron staining" on the roof.

- Never pressure-clean an asphalt shingle roof. The pressure will remove granules and each treatment will take three years off the life of the roof.

Algae Discoloration

The discoloration of asphalt roofing materials due to algae is a common problem in humid parts of the country. You can see evidence of this condition (often called *fungus growth*) in Figure 13-4. If left unchecked, algae can turn a light-colored asphalt shingle roof to dark brown or black. Algae also discolors other roof coverings, including wood shingles, shakes, tile, and built-up roofs. In humid areas, consider using one of the wide variety of algae-resistant roofing products now available.

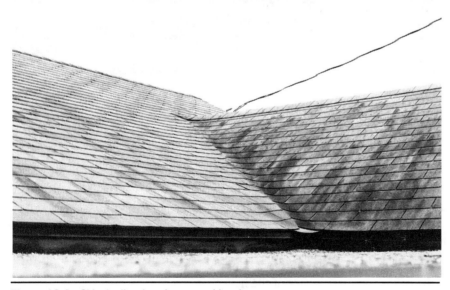

Figure 13-4 Shingle discoloration caused by algae

A discolored roof can be lightened, at least temporarily, by bleaching. First, cover any plants that may be affected by the run-off. Sponge or fog spray a dilute mixture of chlorine bleach (about 50/50) onto the roof surface, then immediately rinse it down with clear water.

Efflorescence on Concrete Tile Roofs

Efflorescence (or *lime bloom*) sometimes occurs on concrete products. The free lime in concrete goes through various chemical reactions during manufacture, resulting in a white chalky deposit on the surface of the tile. This deposit is only temporary. Within a year, rainwater dissolves and carries away the deposit, restoring the tiles to an even appearance. Efflorescence is purely superficial and in no way affects the strength or other functional properties of the roof tile.

Assessing Hail Damage

A common cause of homeowners' insurance claims is hail damage to a roof. If the roof is actually damaged, it should be repaired or replaced. But some homeowners assume that they're entitled to a new roof after any hailstorm, whether the roof has been damaged or not. If the homeowner pushes the issue to the point of filing a lawsuit, the insurance company will often replace the roof to avoid a trial, since juries are seldom sympathetic to an insurance company.

The sad result, however, is that we all help pay to replace the undamaged roof. Insurance companies base their premiums on the cost of doing business in a local area. If there is an inordinate number of claims they have to pay out on, premiums will go up.

Here is a guide to assessing hail damage that's fair to both the homeowner and the insurance company.

Hail Damage to Asphalt Shingles

Hail damage to asphalt shingles leaves such obvious signs that a simple visual inspection of the roof lets you determine whether a roof is damaged or not. Here's what to look for:

- Dark spots on the shingles' surface. These spots mark hail impact points and they look dark because the impact pops the mineral granules loose from the shingles' surface. Spotting is easy to see on light colored shingles, but it's not always obvious on darker color shingles.

- Pitting that's visible on the shingles' surface. The pit locations correspond to the dark spots because they are also a result of the hail impact. Pitting is visible on shingles no matter what color they are.

Roofing Repair & Maintenance

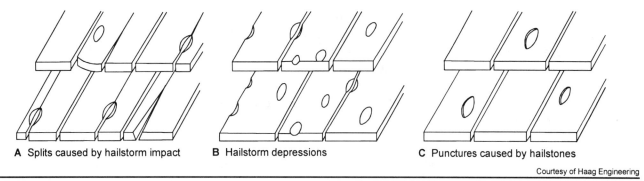

A Splits caused by hailstorm impact **B** Hailstorm depressions **C** Punctures caused by hailstones

Courtesy of Haag Engineering

Figure 13-5 Shingle damage caused by hail

- When touched, the pits and dark spots feel soft, like a bruise or soft spot on an apple.

Hail damage is basically cosmetic, although it can shorten the life of the roof if it's severe. Some insurance companies pay a percentage of the total value of the roof for repair.

Hail Damage to Wood Roofs

Hailstones vary in size, shape and hardness, so impact marks vary in size, shape and depth. Wind-driven hailstones do the most damage, so the windward side of a roof sustains more damage than the leeward side. Examine roof penetrations such as roof vents, soil stacks and heater flues to see an indication of the nature of the hailstones. If thin-sheet components aren't dented, the hail probably didn't damage the shingles or shakes.

Hail damage (versus normal weathering) on a wood shingle or shake roof isn't always easy to identify. To arrive at some uniform way to assess hail damage on wood roofs, Haag Engineering Company of Dallas, Texas has cataloged damage data by observing shingled test panels exposed to various simulated hailstorms. Testing has been conducted since 1963.

One important observation in the Haag study is that flatgrain and slashgrain shingles tend to split and warp more than edgegrain shingles. Curled or warped shingles are more apt to be split by hail impact than flat ones. Hailstones don't significantly affect sound shingles. And edgegrain shingles aren't as likely to split, because they don't warp as much. But when shingles do split, the characteristics of the split are the same for flatgrain, slashgrain, or edgegrain shingles.

When hail causes splits, the splits nearly always match impact marks, and the exposed surface of the split is the color of fresh wood throughout. See Figure 13-5 A. The exception is hailstone impact against a warped shingle, which sometimes causes a split slightly offset from the impact mark. In this case, thumb pressure in the hail-caused dent opens the split to reveal fresh-colored wood.

Assess roof damage as soon as possible after a storm because hail-caused splits are most obvious then. Impact marks fade as surface wood fibers recover. Inspect the roof when it's dry because it's easier to see new splits then. That's because wet wood swells, so the splits tend to close up. Note that splits caused by hail occur at the moment of impact, and hail-caused depressions which don't cause splitting at the time will not cause future splitting (Figure 13-5 B).

Hailstones can sometimes actually puncture a shingle. Punctures are usually rounded and the exposed underlying wood has a fresh wood color (Figure 13-5 C).

■ **Natural Weathering** After 10 years' weather exposure, about a third of all edgegrain shingles, and about two-thirds of slashgrain and flatgrain shingles, split naturally. These splits usually begin, and are widest, at the butt of a shingle, though some splits originate at nail holes.

Natural splits normally have rounded edges and eroded interiors. These splits usually have a V-shaped cross-section and don't extend the full depth of the shingle. Wood near the bottom of the split often has a fresher wood color because it hasn't been exposed to the weather for as long.

The extent of weathering depends on how steep the roof slope is as well as the roof slope direction. South-facing shingles installed on a low roof slope will weather fastest and the most.

■ **Repairing Wood Roofs** Shingles with hail-caused splits which aren't aligned with joints or weathered splits normally need not be repaired. You can repair a wood shingle or shake by replacing it, or by under-shimming the damaged shingle or shake with a galvanized steel sheet, aluminum sheet or a roofing felt shim. A roof is considered beyond economical repair when the repair cost exceeds 80 percent of the replacement cost. This equates roughly to 30 hail-caused splits per square for cedar shingles, and 25 hail-caused splits per square for cedar shakes.

It isn't practical to repair old, badly-weathered roofs. In general, it's better to replace shingle roofs older than 20 years, shake roofs older than 25 years, and roofs damaged by severe weather conditions.

Roofing Demolition

If a tear-off is necessary, strip shingles down to the deck. When you remove any roof, start the tear-off at the ridge and work down the slope, as shown in Figure 13-6. This method is especially necessary when you remove wood shingles or shakes installed over spaced sheathing. Otherwise, broken material will fall through the open sheathing into the attic.

Roofing Repair & Maintenance

Figure 13-6 Remove shingles starting at the ridge

Dislodge wood shingles, shakes, slate and metal roofing panels with a crowbar, removing two to three courses at a time. Remove tile shingles by hand. When you remove asphalt shingles, roll roofing or a built-up roof, use a flat spade shovel to pry up material. File a notch into the blade of the shovel so you can remove shingle nails along with the shingles. If you don't remove the nails, parts of the old shingles will remain under the nails and make the deck surface uneven and bumpy. And the old nails will pop up through new underlayment unless you use roll roofing for underlayment.

If the shingles don't fall apart when you remove them, stack them into bundles for removal. If the shingles are old and brittle, collect and remove them with a scoop shovel and carry them to the edge of the roof with a wheelbarrow.

Some roofers prefer to take off the paper and asphalt shingles all at the same time. Then they can roll the shingles into the paper and roll the whole lot down the roof as a unit. This method reduces the amount of mess left to clean up at the end.

If you have to tear off more than one layer of shingles, remove only one layer at a time. Use a spud or a spud hammer (Figure 13-7) to remove a gravel roof. Spread tarps along the ground under the edges of the roof to help collect debris falling off the roof.

Figure 13-7 Spud hammer

391

If you must remove siding during a tear-off, for instance at a dormer wall, mark and store the pieces in an ordered way so you can reinstall each piece in its original location. Use existing nail holes as a guide when replacing the material.

After you've removed the old shingles and underlayment, level and repair the deck, if necessary. Replace rotted, damaged and warped sheathing, or delaminated plywood. Cover all large cracks, knotholes, loose knots and resinous areas with sheet metal patches nailed to the deck. Remove all loose or protruding nails or hammer them into the deck. Some roofing contractors recommend that gaps between spaced sheathing be filled in with boards, or that the entire roof area be covered with plywood. You don't necessarily need a solid deck under new shakes, and in humid climates it's a good idea to keep the spaced sheathing.

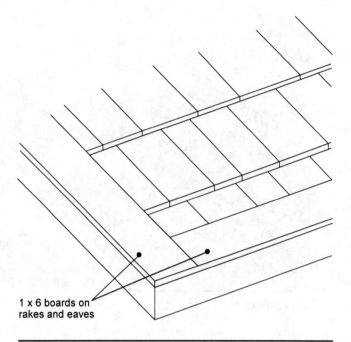

Figure 13-8 New edging strips

If the fascia board or trim is broken or rotted, or if the drip edge is broken or rusted, now is the time to replace them. To install a new drip edge when you're re-roofing over existing shingles, trim the old shingles back from the edge of the roof. If the old roof is wood shingle, cut the shingles back from the eaves and rakes 5½ inches and install a 1 x 6 in their place over the new drip edge. That's shown in Figure 13-8. Also, remove the old hip and ridge units. If there's a concealed metal cover underneath, don't attempt to remove it because it would probably take a lot of shingles with it. Instead, look ahead to Figure 13-13 to see how to install beveled siding over each side of the ridge.

Here's a list of pointers for various re-roofing situations:

- When you re-roof a roof that has ribbon courses, remove the entire top shingle (not just the exposed part) of each ribbon course by hand to avoid humps on the new roof.

- On asphalt shingle re-roofs, old flashings can often be raised. When this isn't possible, cut new shingles to fit around the old flashing, and seal the flashing with plastic cement. Paint the cement and vent with aluminum paint.

- When you tear off a gravel roof, you'll usually have to replace the vent flashings. New metal flashings and valleys are a must when you re-roof with wood shingles or shakes. That's because the bond between wood and plastic cement is only temporary, since wood will soak up water and release the plastic cement.

- Carefully remove any metal flashing demolished during a roof tear-off and use it as a pattern for new flashing.

Roofing Repair & Maintenance

- If metal counterflashing at the chimney and other vertical surfaces hasn't deteriorated, try to temporarily bend it up and out of your way and reuse it.

- Use a flat pry bar to remove a gutter and reassemble it on the ground the same way it was installed at the eaves. When you reinstall the gutter, drive the spikes into fresh wood, instead of into the old hole.

- If the roof deck is warped, but otherwise in good condition, straighten the deck by driving nails through the old shingles into the rafters.

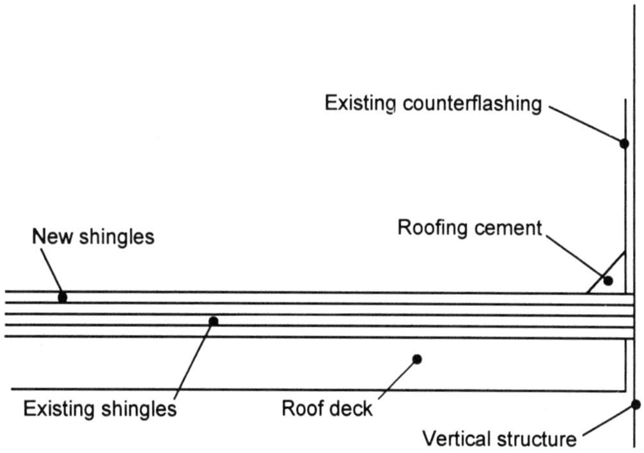

New Flashing

On a re-roof, try to install new flashing under existing metal counterflashing if you can. But sometimes that's impossible. Then you'll have to replace both the flashing and the counterflashing. That adds considerably to the cost. Here's a relatively inexpensive alternative. Seal the top and outside surfaces of the new metal flashing with roofing cement (also referred to as *bull*), then lay in a 4-inch-wide strip of felt and smooth it into the roofing cement. Then coat the felt with a $1/8$-inch-thick layer of roofing cement and spray the coating with a good grade of aluminum paint.

Figure 13-9 Sealing a roof-wall juncture with plastic cement

If you don't install new metal flashing, seal the juncture between the horizontal and vertical surfaces with a $1 1/2$-inch-wide bead of roofing cement applied at a 45-degree angle with respect to both surfaces. See Figure 13-9. Cover the joint with aluminum paint for longer-lasting protection.

When it's not economically feasible to install new counterflashing at a chimney on a wood shingle or shake roof, you must install step flashing. On a shake roof, install metal flashing 13 inches long and extend it 4 inches up the wall and 4 inches under the shingle. On a wood shingle roof, install metal flashing 7 inches long and extend it 2 inches up the wall and 3 inches under the shingle. Reinforce the juncture with a felt strip, then cement and paint the joint as described above.

When you install flashing against a vertical side wall using asphalt shingles over old asphalt shingles, terminate the new shingles within $1/4$ inch of the existing flashing and embed the shingle ends in a 3-inch-wide bed of roofing cement. Then seal the joint with a bead of cement.

When you install flashing against a vertical side wall using asphalt shingles over old wood shingles, nail an 8-inch-wide strip of 50-pound smooth roll roofing over the old roof at the juncture of the shingles and the vertical wall. Drive nails in two rows on 4-inch centers along both edges

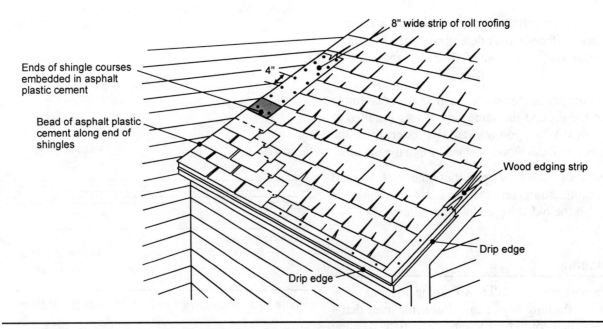

Figure 13-10 Flashing against a vertical wall when re-roofing over wood shingles

of the strip. Always use roofing nails when you re-roof over an existing roof. Embed the ends of new shingles in roofing cement applied to the strip, then seal the end joint with a bead of roofing cement, as in Figure 13-10.

When you install a metal valley on a shingle re-roof, don't use water guards at the metal edges because it won't lie flat enough to produce a smooth roof.

If an existing asphalt shingle roof has an open valley, build up the exposed part of the existing valley to the level of the existing shingles by installing a 90-pound mineral-surfaced roll roofing filler strip. Then you can install a new open valley, woven valley or a closed-cut valley.

Re-Roofing

Whether or not you must tear off the existing roof covering depends on the type of roof or the number of layers or weight of the existing roof covering. Re-roofing with no tear-off obviously takes less time and no underlayment is required between roofs. Also, the existing roof provides additional protection and insulation.

Three layers of shingles is usually the most a roof can support. Most building inspectors as well as many insurance companies follow this rule. However, be cautious about relying too much on that three-layer guideline. There are plenty of roofs that aren't structurally sound enough to carry the load of *two* layers of shingles, let alone three. The load-bearing capacity of a roof depends on many factors, including the rafter size, strength and spacing and the strength of the roof sheathing material. Also, remember that the per square weight of shingles varies not only with shingle type, but also with the exposure amount (this is especially true of wood shingles and shakes). Finally, keep in mind the fact that wood shingles and shakes absorb water and thus weigh far more wet than dry. If you have any doubt about the ability of a particular roof to support another layer, consult a structural engineer. Of course, it might be cheaper to just tear off the existing roof, especially since the engineer might say to tear it off anyway.

Incompatible Substrates

The existing roof covering determines not only whether or not tear-off is necessary, but also what roof covering options are available when no tear-off is required.

You must completely remove certain types of shingle roofs before you install a new roof covering. Because of the irregular surface, never attempt to install new shingles over a shake roof. Never install new shingles over tiles, slates or metal panels because it's too hard to nail down the new roof. A built-up roof with an aggregate surface isn't often resurfaced with shingles due to its rough surface. Also, the roof slope is usually too low for most shingles and the combined weight of aggregate, and new roofing is often too much for the roof frame. *Always* ask a structural engineer before you take on a job like this.

Asphalt Strip Shingles over Asphalt Shingles

You can install new asphalt strip shingles over old ones if you first smooth the substrate by nailing down warped shingles and replacing missing ones. Split a warped shingle and nail down the resulting two shingles. A new asphalt shingle roof will sag over missing shingles.

When you re-roof with asphalt shingles over asphalt shingles, the easiest way is to match the existing shingle pattern. Don't install asphalt strip shingles over dissimilar types of asphalt shingles, such as T-lock, giant individual and hexagonal shingles.

When you re-roof over asphalt shingles, no underlayment is required because the original roof serves that purpose. When you re-roof over existing wood shingles, some building codes require that you install 30-pound felt between the existing and new roofs.

The "Butting-Up" Method

When re-roofing over wood, asphalt shingles or roll roofing, install the new roof using the "butt-up" (or "butt and run") method. This means you install the tops of the new shingles flush against the butts of the old shingles,

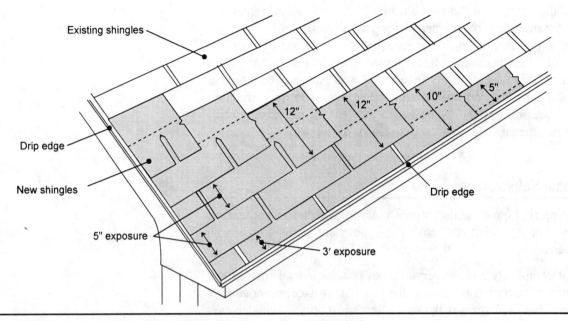

Figure 13-11 Application of new asphalt shingles over existing asphalt shingles

as shown in Figure 13-11. I recommend that you use this method because it's the easiest way to avoid all the problems associated with excessive shingle buildup. A re-roof with excess shingle buildup is unattractive. Furthermore, such a roof is more easily damaged by hail and foot traffic.

For these same reasons, it's equally important to offset overlaps when you re-roof with T-lock shingles over T-lock shingles. (Although T-lock shingles are practically never seen in the western U.S., they're still used in the Southeast.) This also applies to re-roofing with roll roofing over roll roofing.

When re-roofing using three-tab or strip shingles using the butt-up method, you can install the starter course one of two ways. The preferred method is to cut the shingle tabs and tops to make a strip whose width is equal to the exposure of the old shingles (normally 5 inches). Install the resulting strip over the exposed part of the first row of shingles of the old roof. If you're using self-sealing shingles, locate the factory-applied adhesive strips adjacent to the eaves. Remove about 3 inches from the end of the first starter-course shingle to prevent the cutouts and joints of the first course of shingles from being aligned over the joints of the existing starter-course shingles.

Overlay the new starter course by a first course of 10-inch-wide shingles made by cutting 2 inches from the shingle tops. This course will cover the new 5-inch starter course plus the 5-inch exposed part of the second row of shingles of the old roof. That's also shown in Figure 13-11. An easier, but less desirable, method is to cut and install two rows of 10-inch-wide shingles for both the starter course and first shingle course. You'll get a bulge along the bottom of the original second course of shingles. It's also less wind-resistant along the eaves. In either case, apply a spot of roofing cement under each tab of the first course of shingles for added wind resistance.

Install succeeding shingle courses using full-width shingles with their heads butted up against the butts of the old shingles. The full-width shingles will be 2 inches lower than those installed on the old roof. The exposure of the first shingle course is 3 inches and that of the succeeding courses is 5 inches. The difference in exposure isn't apparent, especially if gutters are installed at the eaves.

If the exposure of the old roof is greater than 5 inches or if the old roof is crooked horizontally, remove the old shingles. If the exposure is less than 5 inches, the quantity of new shingles required will be greater than that required for a standard 5-inch exposure roof. Here's Equation 4-2 from Chapter 4:

$$\text{Percentage-of-Increase Factor} = \frac{\text{Recommended Exposure}}{\text{Actual Exposure}}$$

Use this formula to find out how many shingles you'll need. For example, a 4-inch-exposure roof will require:

$$\text{Percentage-of-Increase Factor} = \frac{5 \text{ in.}}{4 \text{ in.}} = 1.25$$

That's 25 percent more shingles. Using the same formula, a 4½-inch-exposure roof will require 11 percent more shingles.

▼ **Example 13-1:** A 30-square re-roof is required over strip shingles with a 4-inch exposure. Assuming that the recommended exposure is 5 inches, find the number of squares of shingles required to cover the roof with strip shingles with a 4-inch exposure.

From the formula above, the Percentage-of-Increase Factor is 25 percent. So you'll need 30 squares x 1.25, or 37.5 squares of shingles.

Asphalt Shingles over Wood Shingles

If wood shingles don't provide a good nailing surface for asphalt strip shingles, but are otherwise in good condition, smooth and improve the surface by installing beveled 1 x 4s or 1 x 6s ("horse feathers" or feathering strips) against the old shingle butts. That's shown in Figure 13-12. Also, put a beveled siding board on each side of the ridge (Figure 13-13).

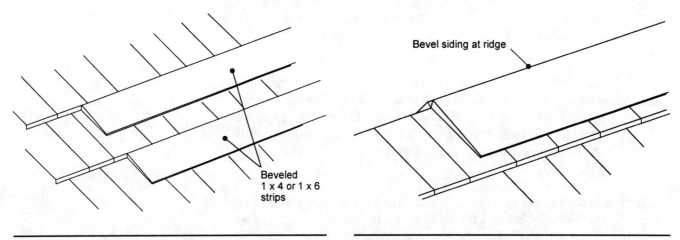

Figure 13-12 "Horse feathers" or feathering strips

Figure 13-13 Beveled siding board at ridge

If shingles and trim at the eaves and rake are badly weathered in areas subject to high winds, you'll also need to trim the wood shingles back from the eaves and rake so you can install 1 x 4 or 1 x 6 edging strips (see Figure 13-8). Install a new drip edge at the same time. This lumber provides a smooth surface as well as a nailing surface for the new roof.

Asphalt Shingles over Built-up or Roll Roofing

You can install asphalt shingles over a built-up roof after you scrape off the aggregate surface, and provided the roof slope is at least 2 in 12. There's no need to reseal the BUR if the felts are in good shape. If there's rigid insulation between the sheathing and the felts, install a plywood nailing substrate over the insulation before you apply new shingles.

You can install asphalt strip shingles over roll roofing, provided the old surface is smooth and the roof slope is at least 2 in 12. Split any buckles or blisters and nail the roll roofing flat against the sheathing. Nail down any lapped joints that have separated.

Installing T-Lock Shingles over an Existing Roof

Use T-lock shingles over a rough roof and you won't have to tear off the old roof. Start the roof by installing a starter strip made from 9 inches of the top part of the shingles, as shown in Figure 13-14, or by installing a 9-inch-wide starter roll (Figure 13-15). Make the first shingle course by cutting the tabs from whole shingles (Figure 13-16). For succeeding courses, install whole shingles (Figure 13-17). You can salvage the tabs you cut off of the starter course to finish the roof directly beneath each side of the ridge.

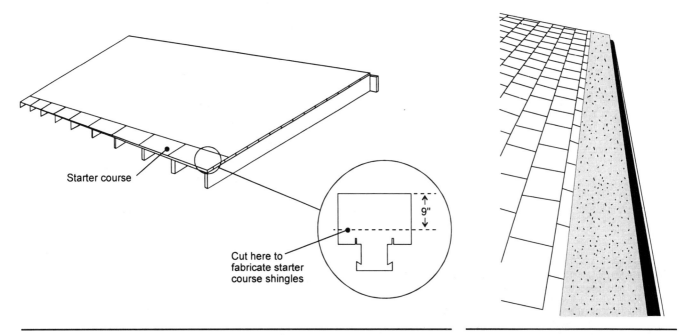

Figure 13-14 Fabricating the starter course from T-lock shingles

Figure 13-15 Prefabricated starter roll

To install a smooth T-lock shingle re-roof over an existing T-lock roof, try to match the new shingles and old shingles. Also, install shingles of the new roof ½ inch lower than the old shingles to avoid an excessive buildup of shingles. The simplest way to do this is to start the first shingle course with an additional ½ inch overhang that you can trim off later.

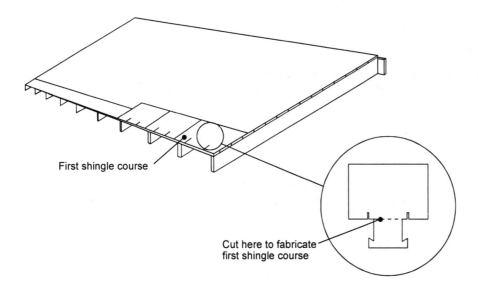

Figure 13-16 Installing the first course

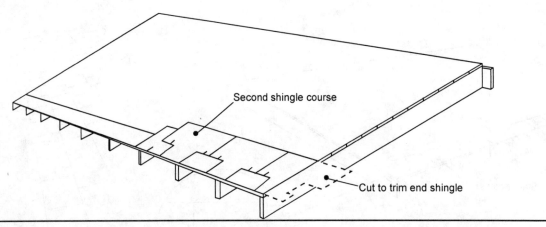

Figure 13-17 Installing subsequent courses

Re-roofing with Wood Shingles and Shakes

You can apply wood shingles over any type of asphalt shingle, roll roofing or smooth-surface built-up roofing, provided the roof has a slope of at least 3 in 12, and the substrate is capable of holding nails. You can install wood shingles over an aggregate-surfaced roof as long as the shingles are installed over spaced sheathing applied to the top of the old roof, as in Figure 13-18. In addition to spaced sheathing, you'll need to install a 1 x 6 along the eaves, rakes and on each side of the ridge. Also install 1 x 4s along the edges of the valleys in order to receive new metal valley flashing. Spaced sheathing has the added advantage of providing good air circulation.

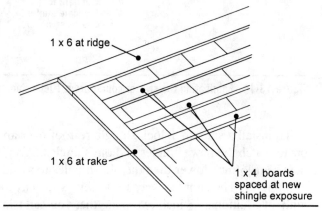

Figure 13-18 Spaced sheathing over an existing roof

You can apply new wood shingles over old wood shingles. If the surface is rough, install feathering strips against the old shingle butts (Figure 13-12).

You can install shakes over any type of asphalt shingles, wood shingles, roll roofing or smooth-surface built-up roofing that's not leaking, provided the roof is slope is at least 4 in 12.

Re-Roofing with Roll Roofing

You can install roll roofing over roll roofing, provided the old surface is smooth. Never install roll roofing over any type of asphalt shingle or over an aggregate-surfaced built-up roof.

Re-Roofing with Metal Panels

You can install metal panels over all types of asphalt shingles, wood shingles, roll roofing and built-up roofs, provided that the roof has an adequate slope. Check the manufacturer's specifications.

Re-Roofing with Tiles

You can install roofing tile over all types of asphalt shingles, roll roofing, and built-up roofs if the old roof is fairly even or can be made even economically. Also, the roof must have an adequate slope, and the roof frame and deck must be designed to carry the load. If in doubt, consult an engineer.

Re-Roofing with Slate

You can install slate over all types of asphalt shingles, roll roofing, wood shingles and built-up roofs. The roof slope must be at least 4 in 12 (unless flat-roof construction is desired) and the roof frame and sheathing must be strong enough to carry the added load.

Slates used to re-roof over wood shingles must be long enough to span two courses of wood shingles in each course. Spanning two courses of wood shingles provides the slates with two points of support. That is, each slate will rest on two wood shingle butts. A shorter slate, spanning a single course of wood shingles, has a single point of support and wouldn't lie well. The slates you use for this re-roof job measure at least 18 inches long. You install them with four punched nail holes instead of the usual two. Although their length makes these slates heavier and more difficult to handle, their only other disadvantage is a tendency to break under foot traffic.

Estimating Re-Roofing

The material quantities required for a re-roof are about the same as those required for a new roof. When you re-roof (as compared to roofing over a new wood deck), there's more labor involved due to the demolition and shingle trimming required. But this may be offset by the fact that it's usually not necessary to install felt under a re-roof.

When you estimate a re-roof, or tear-off and re-roof project, work out your estimate while you're on the roof. That way you can assess the old roof, including the condition and number of layers of shingles. Remember that it's easier to remove nails than staples, so notice how the old roof was attached. You can also check the condition of framing members, sheathing, fascia, drip edge, vents, vent caps and flashing. If the deck has deteriorated to the extent that it will no longer hold a nail, you'll have to replace that as well.

When you're going to tear off a roof, it's sometimes very hard to judge ahead of time how much sheathing you'll have to replace. This is especially true at the eaves, since you can't see that area from inside the attic. Estimate those repairs on a "cost-plus" basis.

You can often spot rotten deck areas by walking over the roof. A rotted deck will sag under your weight and will feel bouncy as you walk over it. That will be most apparent near the ridge, or on the north side of a house where the sun doesn't shine as directly.

Attic Ventilation

Insulation, weatherstripping and caulking make a home more airtight and confine water vapor in the house. Eventually, most of the water vapor passes through the ceiling and accumulates in the attic. In cold weather, in a poorly-ventilated attic, this warm moist air condenses when it reaches the cold underside of the roof sheathing. In hot weather, hot moist air tries to escape through gaps in the roof covering. The end result can be a buckled and rotten deck, deteriorated underlayment and blistered shingles. Many re-roofing jobs result from inadequate attic ventilation.

With proper ventilation, air circulates freely throughout the attic carrying away the water vapor before it can condense. Never cover vent openings during cold weather, and make sure that soffit vents are not blocked by insulation.

Putting a new roof on a building which lacks proper ventilation is like painting over a rotten board. Call the owner's attention to any ventilation problem. That will save you money in the long run because you'll avoid expensive call-backs. It will also establish your reputation for know-how and quality workmanship. And besides that, a ventilation problem may void the shingle manufacturer's warranty.

A rule of thumb for adequate attic ventilation is to allow 1 square foot of net free vent opening for:

- each 150 square feet of horizontal attic surface when the roof has no vapor barrier; or

- each 300 square feet of horizontal attic surface when the roof has a vapor barrier or when half the vent openings are located at the ridge and the other half are located along the eaves.

When you calculate the net free vent opening, be sure that the vent screen is taken into consideration, because screens vastly reduce the free area of the vent.

The gross ventilation area depends on the type of vent openings, as shown in Figure 13-19.

Obstructions in ventilators, louvers and screens[1]	Multiply required net area in square feet by:[2]
1/4 inch mesh hardware cloth	1
1/8 inch mesh screen	1 1/4
No. 16 mesh insect screen (with or without plain metal louvers)	2
Wood louvers and 1/4 inch mesh hardware cloth[3]	2
Wood louvers and 1/8 inch mesh screen[3]	2 1/4
Wood louvers and No. 16 mesh insect screen[3]	3

[1] In crawl-space ventilators, screen openings should be no larger than 1/4 inch; in attic spaces no larger than 1/8 inch.
[2] Net area for attics determined by ratios in Figures 13-22 through 13-24.
[3] If metal louvers have drip edges that reduce the opening, use same ratio as shown for wood louvers.

Figure 13-19 Ventilating area increase required for louvers and screens in crawl spaces and attics

Roofing Repair & Maintenance

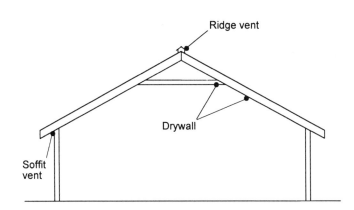

Figure 13-20 Installing a drywall trough over a cathedral ceiling

Continuous Ridge Vent

Drywall installed over a cathedral or attic ceiling forms a trough to collect hot air, as shown in Figure 13-20. Use a continuous vinyl or aluminum ridge vent like the one in Figure 13-21 over the trough. A continuous ridge vent is a good solution because it lets hot air escape, even when there's no wind, through a 2-inch-wide strip cut along the ridge. Run a ridge vent to within 18 inches of each end of the ridge.

Figure 13-21 Continuous ridge vent

Gable, Hip and Flat Roofs

Here's a more detailed approach to calculating proper attic ventilation for gable, hip and flat roofs. The minimum net area required for proper roof ventilation depends on the total ceiling area, the location of the vents and the type of roof design. See Figures 13-22 through 13-24.

You can see in Figures 13-25 and 13-26 that there are a variety of inlet vents for soffits. Make sure that inlet vents are evenly distributed so there won't be dead air spaces. In humid areas like Florida, use perforated aluminum soffit material like that shown in Figure 13-27 to provide inlet ventilation.

In a gable roof, install outlet vents as high as possible (Figure 13-28).

An alternative to tearing out and reconstructing existing ventilation systems is to recommend the installation of attic exhaust fans.

Roofing Construction & Estimating

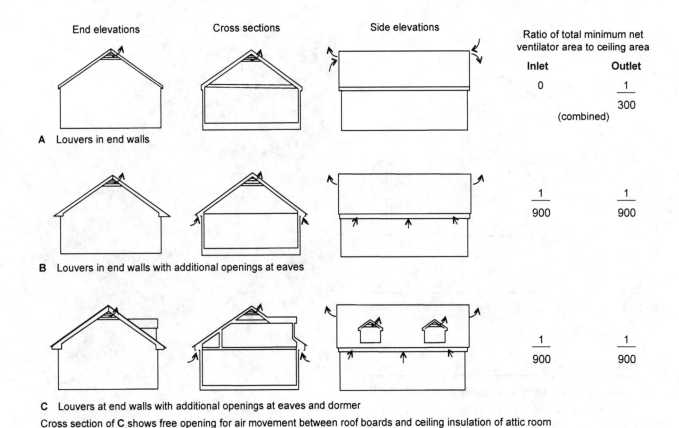

Figure 13-22 Ventilating areas of gable roofs

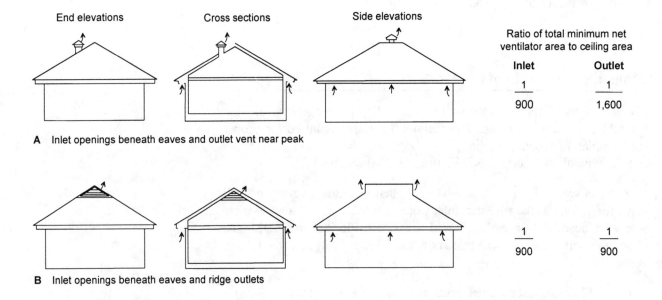

Figure 13-23 Ventilating areas of hip roofs

Roofing Repair & Maintenance

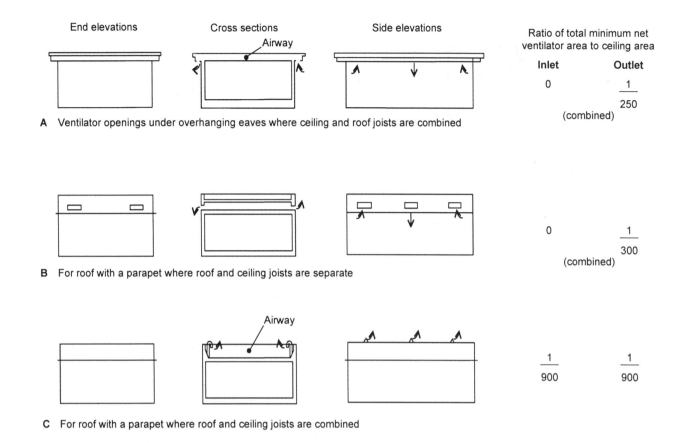

Figure 13-24 Ventilating areas of flat roofs

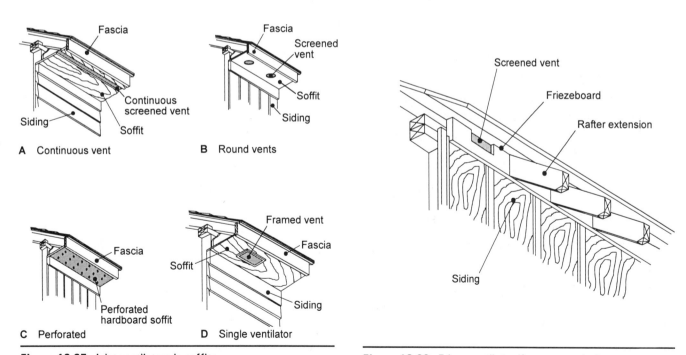

Figure 13-25 Inlet ventilators in soffits

Figure 13-26 Frieze ventilator (for open cornice)

Figure 13-27 Perforated aluminum soffit

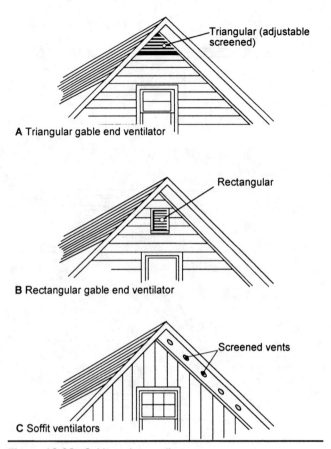

Figure 13-28 Gable outlet ventilators

▼ **Example 17-2:** Assume a ventilation system as shown in part B of Figure 13-22 with six inlet and two outlet ventilators with No. 16 mesh and plain metal louvers. Find the gross ventilation area required for each inlet and outlet ventilator for the building diagrammed in Figure 13-29.

From Figure 13-22 you see the minimum inlet and outlet ratio is 1/900 of the ceiling area. The minimum net area of inlet or outlet ventilation is:

$$\text{Minimum net ventilation area} = \frac{31' \times 26'}{900}$$

$$= 0.90 \text{ square feet}$$

From Figure 13-19 the minimum gross area required is:

Minimum gross ventilation area = 0.90 SF × 2
= 1.8 square feet

For two outlet ventilators, each ventilator requires:

$$\text{Gross ventilation area per outlet vent} = \frac{1.8 \text{ SF}}{2}$$

= 0.9 square feet, or 130 square inches
(0.9 × 144 square inches per SF)

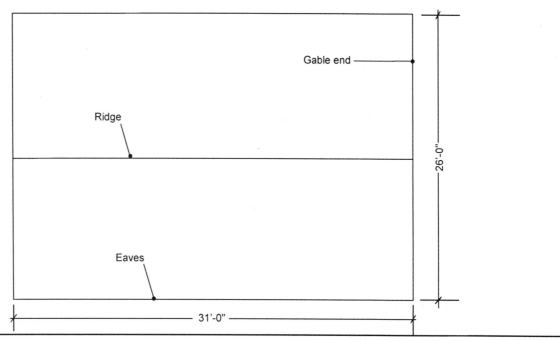

Figure 13-29 Ventilation example

For six inlet ventilators, each ventilator requires:

Gross ventilator area per inlet vent $= \dfrac{1.8 \text{ SF}}{6}$
= 0.3 square feet, or 44 square inches
(0.3 x 144 square inches per SF)

Gutters and Downspouts

Install gutters at the eaves to collect runoff water and divert it away from the building. Without a gutter system, eaves trim is subject to rotting. Also, runoff erodes planting beds, and heavy runoff might beat the plants themselves to death. Even worse, runoff water can saturate soil next to the foundation, leading to moisture penetration through the foundation or basement wall.

The most common materials for gutters and downspouts are aluminum, galvanized steel, and vinyl. Materials less commonly used include copper or zinc. You can buy painted metal gutters, but they also come unfinished so you can paint them to match the building. Use at least 28-gauge metal or 16-ounce copper for drainage systems.

Figure 13-30 shows several gutter shapes. Downspouts (also called *leaders* or *conductors*) are usually rectangular or circular, many with corrugations for added strength. They're shown in Figure 13-31. Corrugations also provide room for expansion if ice forms in the downspout. For the same reason, rectangular downspouts are better than round ones in places where the temperature drops below freezing.

Most gutters come 4, 5 or 6 inches wide. As a rule of thumb, you can use 4-inch-wide gutters on roofs with drainage areas up to 750 square feet. Roof drainage areas between 750 and 1,500 square feet require 5-inch gutters and roof drainage areas greater than 1,500 square feet require 6-inch gutters. In very wet climates, use gutters one size larger than these rules of thumb.

There are several ways to fasten gutters and downspouts to a building. You can use hangers, straps, or spikes, as shown in Figure 13-32. You generally space gutter hangers at 3-foot centers, but in heavy snow areas reduce this spacing to 1½ feet. I don't recommend that you use strap type gutter hangers because they can tear loose under a heavy load of snow or ice, and take the shingles with them.

The spike and spacer tube hanging system has the advantage of allowing the metal gutter to expand and contract freely.

Install downspout straps at 6-foot centers. Use a minimum of two straps, one at the gooseneck and one at the bottom elbow (or shoe) to secure the downspout to the wall. Solder joints in gutters and downspouts, although some specifications only require joints to be caulked with silicone or butyl rubber caulking compound.

Install downspouts large enough to handle all rainwater runoff. A general rule for residential construction is to allow 1 square inch of downspout end area for each 100 square feet of roof area. A 7-square-inch downspout end area is the minimum recommended.

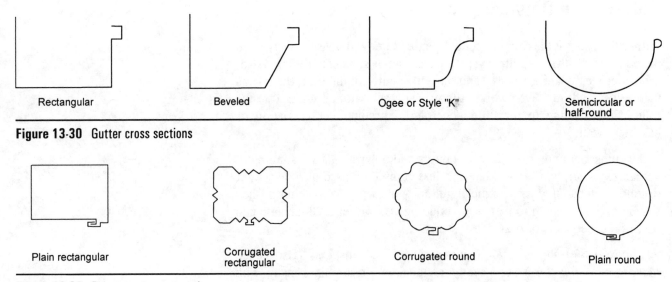

Figure 13-30 Gutter cross sections

Figure 13-31 Downspout cross sections

Roofing Repair & Maintenance

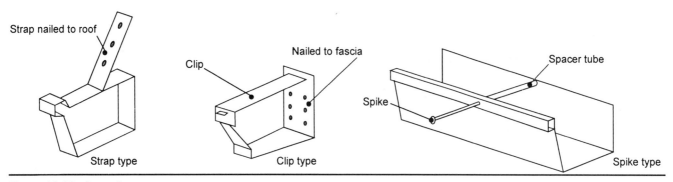

Figure 13-32 Gutter hangers

Slope metal gutters to drain toward downspouts with a slope of at least 1/16 inch of fall per linear foot of gutter run. Allow a maximum run of 10 feet from the high point of the gutter to the downspout if possible, and try to limit runs to 30 feet. If this isn't possible, increase the leader area by 1 square inch for each 20 feet of run between downspouts. On long gutter runs, you can center the high point of the gutter between the downspouts, and slope the gutter toward them in both directions. On straight gutter runs, install expansion joints on 60-foot centers (minimum).

You can also get a molded gutter with a sloping inner lining. You can set a molded gutter true with the eaves line so it doesn't look crooked.

If more than one gutter run empties into a downspout, provide a leader head as shown in Figure 13-33. Don't install downspouts carrying water from a larger and higher roof so they discharge onto a lower and smaller roof. Instead, let each gutter drain to the ground. A concrete splash block at the base of the downspout helps divert water away from the building.

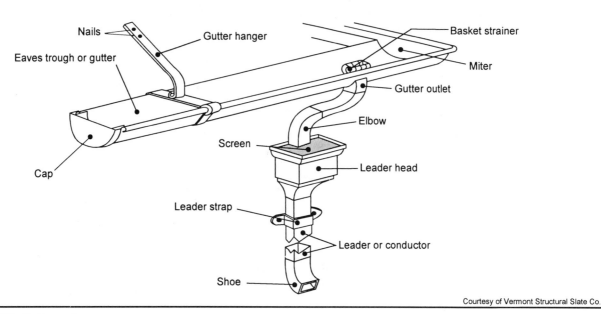

Courtesy of Vermont Structural Slate Co.

Figure 13-33 Gutter parts

409

Estimating Gutter and Downspout Systems

Take off gutter and downspout material by the linear foot. These come in 10-foot lengths. You buy gutter and downspout accessories by the piece. These accessories include inside and outside miters, slip joint connectors, end caps, end pieces, elbows, pipe straps, gutter hangers, basket strainers, and leader heads. Also order splash diverters as in Figure 13-34. The concrete splash blocks placed beneath each downspout are usually installed by the general contractor. If you've agreed to provide them, however, be sure you remember to price and include them in your estimate.

Gutter System Maintenance

Clean debris from gutters and downspouts frequently, especially in the spring and fall. This is very important, because a clogged gutter can fill with water and freeze, causing ice dams. You can keep downspouts from clogging by installing gutter baskets or strainers, as shown in Figure 13-33.

Replace or repair sagging or broken straps with solder, bolts or screws. Fill small holes with epoxy resin and seal larger holes with adhesive-backed aluminum tape. Caulk leaking joints with silicone or butyl rubber caulking compound.

Now you know how to install and repair all types of roofing. That just leaves one very important part of your business to cover in the final chapter — how to estimate (and maximize) production rates.

Figure 13-34 Splash diverter

14 Estimating (and Maximizing) Production Rates

▶ Production rates and unit costs are very hard to predict because of the many variables that affect production. If you don't already have them, start right now keeping good records of your crews' jobs and their production rates. They're the most reliable indicator of your future production rates. Until you have your own rates to go by, you can use published costs. For the examples in this chapter, we'll use figures from the *National Construction Estimator*, published by Craftsman Book Company. There's an order form bound into the back of this book.

Labor Unit Prices

Unit prices based on observation of your own crews are your best predictor of labor costs on your next job. But whether you use your own records or estimating guides, you'll have to consider job complexity and other factors when setting your labor unit prices. You can determine any daily or hourly labor unit price (LUP) by:

$$\text{Labor Unit Price} = \frac{\text{Total Daily Crew Cost}}{\text{Daily Crew Production}}$$

Equation 14-1

or

$$\text{Labor Unit Price} = \frac{\text{Total Hourly Crew Cost}}{\text{Hourly Crew Production}}$$

Equation 14-2

▼ **Example 14-1**: Find the labor unit price in cost per square for installing fiberglass shingles. Assume daily crew production of 6.37 squares per day with a crew of one roofer at $25.00 per hour and one laborer at $10.00 per hour. The hourly crew cost is $35.00 per hour ($25.00 + $10.00).

Total Daily Crew Cost = 8 hours each x $35.00 per hour

= $280.00

From Equation 14-1, the labor unit price is:

$$\text{LUP} = \frac{\$280.00}{6.37 \text{ SQ}} = \$43.96 \text{ per square}$$

Production Based on Labor Unit Prices

If you know the labor unit price, then you can project the production rate as follows:

$$\text{Daily Production} = \frac{\text{Daily Crew Cost}}{\text{LUP}}$$

Equation 14-3

or

$$\text{Hourly Production} = \frac{\text{Hourly Crew Cost}}{\text{LUP}}$$

Equation 14-4

▼ **Example 14-2**: Assume a labor unit price of $43.96 per square with a crew making $35.00 per hour. Find the approximate daily production rate you can expect for installing fiberglass shingles:

Total Daily Crew Cost = 8 hours x $35.00 per hour

= $280.00

From Equation 14-3:

$$\text{Daily Production} = \frac{\$280.00/\text{Day}}{\$43.96/\text{SQ}} = 6.37 \text{ squares per day}$$

Project Duration

You can predict the total project duration for a given item of work with the following formula:

$$\text{Project Duration (Days)} = \frac{\text{Total Material}}{\text{Daily Production}}$$

Equation 14-5

or

Estimating (and Maximizing) Production Rates

$$\text{Project Duration (Hours)} = \frac{\text{Total Material}}{\text{Hourly Production}}$$

Equation 14-6

▼ **Example 14-3**: At a production rate of 6.37 squares per day, find how much time it will take to install 27 squares of fiberglass shingles.

From Equation 14-5:

$$\text{Project Duration} = \frac{27 \text{ SQ}}{6.37 \text{ SQ/Day}} = 4.3 \text{ days}$$

You can find production rates for time required per unit installed as follows:

$$\text{Production (Crew-days/Unit)} = \frac{\text{LUP}}{\text{Daily Crew Cost}}$$

Equation 14-7

or

$$\text{Production (Crew-hours/Unit)} = \frac{\text{LUP}}{\text{Hourly Crew Cost}}$$

Equation 14-8

▼ **Example 14-4**: Find the production time required per square of shingles installed by a crew working at a total wage of $35.00 per hour, and a LUP of $43.96 per square.

From Equation 14-8, you find total hours per square by:

$$\text{Production (Crew-hours/Unit)} = \frac{\$43.96/\text{SQ}}{\$35.00/\text{Hr.}} = 1.26 \text{ hours/square}$$

Use this formula to find production rates in crew-hours per unit:

$$\text{Production (Crew-hours/Unit)} = \frac{\text{Crew-hours/Day}}{\text{Daily Production}}$$

Equation 14-9

▼ **Example 14-5**: Find how many crew-hours per square are required at $43.96 per square for a crew earning $35.00 per hour.

Use Equation 14-9:

$$\text{Production (Crew-hours/Unit)} = \frac{16 \text{ Hr./Day}}{6.37 \text{ SQ/Day}}$$

$$= 2.51 \text{ crew-hours per square}$$

Required Production Rates

Use the following formula to find the production rate required to finish a given work item within its labor budget:

$$\text{Required Daily Production} = \frac{\text{Daily Crew Cost} \times \text{Total Material}}{\text{Total Labor Budget}}$$

Equation 14-10

or

$$\text{Required Hourly Production} = \frac{\text{Hourly Crew Cost} \times \text{Total Material}}{\text{Total Labor Budget}}$$

Equation 14-11

▼ **Example 14-6**: Assuming a crew cost of $280.00 per day, use Equation 14-10 to find the production rate required to install 27 squares of shingles within a labor budget of $1,188.00.

$$\text{Required Production} = \frac{\$280.00/\text{Day} \times 27 \text{ SQ}}{\$1188.00} = 6.36 \text{ squares per day}$$

Prefabrication for Multiple Uses

Some work items such as scaffolding can be prefabricated and reused many times. In those cases, you can calculate the total labor cost as follows:

$$\text{Total Labor Cost} = \frac{\text{Labor Cost to Prefabricate}}{\text{Number of Uses}} + \text{Labor Cost to Install}$$

Equation 14-12

Adjusting Labor Costs

Some labor manuals give only an hourly wage for a particular trade and a labor unit price for a specific task. There are two ways that you can tailor unit prices (the actual wages you pay) to fit your own situation:

Option 1: (Using Equation 14-2)

$$\text{Actual LUP} = \frac{\text{Actual Hourly Wages}}{\text{Production (Units/Hour)}}$$

where (from Equation 14-4):

$$\text{Production (Units/Hour)} = \frac{\text{Published Hourly Wages}}{\text{Published LUP}}$$

Option 2:

$$\text{Actual LUP} = \frac{\text{Actual Hourly Wages}}{\text{Published Hourly Wages}} \times \text{Published LUP}$$

Equation 14-13

▼ **Example 14-7**: Use both options above to find the actual labor unit price to install fiberglass shingles where the published costs are $43.96 per square with a crew making $35.00 per hour when the actual wages you pay a crew are $25.00 per hour.

Solution:

Option 1 (using Equation 14-4):

Production (Units/Hour) = $\dfrac{\$35.00/\text{Hr.}}{\$43.96/\text{SQ}}$ = 0.796 SQ/Hr.

From Equation 14-2:

Actual LUP = $\dfrac{\$25.00/\text{Hr.}}{0.796 \text{ SQ/Hr.}}$

= \$31.41/SQ

Option 2 (using Equation 14-13):

Actual LUP = $\dfrac{\$25.00/\text{Hr.}}{\$35.00/\text{Hr.}}$ x \$43.96/SQ

= \$31.40/SQ

Obviously, Option 2 is the simpler method.

Estimating with Published Prices

The *National Construction Estimator* (NCE) is an encyclopedia of building costs that's updated each year. An estimating program disk called National Estimator is bound into every copy of the book. The program produces an estimate report, using the contents of the book as a database. There's an order blank for the NCE in the back of this book.

The NCE is divided into two parts. The Residential Division contains costs for homes and apartments with a wood or masonry frame. The Industrial and Commercial Division contains other construction costs not covered in the Residential Division. The Residential Division is arranged in alphabetical order by construction trade and type of material. The Industrial and Commercial Division follows the 16-section Construction Specification Institute (CSI) format.

Here's how to understand published costs like those in the NCE and use them to price your jobs.

Published labor costs are an average based on nationwide surveys for the various trades. Hourly wage rates are priced higher for industrial and commercial work than for residential construction. Residential hourly wage rates don't include a markup for overhead or supervision costs. The industrial and commercial wage rates do include a 30 percent markup for supervision expense, contingency allowance, overhead and profit. That's because it's assumed that commercial work is often subcontracted to specialized crafts. There's a 15 percent markup on materials.

Craft Code	Cost Per Manhour	Crew Composition	Craft Code	Cost Per Manhour	Crew Composition
M8	$47.94	1 laborer, 3 pipefitters	S3	$45.35	1 tractor operator, 1 truck driver
M9	51.42	1 electrician, 1 pipefitter			
MR	25.51	1 millwright (Residential)			
MS	46.09	1 marble setter			
					mosaic & terrazzo worker
			T5	44.91	1 sheet metal worker, 1 laborer
		1 plasterer (Industrial & Commercial)	T6	47.16	2 sheet metal workers, 1 laborer
PT	25.14	1 painter (Residential)			
R1	24.75	1 roofer, 1 laborer	TD	41.69	1 truck driver/teamster/oiler (Industrial & Commercial)
R3	46.44	2 roofers, 1 laborer			
RB	53.08	1 reinforcing ironworker (Industrial & Commercial)	TL	24.34	1 tile layer
			TO	49.01	1 tractor operator
RF	50.58	1 roofer (Industrial & Commercial)	TR	22.87	1 truck driver/teamster/oiler (Residential)
RI	27.29	1 reinforcing ironworker (Residential)	U1	44.14	1 plumber, 2 laborers, 1 tractor operator
RR	28.69	1 roofer (Residential)			
S1	43.59	1 laborer, 1 tractor operator	U2	42.51	1 plumber, 2 laborers

Figure 14-1   Craft codes definition

In the NCE, manhours per unit installed and the craft performing the work (craft code) are listed in the "Craft @ Hrs" column. Figure 14-1 is part of the section in the NCE which defines craft codes. Figure 14-2 is part of the residential roofing section from the NCE.

The average manhour cost is calculated by dividing the total hourly crew cost by the number of crew personnel. For example, an R1 crew consists of one roofer and one laborer. These workers make $28.69 and $20.80 per hour, respectively. So the average cost per manhour is:

$$\text{Average Cost per Manhour} = \frac{\$28.69 + \$20.80}{2}$$
$$= \$24.75 \text{ per manhour}$$

The labor unit costs in the NCE are the result of multiplying the installation time (in manhours) by the average cost per manhour. For example, the labor cost listed in Figure 14-2 for installing asphalt shingles is $50.70 per square. This figure is calculated:

Labor Cost Per Unit = 2.05 Manhours/Square x $24.75/Manhour

= $50.737 per square

Estimating (and Maximizing) Production Rates

                                                                                45.00
b coation cement, one coat #107
suita   --   Sq   --   --   115.00
25 lb fiberglass base sheet, 2 plies #184 Rufon polyester fabric embedded in and top
ed with #106 asphalt emulsion, and Henry #120 aluminum coating)
Over suitable substrate   --   Sq   --   --   125.00

Roofing, Composition Shingle. Material costs include 5% waste. Labor costs include roof loading and typical cutting and fitting for roofs of average complexity.

	Craft@Hrs	Unit	Material	Labor	Total
Asphalt shingles					
Celotex Dimensional 4 (355 lb, 25 year)	R1@2.05	Sq	85.00	50.70	135.70
Certainteed Hallmark (340 lb, 30 year)	R1@2.05	Sq	75.00	50.70	125.70
Fiberglass roofing shingles. Add for hip and ridge below					
Celotex Big D (225 lb, 20 year)	R1@1.83	Sq	24.00	45.30	69.30
Celotex Big D 25 (300 lb, 25 year)	R1@1.83	Sq	42.00	45.30	87.30
Celotex Presidential (40 year)	R1@2.60	Sq	75.00	64.30	139.30
GAF Sentinel (225 lb, 20 year class A)	R1@1.83	Sq	24.00	45.30	69.30
GAF Royal Sovereign (25 year)	R1@1.83	Sq	30.00	45.30	75.30
GAF Woodline (25 year)	R1@1.83	Sq	40.00	45.30	85.30
GAF Timberline (30 year)	R1@1.83	Sq	51.00	45.30	96.30
GS Firescreen (20 year)	R1@1.83	Sq	25.00	45.30	70.30
GS Firescreen Plus (25 year)	R1@1.83	Sq	28.00	45.30	73.30
GS Firehalt (25 year)	R1@1.83	Sq	35.00	45.30	80.30
GS Architect 80 (30 year class A)	R1@1.83	Sq	47.00	45.30	92.30
GS High Sierra (40 year)	R1@1.83	Sq	55.00	45.30	100.30
Elk Prestique I (320 lb, 30 year class A)	R1@1.83	Sq	47.00	45.30	92.30
Elk Prestique II (240 lb, 25 year class A)	R1@1.83	Sq	37.00	45.30	82.30
Elk Prestique Plus (40 year, class A)	R1@1.83	Sq	55.00	45.30	100.30
Masonite Woodruf Shingles					
Traditional	R1@1.83	Sq	82.00	45.30	127.30
Add for hip and ridge units	R1@.020	LF	.56	.49	1.05

	Craft@Hrs	Unit	Material	Labor	Total
Allowance for felt, flashing, fasteners, and vents					
Typical price	--	Sq	--	--	3.00

Roofing, Crushed Rock. Standard crush, 1/8" to 7/16", 10' lift and spread in 60 lb per 100 SF roofing asphalt at $11.00 per 100 lbs, typical costs based on rock at $85 per ton for 2 ton quantities.

	Craft@Hrs	Unit	Material		
Arctic white crushed tile, 220 lbs/sq	R1@.340	Sq	18.00		
Canyon red crushed tile, 240 lbs/sq	R1@.340				
Desert bronze, natural rock, 260 lbs/sq					
Desert green, igneous rock, 250 lbs/sq					
Glacier white, crushed					
Lagun					
Cal					

Figure 14-2 Residential roofing cost data

For simplicity, this cost is rounded to $50.70 per square. The loss of accuracy is insignificant.

To find out how many units one worker can finish in an 8-hour day, divide the manhours per unit into 8 hours. To find how many units a crew can complete, multiply the units for one worker by the number of crew members.

For example, the production rate for asphalt shingle installation is 2.05 manhours per square for an R1 crew (one roofer and one laborer). The daily production of each crew member is:

Daily Production (per man) = 8 Hours/Day ÷ 2.05 Manhours/Square

= 3.90 squares per day

The daily crew production is:

Daily Crew Production = 3.90 Squares/Man-Day x 2 Men

= 7.8 squares per day

Adjusting Labor Costs

Figure 14-3 is a section of the NCE chart which defines labor costs. Here's how to customize the labor costs given in a published cost book if your wage rates are different from those in the book.

Start with the taxable benefits you offer. Assume workers on your payroll get one week of vacation and one week of sick leave each year. Convert these benefits into hours. Your figures should look like this:

40 vacation hours + 40 sick leave hours = 80 taxable leave hours

Then add the regular work hours for the year:

80 taxable leave hours + 2,000 regular hours = 2,080 total hours

Multiply these hours by the base wage per hour. If you pay roofers $10.00 per hour, the calculation would be:

2,080 hours x $10.00 per hour = $20,800 per year

Next calculate the payroll tax and insurance rate for each trade. If you know the rates that apply to your employees, use those rates. If not, use the rates from a published price guide. For this example, we'll use 47.3% (the rate for roofers in the NCE, Figure 14-3). To increase the annual taxable wage by 47.3%, multiply by 1.473:

$20,800 per year x 1.473 tax & insurance rate = $30,638 annual cost

Then add the cost of non-taxable benefits. Suppose your company has no pension or profit sharing plan but does pay for medical insurance for employees. Assume that the cost for your roofer is $100 per month or $1,200 per year.

Estimating (and Maximizing) Production Rates

Residential Division

Hourly Labor Cost

	1	2	3	4	5	6
Craft	Base wage per hour	Taxable fringe benefits (@5.15% of base wage)	Insurance and employer taxes(%)	Insurance and employer taxes($)	Non-taxable fringe benefits (@4.55% of base wage)	Total hourly cost used in this book
Bricklayer	$17.75	$0.91	32.8%	$6.12	$0.81	$25.59
Bricklayer Helper	13.65	0.70	32.4	4.65	0.62	19.62
Building Laborer	14.15	0.73	35.5	5.28	0.64	20.80
Building Carpenter	17.25	0.89	36.1	6.54	0.78	25.46
Cement Mason	17.25	0.89	29.4	5.34	0.78	24.26
Drywall Installer	17.60	0.91	38.8	7.19	0.80	26.50
Drywall Taper	17.60	0.91	31.7	5.86	0.80	25.17
Electrician	20.00	1.03	27.3	5.73	0.91	27.67
Floor Layer	17.45	0.90	28.9	5.30	0.79	24.44
Glazier	16.85	0.87	32.4	5.74	0.77	24.23
Lather	18.50	0.95	25.3	4.92	0.84	25.21
Marble Setter	16.00	0.82	32.4	5.45	0.73	23.00
Millwright	18.00	0.93	30.4	5.76	0.82	25.51
Mosaic & Terrazzo Worker	16.75	0.86	29.1	5.13	0.76	23.50
Operating Engineer	20.00	1.03	31.6	6.64	0.91	28.58
Painter	17.50	0.90	32.3	5.94	0.80	25.14
Plasterer	17.60	0.91	32.4	5.99	0.80	25.30
Plasterer's Helper	13.50	0.70	32.4	4.60	0.61	19.41
Plumber	20.00	1.03	28.8	6.06	0.91	28.00
Reinforcing Ironworker	17.50	0.90	44.0	8.09	0.80	27.29
Roofer	18.00	0.93	47.3	8.94	0.82	28.69
Sheet Metal Worker	20.00	1.03	30.5	6.42	0.91	28.36
Tile Layer	17.35	0.89	29.1	5.31	0.79	24.34
Truck Driver	15.70	0.81	33.6	5.55	0.71	22.87

The labor costs shown in Column 6 were used to compute the manhour costs for crews on pages 5 to 7 and the figures in the "Labor" column of the Residential Division of this manual. Figures in the "Labor" column of the Industrial and Commercial Division of this book were computed using the hourly costs shown on page 230. All labor costs are in U.S. dollars per manhour.

It's important that you understand what's the figures in each of the six columns above explana...

benefits average 5.15% o... construction contractors. the base wage.

Figure 14-3 Labor costs

$30,638 annual cost + $1,200 medical plan = $31,838 total annual cost

Divide this total annual cost by the actual hours worked in a year. This is your total hourly labor cost including all benefits, taxes and insurance. Assume your roofer will work 2,000 hours a year:

$31,586 ÷ 2000 = $15.92 per hour

Finally, find your modification factor for the labor costs. Divide your total hourly labor cost by the total hourly cost shown in the published costs. For the roofer in our example, the figure in the NCE is $28.69:

$15.92 ÷ $28.69 = 0.55

Your modification factor is 55 percent. Multiply labor costs for roofers in the cost book by 0.55 to find your estimated cost.

If You Don't Know the Labor Rate

On some estimates you may not know what actual labor rates will apply. In that case, use both labor and material figures in the cost book without making any adjustment. But to make them more accurate, apply the area modification factor for your region. The NCE lists percentage guidelines you use to adjust the prices to fit the part of the country where you live. When you've compiled all labor, equipment and material costs, multiply the totals by the appropriate factor in the area modification table. Figure 14-4 shows part of the NCE's area modification table.

Roofing Labor Tips

Study each job carefully before you assign costs. The number and sizes of roof penetrations, the number of hips and especially the number of valleys, all affect production rates. Structures such as dormers which interrupt the shingle pattern require you to project the shingle pattern and tie in remote shingle courses. That increases your time. When tie-ins are required, the shingle pattern affects production. For example, with 3-tab shingles the 6-inch pattern is the easiest to use when making a tie-in.

You're the only one who can judge how efficient and dependable your workers are, and how well they work together. To a certain point, a larger crew will produce more than a smaller crew. But if the crew is too large, management problems and workspace limitations will actually hinder production.

The strength and stamina of individual crew members can also affect production rates. But strength isn't as important as basic common sense and a positive attitude toward the work. Workers' attitudes can change daily, or even hourly. A disgruntled or frustrated worker slows everyone down, and bad attitudes can be contagious.

Maximizing Production Rates

Job complexity and the skill and experience of your workers probably influence productivity more than anything else. But there are many other factors that can affect your production costs.

■ **Work Habits** Place as many shingles as possible without changing position — but don't try to shingle the whole roof from one position. Try to find a happy medium. Also, lay out partial bundles of shingles at strategic

Area Modification Factors

Construction costs are higher in some cities than in other cities. Add or deduct the percentage shown on this page or page 11 to adapt the costs listed in this book to your job site. Adjust your estimated total project cost by the percentage shown for the appropriate city in this table to find your total estimated cost. Where 0% is shown it means no modification is required.

These percentages were compiled by comparing the actual construction cost of residential, institutional and commercial buildings in 402 communities throughout the United States. Because these percentages are based on completed project costs, they consider all construction cost variables, including labor, equipment and material cost, labor productivity, climate, job conditions and markup.

Use the percentage shown for the nearest or most comparable city. If the city you need is not listed in the table, use the percentage shown for the appropriate state.

These percentages are composites of many costs and will not necessarily be accurate when estimating the cost of any particular part of a building. But when used to modify all estimated costs on a job, they should improve the accuracy of your estimates.

Alabama -18%	Sacramento +3%	**Hawaii** +46%	**Louisiana** -13%	
Birmingham -16%	San Bernardino +17%	Hilo +47%	Alexandria -18%	
Dothan -21%	San Diego +13%	Honolulu +40%	Baton Rouge -9%	
Florence -20%	San Francisco +39%	Kauai +50%	Houma -11%	
Huntsville -18%	San Jose +25%	Kona +49%	Lafayette -13%	
Mobile -15%	Santa Ana +23%	Maui +46%	Lake Charles -10%	
Montgomery -17%	Santa Barbara +20%		Marshall -10%	
Tuscaloosa -19%	Santa Cruz +24%	**Idaho** -5%	Monroe -17%	
	Santa Maria +18%	Boise -7%	New Orleans -10%	
Alaska +45%	Santa Rosa +13%	Coeur D'Alene -2%	Shreveport -15%	
Anchorage +43%	South Lake Tahoe +5%	Idaho Falls -7%		
Fairbanks +53%	Tehachapi +16%	Pocatello -3%	**Maine** +2%	
Juneau +47%	Ventura +19%		Augusta +3%	
Kenai +31%	Victorville +16%	**Illinois** 0%	Bangor -3%	
Ketchikan +47%		Bellville -2%	Brunswick +6%	
Sitka +48%	**Colorado** -5%	Chicago +7%	Lewiston +4%	
	Aspen-Vail +3%	East St. Louis 0%	Portland +4%	
Arizona -3%	Denver -10%	Moline 0%	Waterville -1%	
Casa Grande +1%	Grand Junction -8%	Springfield -3%		
Flagstaff -6%			**Maryland** -1%	
Kingman -6%	**Connecticut** +9%	**Indiana** -4%	Baltimore +1%	
Phoenix +3%	Bridgeport +12%	Bloomington -5%	Baltimore City +4%	
Sierra Vista -8%	Hartford +10%	Evansville -7%	Hagerstown -6%	
Tucson 0%	New Haven +8%	Fort Wayne -7%	Salisbury -3%	
Yuma -6%	New London +6%	Gary +3%	Waldorf -1%	
	New Milford +9%	Hammond +1%		
Arkansas -16%	Norwich +5%	Indianapolis -6%	**Massachusetts** +14%	
Fayetteville -17%	Ridgefield +13%	Lafayette -5%	Boston +21%	
Fort Smith -17%	Windam +5%	South Bend -4%	Fall River +11%	
Jonesboro -18%		Terre Haute -3%	Worcester +9%	
Little Rock -13%	**Delaware** +3%			
Texarkana -17%	Dover +2%	**Iowa** -8%	**Michigan** -2%	
	Wilmington +4%	Bettendorf -3%	Ann Arbor +2%	
California +16%		Cedar Rapids -7%	Battle Creek -6%	
Arrowhead +17%	**District of Columbia** +4%	Council Bluffs -11%	Benton Harbor -3%	
Bakersfield +12%	Washington D.C. +4%	Davenport -5%	Detroit +4%	
Barstow +16%		Des Moines -8%	Flint +1%	
Big Bear +18%	**Florida** -13%	Dubuque -8%	Grand Rapids -7%	
Desert Center +23%	Jacksonville -19%	Mason City -9%	Jackson -1%	
El Cajon +14%	Key West -10%	Sioux City -10%	Lansing +2%	
Eureka +11%	Miami -11%	Waterloo -8%	Marquette -2%	
Fresno +4%	Orlando -5%		Mt. Pleasant -3%	
Inyokern +14%	Pensacola -20%	**Kansas** -9%	Muskegon -8%	
Lancaster +22%	Tampa -13%	Garden City -13%	Saginaw -4%	
Los Angeles +26%		Kansas City +3%	Ypsilanti +5%	
Modesto +2%	**Georgia** -18%	Pittsburg -14%		
Monterey +22%	Albany -18%	Salina -12%	**Minnesota** -2%	
Mojave +17%	Atlanta -15%	Topeka -6%	Duluth 0%	
Needles +16%	Augusta -18%	Wichita -12%	Mankato -4%	
Oakland +25%	Brunswick -19%		Minneapolis +3%	
Ojai +16%	Columbus -19%	**Kentucky** -8%	Rochester -2%	
Oxnard +19%	Macon -19%	Ashland -5%	St. Cloud -3%	
Paso Robles +16%	Rome -20%	Covington -3%	Worthington -9%	
Piru +15%	Savannah -19%	Louisville -12%		
Placerville +4%	Valdosta -19%	Middlesboro -8%	**Mississippi** -16%	
Redding 0%		Owensboro -11%	Corinth -16%	
Ridgecrest +11%	**Guam** +45%	Paducah -12%	Greenville -15%	
	Agana +45%		Greenwood -16%	

Figure 14-4 Area modification table

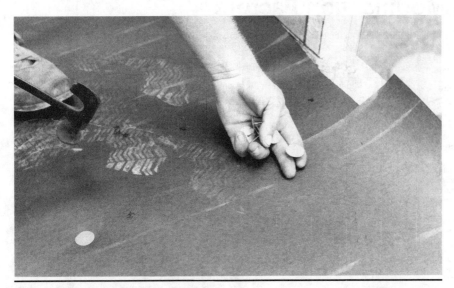

Figure 14-5 How to hold a roofing nail

locations along the roof so that they can be easily reached when needed. When you install any type of shingle, apply four or five courses at a time without changing body position. And most roofers I work with think it's more comfortable to sit on your hip than to kneel.

Hold roofing nails between your index finger and middle finger as shown in Figure 14-5. If you're right-handed, start shingling in the lower left corner of the roof. Work up and out toward the right with your left leg curled under you. Also, nail each shingle from left to right. Use the exposure gauge on your roofer's hatchet to set shingle exposure. Unless the roof has been lined out with horizontal and vertical chalk lines, check both ends of a long shingle for proper alignment. Shingle high-traffic areas (like the area immediately above a ladder) last.

In hard-to-nail places, dab the head of the hatchet in a bit of roofing cement, then embed the nail head in the cement. The cement will hold the nail in place as though it were magnetized to the hatchet.

When installing T-locks, pick up about ⅓ bundle at a time and drape them over your right leg while you're sitting on the roof. This eliminates unnecessary movement and increases your traction.

Place a stack of 3-tab shingles to your left (if you are right-handed) with the tabs facing you. This way you can pick up and place each shingle in one motion and it keeps your right hand on the hatchet at all times. Notice the worker in Figure 14-6.

■ **Tools** Use proper roofing tools and keep all cutting tools sharp. Don't use a straightedge for a guide when you cut shingles. This takes too much time. Learn to make straight, accurate cuts freehand. Cut asphalt shingles by scoring from the back side and then bend the shingle until it breaks. Cutting through the mineral surface will quickly dull a cutting tool.

Figure 14-6 Positioning the body and shingles

It may seem obvious that power tools like pneumatic nailers and electric saws and drills will increase production. But that's only true if the job is big enough to justify moving and setting up the power equipment. Also, equipment such as mobile placers aren't cost-effective unless they're used enough to justify their purchase and maintenance costs. It's often wiser to rent than to own.

■ **Materials** All things being equal, a large, heavy shingle is harder to install than a light one. But, the greater the exposure, the faster you'll cover the roof. Lower grades of wood shingles are narrower than the higher grades, reducing productivity and raising labor costs.

■ **Roof Slope** The steeper the roof or the higher the roof above the ground, the harder and more expensive it will be to shingle. On roofs steeper than 6 in 12, you'll need toe boards (Figure 14-7). Also, scaffold rental adds significantly to your job cost.

■ **Job Organization** When a job is poorly planned, the specs are unclear, or the plans themselves are sloppy, it's hard for workers to follow a smooth routine. There's often a lot of debate, and the workers design the job as they go along. Change orders, especially those that require work to be demolished and rebuilt or relocated, delay the job momentum. And if the building owner can't make timely decisions, the job will suffer.

Production is always lower on remodeling projects where tenants remain in the property. Accident prevention, noise and dust control, and job scheduling are problems that must be anticipated under those conditions.

■ **Other Contractors** The work of subcontractors who precede you on the job is beyond your control, but can still make your life miserable. An uneven or out-of-square slab adds time and headaches to the framer's job. If the framer installs rafters off-center, that makes it difficult to attach sheathing properly. As a result the eaves and rake won't be square and that makes your job difficult.

■ **Weather** High winds hamper installation of sheet material, and wet weather prohibits the use of power tools. During hot weather, try to shingle areas exposed to the sun early in the day and work in shaded areas during the hotter time of day. Sit on freshly installed shingles which are still cool, not on the part of the roof where the sun's been beating down for hours. During cold weather, try to follow the sun throughout the day.

■ **Protect the Surroundings** Protect existing siding by laying 4 x 8 sheets of plywood or waferboard against the building. You can also lay this material over shrubs and the lawn for protection. If bitumen drips onto siding, clean it with kerosene or tar remover sold by auto supply shops.

Accident Prevention

Accidents and their prevention are major labor cost factors for roofers. Remember that roofing contractors pay the highest workers' compensation insurance rates of all the construction trades in the nation because of their high accident rate. Help keep the lid on insurance premiums by making sure your people know — and follow — the safety rules. Accidents will happen, but don't contribute to them by your own negligence, carelessness, or poor supervision.

Here's a mini-manual of safety tips you should follow yourself, and pass on to your crews. These procedures will make the work go faster and safer.

■ **Apparel and Protective Gear** Wear loose and non-binding clothing, but be watchful of loose shirttails or baggy clothes around power equipment and ladders. Some roofers buy their work clothes from a second-hand store and discard them after they get too dirty to tolerate. Clothes worn while roofing will eventually ruin a washing machine.

Wear protective eye goggles when you use power equipment for cutting, or around hot or caustic materials. Wear protective gloves when you're working with hot bitumen. It's also a good idea to wear gloves when you work with sheet metal.

Wear soft-soled shoes to prevent slipping. That also helps prevent damage to shingles. Never step on loose granules or on a shingle that isn't nailed down. To prevent slipping and to protect your pants, cut a piece of inner tube long enough to reach from the hip to the knee and slip it around your leg. Use a couple of old belts to hold it on. Or sit on a thick foam rubber pad to keep you from sliding off the roof.

■ **Ladders and Scaffolding** Regularly inspect ladders to see that they are solidly put together, including rungs and locking devices. When you use an extension ladder, be sure at least one rung is above the edge of the roof, and that the top of the ladder is securely tied to a support. If you're lucky, there's a tree handy. Whenever you can, rest the ladder against a horizontal part of the roof like the eaves. Avoid leaning it against sloping surfaces like the gable ends.

Put a bundle of shingles at the eaves to support the upper end of the ladder. This keeps the ladder from slipping, and prevents it from denting the drip edge.

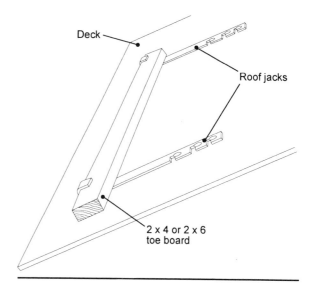

Figure 14-7 Roof jacks and toe board

Set the base of the ladder away from the roof's edge a distance equal to ¼ the working height of the ladder. Set the ladder on level ground, and if shimming is required to level a leg, use plywood sheets, not blocks. Glue rubber or carpet to the ladder feet to prevent the ladder from slipping on a smooth surface. Or use a sheet of plywood or waferboard between the ladder feet and a slick surface. Don't lean out on the ladder, and don't overload yourself or the ladder.

When you use wood board planking over ladders or tube scaffolding, make sure the boards extend beyond the supports in the event that the boards bend under weight. You don't want the boards falling through the supports. Rope off the outside of scaffolds to prevent falls. Keep metal scaffolding away from power lines.

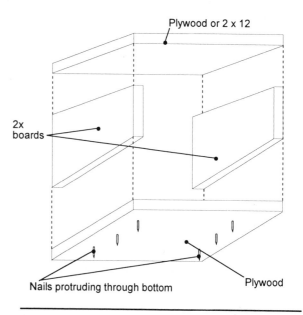

Figure 14-8 Wooden roofer's seat

Especially on steep roofs, apply shingles along the eaves while standing on some type of scaffolding until the roof covering has progressed to the point where you can install roof jacks (Figure 14-7) or use a roofer's seat (Figure 14-8). If it isn't possible to set up scaffolding while you work along the eaves, tie yourself to a rope that is securely fastened to a fixed object such as a tree located on the opposite side of the building. Don't make the mistake of one poor do-it-yourselfer I heard of. There wasn't a tree in his front yard, so he tied the rope to the rear bumper of his car in the driveway. His second mistake was to not tell his wife!

■ **Keep Your Eyes Open** Always watch for loose shingles and debris. Be wary of release paper that's applied to certain roll roofing products such as ice shield material; the release paper is very slippery. And fiberglass felts have very little tear strength. If you walk over them, they tear away from the nails easily. Look out for slippery areas, moss and wet leaves. Don't walk on a wet roof.

When you remove an old roof, work from the top down and periodically clear the debris. Rope off the dump zone, and avoid stepping on exposed nails.

Keep off a roof during a lightning storm. Don't touch wires crossing over the roof. If you use cranes or forklifts to raise roofing materials, watch out for overhead power lines.

■ **Watch That Pot** Keep an eye on kettle temperature since bitumens can explode if they're overheated.

■ **"High Anxiety"** Be patient with apprentices, especially when it comes to fear of heights. Don't encourage anyone to do anything they feel is dangerous. People can usually overcome this fear, but for some it takes longer than for others. A friend of mine had a rough time qualifying for a position with the local volunteer fire department because heights panicked him. This same man now teaches firefighting to new recruits in one of the largest cities in Florida.

That's a real success story. And I hope this book will contribute to your success, by opening up new horizons and helping you make your roofing business more efficient and profitable.

Appendix

A Roof-slope factors for determining rafter lengths (common, jack, hips and valleys)

B Valley length factors

C Roofing equations used in this book

Appendix A Roof-slope factors for determining rafter lengths (common, jack, hips and valleys)

(1) Roof slope	(2) Common or jack rafters (factor x run = actual length)	(3) Hips or valleys (factor x run = actual length)	(4) Hips or valleys (factor x plan length = actual length)
1 in 12	1.004	1.417	1.002
2 in 12	1.014	1.424	1.007
3 in 12	1.031	1.436	1.015
4 in 12	1.054	1.453	1.027
5 in 12	1.083	1.474	1.042
6 in 12	1.118	1.500	1.061
7 in 12	1.158	1.530	1.082
8 in 12	1.202	1.564	1.106
9 in 12	1.250	1.601	1.132
10 in 12	1.302	1.642	1.161
11 in 12	1.357	1.685	1.191
12 in 12	1.414	1.732	1.225
13 in 12	1.474	1.782	1.260
14 in 12	1.537	1.833	1.296
15 in 12	1.601	1.888	1.335
16 in 12	1.667	1.944	1.375
17 in 12	1.734	2.002	1.416
18 in 12	1.803	2.062	1.458
19 in 12	1.873	2.123	1.501
20 in 12	1.944	2.186	1.546
21 in 12	2.016	2.250	1.591
22 in 12	2.088	2.315	1.637
23 in 12	2.162	2.382	1.684
24 in 12	2.236	2.450	1.732

Appendix B   Valley length factors

Factor × Run of Low-Sloped Roof = Actual Length of Valley

Slope of Steep Roof

Low Roof Slope	1/12	2/12	3/12	4/12	5/12	6/12	7/12	8/12	9/12	10/12	11/12	12/12	13/12	14/12	15/12	16/12	17/12	18/12	19/12	20/12	21/12	22/12	23/12	24/12
1 in 12	1.417	1.123	1.058	1.035	1.023	1.017	1.014	1.011	1.010	1.009	1.008	1.007	1.006	1.006	1.006	1.005	1.005	1.005	1.005	1.005	1.004	1.004	1.004	1.004
2 in 12		1.424	1.213	1.134	1.090	1.067	1.053	1.044	1.038	1.033	1.030	1.027	1.025	1.024	1.023	1.021	1.021	1.020	1.019	1.019	1.018	1.018	1.018	1.017
3 "			1.436	1.275	1.193	1.146	1.116	1.097	1.083	1.074	1.066	1.061	1.056	1.053	1.050	1.048	1.046	1.044	1.043	1.042	1.041	1.040	1.039	1.038
4 "				1.453	1.323	1.247	1.199	1.167	1.144	1.127	1.115	1.106	1.098	1.092	1.087	1.083	1.080	1.077	1.075	1.073	1.071	1.070	1.068	1.067
5 "					1.474	1.367	1.298	1.251	1.217	1.193	1.175	1.161	1.150	1.141	1.133	1.128	1.123	1.118	1.115	1.112	1.109	1.107	1.105	1.103
6 "						1.500	1.409	1.346	1.302	1.269	1.244	1.225	1.210	1.197	1.187	1.179	1.172	1.167	1.162	1.158	1.154	1.151	1.148	1.146
7 "							1.530	1.451	1.395	1.353	1.321	1.296	1.277	1.261	1.248	1.238	1.229	1.221	1.215	1.209	1.205	1.201	1.197	1.194
8 "								1.564	1.495	1.444	1.405	1.374	1.350	1.331	1.315	1.302	1.291	1.281	1.273	1.267	1.261	1.256	1.251	1.247
9 "									1.601	1.540	1.494	1.458	1.429	1.406	1.387	1.371	1.357	1.346	1.337	1.328	1.321	1.315	1.310	1.305
10 "										1.642	1.588	1.546	1.512	1.485	1.462	1.444	1.428	1.415	1.404	1.394	1.386	1.379	1.372	1.367
11 "											1.685	1.637	1.599	1.568	1.542	1.521	1.503	1.488	1.475	1.464	1.455	1.446	1.438	1.432
12 "												1.732	1.689	1.654	1.625	1.601	1.581	1.563	1.549	1.536	1.525	1.516	1.507	1.500
13 "													1.782	1.742	1.710	1.683	1.661	1.642	1.625	1.611	1.599	1.588	1.579	1.571
14 "														1.833	1.798	1.768	1.743	1.722	1.704	1.688	1.675	1.663	1.653	1.644
15 "															1.888	1.855	1.828	1.805	1.785	1.768	1.753	1.740	1.729	1.718
16 "																1.944	1.914	1.889	1.867	1.849	1.833	1.818	1.806	1.795
17 "																	2.002	1.975	1.951	1.931	1.914	1.898	1.885	1.873
18 "																		2.062	2.037	2.015	1.996	1.980	1.965	1.953
19 "																			2.123	2.100	2.080	2.062	2.047	2.033
20 "																				2.186	2.164	2.146	2.129	2.115
21 "																					2.250	2.230	2.213	2.197
22 "																						2.315	2.297	2.281
23 "																							2.382	2.365
24 "																								2.450

Appendix C Roofing equations used in this book

Key to abbreviations:

E = exposure (Chapter 5)	N = number of ribbon courses
E = coefficient of linear expansion (Chapter 10)	R = recess (Chapter 1)
FS = factory squares	R = ridge length (Chapter 5)
HL = head lap	SF = square feet
L = length	T = change in temperature
LF = linear feet	TL = top lap
LSW = last strip width	W = width
LUP = labor unit price	

Chapter 1 Measuring and Calculating Roofs

Equation 1-1: Area of a level rectangular roof = L x W

Equation 1-2: Perimeter of an irregular roof = L + W + L + W + R + R
or 2L + 2W + 2R,
or 2(L+W+R)

Equation 1-3: Perimeter of a rectangular roof = 2(L+W)

Equation 1-4: $\text{Slope} = \dfrac{\text{Total Rise}}{\text{Total Run}} = \dfrac{\text{Rise (in inches)}}{12 \text{ (inches per foot of run)}}$

Equation 1-5: $\text{Pitch} = \dfrac{\text{Total Rise}}{\text{Total Span}}$

Equation 1-6: Slope = 2 x Pitch

Equation 1-7: Perimeter of an irregular sloped roof = P = 2(L + W + R)

Equation 1-8: Perimeter of a rectangular sloped roof = P = 2(L + W)

Equation 1-9: Perimeter of a gable roof = 2(Length + Actual Width)

Equation 1-10: Actual Width = 2(Run x Roof-Slope Factor)

Appendix C

Equation 1-11: Perimeter of a gable roof = 2[L + (W x Roof-Slope Factor)]

Equation 1-12: Allowance factor = $\dfrac{\text{Area Covered (including allowances)}}{\text{Net Roof Area}}$

Equation 1-13: Actual (Net) Roof Area = Roof Plan Area x Roof-Slope Factor

Equation 1-14: Plan Length (Hip or Valley) = 1.414 x Run

Equation 1-15: Ridge = L - [2 x ($\dfrac{W}{2}$)] = L - W

Chapter 3   Underlayment on Sloping Roofs

Equation 3-1: Lap Area (ridge, hip and valley) = Total Length x 1 SF/LF

Equation 3-2: Overcut Area (gables) = LF of Rake x 0.34 SF/LF

Equation 3-3: Coverage (SF/FS) = $\dfrac{\text{(Roll Length–End Lap) x Exposure}}{\text{Number of FS per Roll}}$

Equation 3-4: Exposure = $\dfrac{\text{Roll Width–Top Lap}}{\text{Number of Plies}}$

Equation 3-5: FS/Sq. = $\dfrac{100\ \text{SF/Square}}{\text{Coverage/FS}}$

Chapter 4   Asphalt Shingles

Equation 4-1: Shingles/Square = $\dfrac{100\ \text{SF}}{\text{Shingle Length (in.) x Exposure (in.)}}$ x 144 sq. in./SF

Equation 4-2: Percentage-of-Increase Factor = $\dfrac{\text{Recommended Exposure}}{\text{Actual Exposure}}$

Equation 4-3: Courses = $\dfrac{\text{Dimension of Structure}}{\text{Exposure}}$

Equation 4-4: $\text{Exposure} = \dfrac{\text{Dimension of Structure}}{\text{Number of Courses}}$

Equation 4-5: Top Lap = W - E

Equation 4-6: Top Lap = E + HL

Equation 4-7: Head Lap = TL - E

Equation 4-8: Head Lap = W - 2E

Equation 4-9: $\text{Exposure} = \dfrac{W - HL}{2}$

Equation 4-10: Starter Course (SF) = Eaves (LF) x Exposed Area (SF/LF)

Equation 4-11: $\text{Area (SF/LF)} = \dfrac{\text{Exposure (in.)}}{12}$

Equation 4-12: $\text{Number of Eaves Shingles} = \dfrac{\text{Eaves Length (ft.)}}{\text{Shingle Length (ft.)}}$

Equation 4-13: Net Allowance (ridge and hip units) = SF Required - SF Salvaged

Equation 4-14: $\text{Number of ribbon courses (Gable)} = \dfrac{2 \times \text{Roof Dimension}}{\text{Ribbon Spacing}} - 2$

Equation 4-15: $\text{Number of ribbon courses (Hip)} = \dfrac{\text{Roof Dimension}}{\text{Ribbon Spacing}} - 1$

Equation 4-16: Total Length of ribbon courses (Gable) = N x L

Equation 4-17: Total Length of ribbon courses (Hip)
$= 2N\,[(L + W) - 2(N + 1) \times \dfrac{\text{Ribbon Spacing}}{\text{Roof–Slope Factor}}]$

Chapter 5   Mineral-Surfaced Roll Roofing

Equation 5-1: Starter-Strip Waste at Ridge (single coverage) = 4 x E x (L - R - E)

Equation 5-2: Starter-Strip Waste at Ridge (double coverage) = 5.67' x (L - R - 1.417')

Appendix C

Equation 5-3: Starter Strip Waste at Hips = 4 x E x (Run - E) x Roof-Slope Factor

Equation 5-4: Starter Strip Waste at Hips (17" exposure)
= 5.67' x (Run - 1.417') x Roof-Slope Factor

Equation 5-5: Waste at Starter Strip & Hips
= 4 x E x [(L - R - E) + (Run - E)] x Roof-Slope Factor

Equation 5-6: Waste at Starter Strip & Hips (17" exposure)
= 5.67' x [(L - R - 1.417') + (Run - 1.417')] x Roof-Slope Factor

Equation 5-7: Waste at Starter Strip & Hips (double coverage, perpendicular to eaves)
= 4 x E x (Run - E) x Roof-Slope factor

Equation 5-8: Last strip width (LSW) = E x the decimal part (numbers to the right of the decimal point) of the expression: $\dfrac{W/2 \times \text{Roof-slope factor}}{E}$

Equation 5-9: LSW = E x the decimal part of the expression: $\dfrac{L}{E}$

Equation 5-10: Waste (SF/LF of ridge) = E - (2 x LSW)
when LSW is equal to or less than half the exposure

Equation 5-11: Waste (SF/LF of ridge) = 2 x (E - LSW)
when LSW is greater than half the exposure

Chapter 6 Wood Shingles and Shakes

Equation 6-1: Hip and Ridge Units = $\dfrac{\text{Total LF Ridge and Hips}}{\text{Exposure (ft.)}}$

Equation 6-2: Total Units = $\dfrac{\text{Total LF Ridge and Hips}}{\text{Exposure (ft.)}}$ + Number of Hips

Chapter 8 Slate Roofing

Equation 8-1: Squares Ordered per Square Covered equal:
$$\frac{\text{Slates per 100 SF at your head lap}}{\text{Slates per 100 SF at 3-inch head lap}} \times \text{Roof area (in squares)}$$

Equation 8-2: Hip and ridge slates $= 2 \times \dfrac{\text{LF of ridge and hip}}{\text{Exposure}}$

Chapter 9 Metal Roofing and Siding

Equation 9-1: Waste Factor $= \dfrac{\text{Gross Panel Area}}{\text{Net Panel Area}}$

Equation 9-2: Panels per Square $= \dfrac{100 \text{ SF}}{\text{Net Panel Area (SF)}}$

Equation 9-3: Panels per Square $= \dfrac{14{,}400 \text{ sq. in.}}{\text{Net Panel Area (sq. in.)}}$

Chapter 10 Built-up Roofing

Equation 10-1: Exposure $= \dfrac{34 \text{ inches}}{\text{No. of Plies}}$

Equation 10-2: Change of Length $= E \times T \times L$

Chapter 12 Insulation, Vapor Retarders and Waterproofing

Equation 12-1: R value $= \dfrac{1}{K} \times$ Material Thickness (inches)

Equation 12-2: U (coefficient of transmission) $= \dfrac{1}{R}$

Appendix C

Chapter 14 Estimating (and Maximizing) Production Rates

Equation 14-1: $\text{Labor Unit Price} = \dfrac{\text{Total Daily Crew Cost}}{\text{Daily Crew Production}}$

Equation 14-2: $\text{Labor Unit Price} = \dfrac{\text{Total Hourly Crew Cost}}{\text{Hourly Crew Production}}$

Equation 14-3: $\text{Daily Production} = \dfrac{\text{Daily Crew Cost}}{\text{LUP}}$

Equation 14-4: $\text{Hourly Production} = \dfrac{\text{Hourly Crew Cost}}{\text{LUP}}$

Equation 14-5: $\text{Project Duration (Days)} = \dfrac{\text{Total Material}}{\text{Daily Production}}$

Equation 14-6: $\text{Project Duration (Hours)} = \dfrac{\text{Total Material}}{\text{Hourly Production}}$

Equation 14-7: $\text{Production (Crew-days/Unit)} = \dfrac{\text{LUP}}{\text{Daily Crew Cost}}$

Equation 14-8: $\text{Production (Crew-hours/Unit)} = \dfrac{\text{LUP}}{\text{Hourly Crew Cost}}$

Equation 14-9: $\text{Production (Crew-hours/Unit)} = \dfrac{\text{Crew-hours/Day}}{\text{Daily Production}}$

Equation 14-10: $\text{Required Daily Production} = \dfrac{\text{Daily Crew Cost} \times \text{Total Material}}{\text{Total Labor Budget}}$

Equation 14-11: $\text{Required Hourly Production} = \dfrac{\text{Hourly Crew Cost} \times \text{Total Material}}{\text{Total Labor Budget}}$

Equation 14-12: $\text{Total Labor Cost} = \dfrac{\text{Labor Cost to Prefabricate}}{\text{Number of Uses}} + \text{Labor Cost to Install}$

Equation 14-13: $\text{Actual LUP} = \dfrac{\text{Actual Hourly Wages}}{\text{Published Hourly Wages}} \times \text{Published LUP}$

Index

A

Accident prevention..... 424-426
Acid-flux solder......... 249
Acrylic
  carbide sealers......... 379
  caulk................ 365
  emulsion............. 310
  polymer binder....... 226
Acrylic-based adhesive..... 306
Actual length, common rafters. 15
Actual width, roofs........ 17
Adhesion, caulk.......... 366
Adhesive
  acrylic-based......... 306
  bonding.......... 339-342
  cold-process......... 337
  heat activated........ 74
  neoprene............ 306
  strips....... 75, 106, 396
Adjustable vent.......... 406
Aggregate
  built-up roof.... 291, 314, 317
  crushed stone....... 307, 339
  elastomeric roofing... 334, 337
  grading........... 305, 307
  moisture content..... 307
  underlayment....... 37
Aggregate-surfaced roofs.. 41, 89
Air circulation.. 29, 37, 172, 337
  .... 362, 402
Algae-resistant roofing..... 387
Alligatoring......... 329, 330
Allowance factor........ 18, 42
Alloy 255, 268, 277, 279, 285, 287
Aluminum
  coatings 291, 308, 312, 316, 319
  composites.......... 343
  flange.............. 321
  flashing............ 65
  foil vapor barrier...... 362
  paint........ 382, 392-393
  pigment flakes....... 309
  ridge............... 403
  roof coating....... 309-310
  shim................ 390
  shingles............ 288
  soffit.............. 403
  solder.............. 268
  tape............ 383, 410
Aluminum-core shingles.... 288
Aluminum-faced insulation
  .... 350, 353
American method..... 128-129
Anchor
  bar................ 340
  clips............... 258
Anodic metal........... 279
Anti-ponding metal..... 202-203
Antimony.............. 286
Apex tile........... 212, 226
  trim................ 213
Apprentices............ 426
Apron flashing. 168-169, 246, 262
Area
  level roof............ 7-8
  sloped roof......... 17-21

Asbestos.............. 306
Asphalt
  bitumen....... 297, 333, 335
  .... 342, 355, 360
  cement............. 305
  dead-level..... 310, 320, 337
  emulsion 305, 307-310, 374, 378
  flux................. 36
  mastic........... 305, 378
  plastic cement......... 98
  primer........ 97, 296, 300
Asphalt shingles....... 73-130
  buckling............ 106
  bundles......... 108, 111
  cap................ 92-93
  colors.............. 76-77
  components......... 74
  coursing....... 82-84, 109
  decking............. 76
  dimensional........ 78
  estimating costs...... 130
  estimating quantities... 113-127
  fiberglass.......... 74, 78
  flashing........ 96-105
  free-tab............ 81
  giant............ 127-129
  hexagonal...... 78, 127-129
  high-slope applications..... 88
  hip units......... 114-115
  individual.... 78, 80, 127-129
  installation......... 80-96
  interlocking....... 78-80, 127
  laminated.......... 78
  life expectancy....... 73
  low-slope applications..... 88
  organic............. 88
  patterns........ 82-84, 87
  per square requirements. 108-109
  plies................ 111
  random-tab......... 78
  removing........... 391
  repairing........... 384
  re-roof, over........ 395
  re-roofing with....... 395-398
  ribbon coursed...... 124-127
  ridge units........ 114-115
  self-sealing......... 79
  slope requirements.... 76-77
  square-butt......... 78
  square-tab.......... 79
  stapling.......... 107-108
  starter course.... 80-81, 113-114
  storage............. 76
  stretch............. 24
  strip............. 78, 80
  tabs................ 78
  thick-butt.......... 78
  three-tab...... 74, 78-79
  triple-tab........... 78
  two-tab......... 78, 106
  UL ratings....... 75, 79
  underlayment....... 38
  waste factors...... 118-120
  weights........... 78-79
  wind-resistant..... 75, 88
Asphalt-faced insulation.... 337
Asphalt-fibrated emulsion... 305

Asphalt-laminated paper.... 362
Asphalt-saturated cotton. 370, 382
Asphalt-treated lumber..... 341
Asphaltic
  paint.............. 377
  primer......... 311, 373, 374
Attic
  inspection....... 381, 383
  insulation.......... 356
  ventilation.... 29, 35, 358, 402
Average manhour costs.... 416

B

Back-nailing....... 48, 297-278
Backer rod............. 369
Bacteria........... 328, 376
Baffles............ 57, 198
Ballast......... 307, 336-337
Ballasted elastomeric system....
  .... 339, 342-344
Bar-anchored system...... 343
Barge board........... 211
Barge tiles............ 208
Base flashing
  asphalt shingles..... 96-97, 99
  built-up roof..... 298, 312, 316
  elastomeric roofing..... 344
  slate roofing........ 246
Base mat.............. 73
Base sheet
  built-up roof. 297-298, 302-303
  .... 308-309, 312
  coated............. 298
  fiberglass........... 36
  roll roofing......... 131
  underlayment...... 37, 47
  vapor retarder... 293, 298-299
  vented....... 298, 331, 363
  waterproofing...... 375
Basement walls......... 371
Basket strainer....... 409-410
Batt insulation. 348-350, 357-358
Battens
  counter....... 200, 202-203
  expansion.......... 264
  integral............ 259
  metal roofing....... 260
  seams... 266-267, 283, 285-286
  slotted............. 201
  spaced.......... 171-172
  strip............... 341
  tile roofing..... 200-202, 204
  .... 206-207, 222, 224
  wood.......... 264, 266
Bauxite.............. 307
Bell eaves......... 171-172
Bellows........... 321-322
Bentonite clay......... 305
Bermuda roof..... 281-282, 284
Beveled
  gutter.............. 408
  siding......... 392, 397-398
Binders, plastic......... 36
Bird holes......... 204-205
Birdstop.............. 203

Bitumen
  asphalt........... 303, 373
  built-up roof. 292-293, 295-310
  .. 313, 315, 319, 324, 329, 331
  coal-tar..... 302-304, 315, 373
  cold-applied......... 305
  cutback in........ 305, 374
  storage............. 305
  types.............. 304
Bitumen trap.......... 318
Bitumen-saturated cotton. 303, 372
Bituminous fiber pipe...... 372
Bituminous paint... 66, 192, 245
Black steel............ 276
Blanket insulation..... 348, 357
Bleaching............. 388
Blind nail method........ 137
Blisters...... 298, 304, 328-330
  .... 333, 398, 402
Blocking............ 28, 29
Blown insulation........ 351
Body-colored tile........ 198
Bond lines............ 78
Bonded roof........... 377
Bonding adhesive.... 339, 340-342
Boston hip....... 240-241, 252
Brazing............. 245
Bright plate........... 270
British thermal units (Btu)... 361
Broken slates.......... 386
Bubble level........... 12
Bubbles.............. 329
Buckled deck.......... 29
Buckling underlayment.... 62, 133
Building code.. 23, 29, 37, 45, 62
 . 65, 70, 167, 194, 200, 204, 260
Building insulation....... 347
Building paper....... 188-191
Building protection...... 424
Built-up roofs......... 291-331
  aggregate-surfaced 291, 306-308
  base sheets....... 298-300
  cold-process...... 305-306
  estimating costs..... 325-326
  estimating quantities.. 323-325
  fasteners........... 294
  flashing..... 311-312, 315-320
  four-ply......... 303, 309
  insulation...... 293, 355, 358
  life expectancy....... 291
  membranes...... 300-303
  metal roofing over...... 288
  mineral-surfaced..... 291, 300
  repairs.......... 329-331
  roll-roof........... 308
  slope.............. 292
  smooth-surfaced..... 291, 308
  substrates.. 24, 27, 36, 292-297
  testing............. 327
  three-ply........ 302, 309
  vapor barriers...... 363-364
  warranties........ 327-328
Bull................. 393
Butt thickness.......... 161
Butted corner.......... 187
Butting-up method..... 395-397
Butyl caulk...... 365, 408, 410
Butyl rubber....... 336, 376

C

Caliche 307
Canoe valley 243-244
Cant strip
  built-up roof 299, 311-313
  .... 316, 319
  slate roofing 236, 240
  tile roofing 202-203
  wood shingles & shakes ... 192
Cap bead 365
Cap flashing
  Asphalt shingles 96, 98-100
  built-up roof 299, 316
  slate roofing 245, 246
Cap sheet
  built-up roof 299, 311-313
  .... 316, 319, 329
  modified asphalt 309, 316
  roll roofing 131
  tile roofing 200
  underlayment 37
Cap shingles 92-95
Cap strip 340
Capillary action 377
Capping in 35
Caps, tin 44, 360
Cathedral ceiling ... 202, 364, 403
Caulk 364-369
  adhesion 366
  acrylic 366
  butyl 365, 408, 410
  doors 357
  epoxy 365
  estimating quantities 367
  fireproof 369
  latex 365
  life expectancy 365-366
  masonry 99
  polysulfide 366
  polyurethane 366
  roof repair 383, 402, 410
  silicone 365-366
  solvent acrylic 366
  vinyl acrylic 365
  windows 357
Caulking tubes 367, 369
Ceiling
  insulation 61-63, 351, 356
  .... 402, 404
  joists 25
  ventilation 364
Cellular glass insulation . 360, 363
Cement
  hydrolithic 379
  lightweight insulating . 297, 333
  .... 338, 355
  roofing 41, 43, 46, 62
  .... 67-69, 81, 88-89
Cement smudges 144
Cement-fiber boards 295
Centipoise 305
Ceramic granules 314, 335
Certi-guard
  shakes 159
  shingles 159
Certi-last shingles 159
Certi-sawn shakes 163
Change orders 423
Chimney flashing
  asphalt shingles 96-100
  metal roofing 261-262
  roofing repair 382, 393
  slate roofing 246-247
  tile roofing 220, 222
  wood shingles & shakes . 168-169

Chipper 211
Chlorinated polyethylene . 336, 342
Chlorine bleach 388
Chlorophane 335, 376
Chlorosulfonated polyethylene
  .... 335, 336
Chromium 285
Cleats 245, 247, 249
  .... 268, 285, 287
Clip hangers 409
Clip-down shingles 127, 128
Clips 249, 285
  anchor 258
  edge 29
  panel 258
  storm 206-207, 210, 213
Closed valley 242, 243
Closed-cut valley 91, 92
Closure caps 258
Closure strip
  metal roofing 257, 258, 260, 263
  tile roofing 203
Coal-tar bitumen 303-306
  elastomeric membranes ... 336
  epoxy coatings 379
Coated
  base sheets 298
  metals 377
  roll 62
  sheet 131, 300, 313, 314
Coefficient of expansion
  ... 264, 266-279, 286, 321, 323
Coefficient of transmission 361-362
Coil stock 289
Coke 303
Cold asphalt emulsion 305
Cold-applied bitumen .. 132, 133
  ..... 300, 359, 372, 373, 376
Cold-process adhesive 337
Cold-process roofing . 48, 305-306
Cold-rolled copper 287
Collar beam 25
Color patterning 34
Colorless sealers 377
Comb ridge 238-239
Combing slate 238-239, 240
Commercial standard slate ... 232
Common rafter 13
Compressed insulation 312
Compressed particleboard ... 353
Concealed clip panels 278
Concealed-nail method 144, 148
Concentrated roof loads 32
Concrete cant 313
Concrete form lumber 26
Condensation 29, 69, 70
  .... 362, 363, 364
Conductor 409
Contingency allowance 415
Continuous
  cleat 299
  clip 261
  flashing 100, 143, 217-220
  261-262
  vent 405
Controlled-flow drainage system . .
  315
Copper
  composites 343
  cornice temper 286
  flashing 65, 70, 242
  .... 245-247, 287
  roofing 39, 40, 263, 264, 286-288
  seams 245
  sheet metal 242, 286-287
  shingles 288
  waste factors 287

Copper-bearing steel ... 279, 283
Cork board 353
Corner
  laced 187, 194
  woven 187
Corner flashing 169, 186, 219, 370
Cornice, open 30, 405
Cornice temper copper 286
Corrugated metal
  downspouts 408
  panels 29, 257, 274, 276
  roofs 24, 29, 271-276
Cotton fabric 372
Counter battens 200, 202-203
Counterflashing
  asphalt shingles 96
  built-up roof 298, 312
  .... 316-319, 322
  metal roofing 262
  roofing repair 383, 393
  slate roofing 246
  tile roofing 218-222
Coursing roll roofing ... 136-142
Coxcomb ridge 240
Craft codes 416
Cranes 426
Creosote 373
Creosote-treated lumber 341
Crickets 96-98, 168, 220
  .... 222, 246, 262
  flashing 96, 168, 220, 222
  valley 98
Cross seams 269, 286
Cross-bond method 208
Crossover 321, 322
Crushed file 307
Crushed stone 307, 339
Cupped shingles 386
Curb
  flange 321, 322
  flashing 318-319, 341
  roof 269
Curled shingles 385, 389
Cushion strips 374
Cutback bitumen ... 303, 305-307
  .... 308, 373-374, 378
Cutouts 75, 78, 89, 106, 396
Cutting allowance 18, 42

D

Damaged sheathing 392
Damaged tabs 385
Dampproofing 371, 377
  bituminous 378
  cavity wall 378
  existing wall 378-379
  hydrolithic 377, 379
Dead load 27-28, 315, 337
Dead soft copper 286
Dead-level asphalt . 310, 320, 337
Dead-level roof 303
Decay-resistant wood shingles 162
Decking
  concrete 296
  defects 292
  false 172-173
  fasteners 294
  fluted metal 257
  grades 32
  gypsum . 295, 298, 300, 333, 355
  gypsum concrete 294
  heavy-load bearing 315
  lightweight insulating . 293-294
  metal 173-174, 293
  metal roofing 310, 359
  .... 382, 391, 395

  nailable 297, 300, 308
  non-nailable 296-297, 300
  ... 308-309, 313, 315, 331, 339
  perforated steel 295
  precast concrete 315
  roof 23-24, 32, 76
  staggered joints 26, 29
  structural cement fiber 300
  structural wood fiber 296
  tongue and groove 26
  ventilation 295
  wet-fill 316
  wood 296, 300
Delays 423
Demolition, roofs 390-394
  asphalt shingles 390
  gravel 392
  flashings 392
  metal 391
  roll 391
  shake 390-391
  slate 391
  tile 391
  wood shingles & shakes 390, 391
Demolition, siding 392
Daily production 412
Deterioration
  deck 29, 35, 402
  flashing 401
  metal roofing 255
  underlayment 29
Diagonal
  method 83
  pattern 87
  trails 87
Dimensional shingles 87
Direct-nail method 46
Discoloration 387
  asphalt shingles 76, 387-388
  concrete tiles 388
Diverters
  splash 64, 66, 410
  water 213
Dolomite 307
Door flashing 186, 279, 370
Doors
  fireproof 279
  insulating 357
  storm 357
Dormer ridge 89
Dormers
  asphalt shingles 84-86
  .... 89-90, 95-96, 100
  estimating labor 420
  flashing 219-220, 246-247
  siding, removing 392
  slate roofing 246-247
  underlayment 64
  ventilating 404
  wood shingles & shakes 168-169
Double lock seam 260
Double-coursing ... 185, 188-189
Double-coverage
  roll roofing . 132, 142, 144, 149
  underlayment . 38-39, 43, 51, 62
Downspouts 387, 407-409
Drain slots 201
Drainage 292, 314-315, 408
Drip edge
  asphalt shingles 80, 88
  estimating quantities 42
  metal roofing 258, 269
  roll roofing 137, 141
  roofing repair ... 392, 394, 401
  tile roofing ... 200, 202-203
  underlayment 35, 40-41, 43
  wood shingles & shakes 164-165

Dry hip............ 212, 213
Dryer vents 362
Drying in 35
Dubbing 89
Dump zones 426
Duration of project 412
Dutch lap method 128
Dutch weave 185

E

Eaves
  bell............. 171, 172
  estimating 18
  flashing 35, 61-62, 318
  length 59
  protection 30, 31, 57, 70, 164-165
  roll roofing 133, 137, 146
  roofing repair 383
  swept............. 171-172
  trough 409
  underlayment......... 40, 44
  units 111
  ventilation.......... 358
Eaves closures
  EPDM............. 203
  metal 203, 206
  metal roofing 257
  rubber 203, 206
  tile roofing 202-204, 206
Edge clips............. 29
Edge strips........ 37, 138
  .... 298-299, 313, 398
Edge grain wood shingles 160, 162
  .... 389, 390
Edging, roof 40
Efflorescence, on tile...... 379
Elastomeric roofing..... 333-346
  ballasted.......... 336-339
  composite 343-344
  CPE 336, 342
  CSPE........ 336, 342
  elasticity......... 333
  EPDM 336-340
  estimating costs 346
  estimating quantities 345-346
  flashing 340-342, 344-345
  fully-adhered.... 336-337, 339
  hypalon 335, 343
  insulation 339
  life expectancy... 338, 342-343
  liquid........... 335
  mechanically-fastened . 336, 340
  neoprene.......... 335
  PIB 336
  PVC 333, 336
  silicone........... 335
  traffic pads 314
  urethane 335
  waste factors 345
Elastomeric waterproofing ... 373
Elbow 408-410
Electric thermal wire....... 63
Electrolysis............ 245
Emery 379
Emulsified asphalt..... 373, 374
Emulsifier 305
Enamel-coated shingles 288
End caps............ 410
End covers 258
End laps 45, 46, 274, 309
End-wall flashing 288-289
Envelope strip 299, 313
EPDM elastomeric systems 337-342
  flashing 334
  waterproofing 376

Epoxy caulk........... 365
Epoxy resin 410
Epoxy sealant, life expectancy 367
Equations, roofing..... 430-435
Equiviscous temperature.... 305
Estimating costs, labor... 411-426
  asphalt shingles 130
  built-up roofs......... 326
  elastomeric roofing 346
  insulation 351
  metal roofing........ 289
  roll roofing 157
  slate roofing 254
  wood shingles & shakes... 196
Estimating costs, materials
  asphalt shingles 130
  built-up roofs....... 325-326
  elastomeric roofing 346
  metal roofing........ 289
  roll roofing 157
  slate roofing 254
  wood shingles & shakes... 196
Estimating material quantities
  asphalt shingles 113-127
  built-up roofs....... 323-325
  caulk 367
  drip edge 42
  elastomeric roofing.... 345-346
  flashing 70-72
  gutters 410
  hip units 114-115, 226, 228
  interlayment 58-61
  metal roofing & siding . 269-275
  280-285
  ridge units......... 114-115
  roll roofing 144-156
  roofing repair 401-402
  shakes 176-185
  sheathing 34
  slate roofing 249-254
  tile roofing 225-226
  underlayment......... 49-56
  valley flashing 70-72
  waterproofing 375-376
  wood shingles & shakes . 176-185
Ethyl alcohol 379
Exceeding dead load 27
Exhaust fans 362, 403
Expanded insulation
  polystyrene 359
  urethane 354
Expansion 310, 337, 359, 366, 408
Expansion batten....... 264
Expansion cleat ... 258-259, 266
Expansion coefficient 264, 266-267
  .... 286, 321, 323
Expansion joint sealers..... 369
Expansion joints 315, 319
  .... 321-322, 375
Exposed insulation 354
Exposed-frame ceiling 360
Exposed-nail method 135, 139
  .... 141, 144, 147
Exposure gauge 422
Exposure, weather
  asphalt shingles 82, 106, 107, 111
  built-up roof 301-303
  decking 30-31
  re-roofing 396-397
  roll roofing 132
  sheathing 30-32
  underlayment...... 45, 57, 58
  wood shingles & shakes ... 174

F

Fabrics........... 300, 303
  composites 343, 377

polyester...... 303, 335, 342
  waterproofing 372, 374
Face grain, plywood 27
Factors
  allowance 18, 42
  hip-slope.......... 14-15
  percentage-of-increase . 225, 397
  roof-slope......... 428
  valley length 429
  valley-slope........ 14-15
  waste. See specific material type
Factory square 36-37, 51
Fancy-butt wood shingles ... 186
Fantail hip......... 240-241
Farm tile 372
Fascia 44
  cover 317
  raised......... 203, 206
  rake 24
  roofing repair 392, 401, 405, 408
  sloping 14
  tile roofing .. 202-203, 206, 207
  wood shingles & shakes 164-165
Fasteners
  asphalt shingles 106-107
  built-up roof 293-297
  decking 294
  elastomeric roofing ... 336, 338
  .... 340, 342-344
  mechanical .. 293-297, 342-344
  359-360
  metal roofs and siding ... 263
  slate roofing 244-245
  staggered nailing . 80, 106, 138
  tile roofing 204-206
  wood shingles & shakes ... 167
Feathering strips ... 397, 398, 400
Felt 36
  asphalt-saturated.. 36, 131, 291
  buckling 37
  coated 36, 291, 298, 300
  exposure 302
  organic 37, 301
  roofing 26, 31
  shim 390
  slate-faced......... 309
  strips 311, 317
  tar-saturated 36, 291, 315
  tearing 37
  warped........... 26
  waterproofing 373-375
Fiberboard
  high-density......... 339
  insulation 297, 298, 359
  rigid 313
Fibered asphalt mastic 373
Fiberglass
  asphalt-saturated.... 36-37, 301
  corrugated sheets 275
  fabric......... 303, 372, 379
  felts, built-up roof.... 300, 301
  .... 310, 329
  felts, waterproofing 360, 373, 375
  flashing 316, 317
  insulation ... 339, 348-350, 360
  insulation, built-up roof 295
  shingles 24, 38-40
  underlayment........ 36-37
Fiberglass shingles. See also
  Asphalt shingles
  laminated 37
  life expectancy....... 40, 73
Fibrated emulsion 378
Field slates 236, 252-254
Fire-resistant
  asphalt shingles 75

elastomeric roofs 342
  felts........... 301, 307
Fire-retardant,
  wood shingles & shakes.... 159
Fireproof doors 279
Fishmouths 304, 330
Flame-resistant insulation ... 350
Flange
  curb............ 321-322
  low-profile 321
  rubber 102
  straight........ 321, 322
  vent... 101-103, 105, 168, 220
Flash point........... 304
Flashing
  aluminum 65
  apron...... 168-169, 246, 262
  asphalt shingles 96-105
  built-up roof . 311-312, 315-320
  cement........ 47, 344
  continuous........ 100, 143
  .... 217-220, 261-262
  copper 65, 70, 242
  .... 245, 247, 287
  corners. 169, 186, 218-219, 370
  cricket 96, 168, 220, 222
  curb............ 318-319, 341
  demolition......... 392
  deteriorated 401
  door......... 186, 279, 370
  dormer..... 219-220, 246-247
  eaves 35, 61-62, 318
  edge............ 142
  elastomeric roofing ... 340-342
  344-345
  end-wall 288-289
  EPDM............ 334
  equipment stand 319-320
  estimating 70-72
  failure 330
  fiberglass 316-317, 370
  flat roof penetration... 318-320
  foundation 370
  front-wall 101
  gable 289
  gauge 65
  head............ 370
  hip 170
  J-bead 219
  lead........... 65, 70
  leaks 384
  metal roofing 261-262
  non-metal 67-68
  pan 210, 217-219, 222
  parapet wall....... 315-318
  pipe............ 318-319
  PVC 317
  rake 210-211
  ridge 170, 210
  roof edge 315, 317, 340
  saddle 70, 96, 168-169
  sealers 305-306
  shake roof........ 65, 70
  sidewall shingles........ 288
  skylights 315-316
  slate roofing 245-249
  soil stack 101-105, 220, 223, 344
  spandrel 370
  step.. 97, 99-100, 168, 169, 393
  through-wall 370
  tile roofing 210, 213-223
  tin............ 65, 245
  upper roof edge flashing ... 142
  vertical wall 99-101
  vinyl 317
  wall 194
  window 186, 279

wood shingles & shakes . . 65, 70
. . . . 192-194
   Z-bar 218-219
   zinc 65, 286
Flashing, base
   asphalt shingles 96-97, 99
   built-up roof 298, 312, 316
   elastomeric roofing 344
   slate roofing 246
Flashing, cap
   asphalt shingles 96, 98-100
   built-up roof 299, 316
   slate roofing 245, 246
Flashing, chimney
   asphalt shingles 96-100
   metal roofing 261-262
   roofing repair 382, 393
   slate roofing 246-247
   tile roofing . . 168-169, 220, 222
Flashing, counter
   asphalt shingles 96
   built-up roof 298, 312
   . . . . 316-319, 322
   metal roofing 262
   roofing repair 383, 393
   slate roofing 246
   tile roofing 218-222
Flashing, roll roofing 134, 142-143
   built-up roof 305
   repair 393
   underlayment 65, 68
Flashing, valley
   asphalt shingles 80
   metal roofing 261
   roll roofing 134
   tile roofing 200, 213-216
   underlayment 35, 50, 64-72
   wood shingles & shakes . . . 164
Flashing, vent
   asphalt shingles 101-105
   elastomeric roofing . . . 344-345
   roll roofing 142-143
   slate roofing 248
   tile roofing 220, 223-224
Flat asphalt 373
Flat seam, metal roofing 260
. . . . 285-287
Flat slate 232, 235
Flat spade 391
Flatgrain wood shingles . 160, 162
. . . . 389, 390
Flexible asphalt 73
Flood coat 305, 306, 307
Floor insulation . . . 356, 362, 365
Flue 382
Fluted metal deck 257
Fly rafter 24
Foam insulation . 292, 337, 348
. . . . 355, 360
   glass 343, 353, 359
Foil-faced insulation . . . 349, 350
Foot traffic
   built-up roof 292, 307
   elastomeric roofing 335
   metal roofing 257, 264
   roofing repair . . . 381, 396, 401
   slate roofing 234
   underlayment 35, 62
Forklifts 426
Foundation flashing 370
Foundation, pier and beam . . 364
Four-ply, built-up roof . . 302, 309
Four-ply, waterproofing . . 373-374
Framing, ladder 23
Free lime 388
Free-tab shingles 80
Freezebacks 61, 62

Friezeboard 405
Full-lace valley 382
Fully-adhered system 338, 343, 344
Fungus
   built-up roof 328
   roofing repair 383, 387
   waterproofing 376
   wood shingles & shakes 159-160
Furring strips 192
Fusion 303

G

Gable
   flashing 289
   molding 164
   overhang 30
   roof landing 33
Gable-end vent 406
Galvanic corrosion 269
Galvanized metal
   drip edges 40
   flashing 65
   metal roofing 272, 276
   roofing repair 385-386, 390
Galvanized steel shingles . . 288
Gauged shingle patterns . . . 185
Glass bead binder 355
Glass fiber mats 36
Glass, insulating 357
Glaze coat, built-up roof . 303, 307
. . . . 308, 310
Glazed tile 198
Gooseneck 408
Graded mineral aggregate 305, 307
Graduated roof, slate . . . 232-234
. . . . 237, 242
Gravel 307, 338, 339, 392
Gravel stop 41
   built-up roof 299, 317-318
   elastomeric roofing 340
Green concrete 378
Green lumber 26
Gross roof area 17-18
Gross vent area 402
Gutters
   basket 410
   built-up roof 292
   debris 383, 410
   estimating quantities 410
   half-round 408
   hangers 408-410
   molded 409
   ogee 408
   outlet 409
   rectangular 408
   round 408
   roofing repair . 386, 393, 407-410
   run 409
   spikes 408-409
   straps 408-410
   wood shingles & shakes . . . 165
Gypsum
   decks 295, 298, 300, 333
   insulation 307, 354, 355

H

Hail damage
   built-up roof 308
   elastomeric roofing 335
   metal roofing 288
   repairing 384, 388-390
Half pattern 82
Half-round gutter 408

Hand lugs 206
Handsplit wood shakes 162
. . . . 163, 175
Hard copper 286
Hard lead 245, 286
Head flashing 370
Head lap
   asphalt shingles 111-112
   slate roofing 234, 235
   . . . . 238, 249, 250
Head lugs 199
Headers 62, 63
Heat aging 343
Heat loss, interior 62
Heat transmission 361
Heat-activated adhesive 74
Heavy-load bearing decks . . . 315
Hexagonal shingles 395
High-density fiberboard 339
High-tin solder 268
High-wind precautions
   asphalt shingles 81, 93
   built-up roof 307
   elastomeric roofing . . . 337, 340
   roofing repair 384
   underlayment 48
Hip
   caps 288
   covers 136, 137, 139
   . . . . 140, 142, 269
   flashing 170, 240, 392
   underlayment 45, 51
Hip (jack) rafter 13-14, 21
Hip roof landing 34
Hip units
   asphalt shingles . . . 92-95, 109
   . . . . 111, 114-115
   gross roof area 18, 51
   roll roofing 145
   roofing repair 392
   slate roofing 241, 251
   tile roofing . . 211-213, 226, 228
   wood shingles & shakes . . . . .
   . . . . 169-170, 178
Hip-ridge junctures 93, 94
Hip-slope factors 14-15
Hooded pitch pan 319, 320
Horizontal supports 25
Horse feathers 192
Hot-dip process 255, 279
Hourly production 412
Humidity, relative 29
Hurricane precautions . . . 48, 206
Hydration 372
Hydrolithic cement 379
Hydrostatic head . . 370, 376, 377
Hypalon roofing 338

I

Ice damage precautions
   asphalt shingles 88, 96
   gutters 408, 410
   ice dams 61-63
   ice shields 62
   insulation 358
   sheathing 24
   underlayment 61-63
Impact marks 389
Inert fillers 314
Infrared radiation 342, 350
Injuries, roofing 334
Inlet vents 403, 404
Insulated doors 357
Insulating glass 357

Insulation 347-362
   aluminum-faced 350, 353
   asphalt-faced 337
   attic 356
   batt 348-350, 357-358
   benefits 347-348
   blanket 348, 357
   blown 352, 364
   building 347
   cellular glass 360, 363
   ceiling 61-63, 351, 356-358, 402
   compressed 312
   fiberboard 297-298, 359
   flame resistant 350
   floor 355, 362, 364
   foam 337, 343, 348
   . . . . 353-354, 359-360
   foil-faced 350, 354
   installation 351
   lightweight 293-294
   lightweight concrete fill . . . 355
   loose fill 350-351
   masonry wall 350-352
   mineral fiber 353
   mobile home 350
   paper-faced 349
   perimeter 353-355
   perlite 351
   phenol formaldehyde 339
   polystyrene 334, 339
   . . . . 343, 354, 360
   polyurethane 335, 353-354
   poured 351
   rigid 171, 173, 296, 334
   . . . 337, 346, 353, 359-360, 398
   rigid-fit 349
   rock wool 348, 350, 361
   roof . . 292, 298-301, 309-311
   315, 321, 337-338, 340, 342-343
   . . . . 356, 358-360
   slab 355
   sound-control batts 349
   sprayed 359-360
   sprayed foam 353-355
   stops 297, 317
   tapered 320, 354
   unfaced 349
   vermiculite 351-352
   wall 336, 353, 356-357
Insulation, R-values
   blown fiberglass 352
   blown rock wool 352
   calculating 361-362
   fiberglass batt 349
   loose fill 350
   recommendations 356
   rigid 353
   rock wool 349
Insurance
   homeowner's 24, 197
   . . . . 231, 388-389
   workers' compensation . . . 424
Integral battens 259
Integral standing seam 269
Interior heat loss 62
Interior partitions 363
Interlayment 32, 38-39
   . . . . 57-58, 162, 164
   coverage 58
   estimating quantities . . . 58-61
Interply bitumen 327, 373
Iron oxide 377
Iron shingles 288
Iron staining 387
Isbutylene-isoprene elastomers 376
Isocyanurate foam insulation . . . .
. . . . 354, 359

J

J channel............ 289
Jack rafter........... 13-14
   run............... 21
Job complexity...... 411, 420
Job delays........... 423
Job organization..... 423
Joints
   expansion 315, 319, 321-322, 375
   malleted............ 265
   mitered............. 169
   staggered.......... 26, 29
Joint spacing, plywood.... 29
Joint tape............ 341
Joists................ 25
Junctures............ 192
   apex............... 194
   concave............ 193
   convex............. 193
Jute.............. 303, 372

K

K-values.............. 361
Kaolin................ 307
Kerosene.............. 424
Kettle........ 304, 334, 376
Keys.................. 78
Kiln-dried lumber...... 26
Knots................. 165

L

L-brackets............ 173
Labor costs........... 411-426
   adjusting...... 414, 418-421
   asphalt shingles...... 130
   built-up roof......... 326
   elastomeric roofing... 346
   estimating......... 411-426
   metal roofing....... 289
   per unit........... 411-412
   production rates.... 411-412
   published......... 415-418
   reducing........... 420
   roll roofing......... 157
   slate roofing........ 254
   wood shingles & shakes... 196
Laced corner....... 187, 194
Lacing............... 57-58
Ladder framing........ 23
Ladders.......... 41, 425
Laminated
   fiberglass shingles..... 37
   roof decking......... 32
Lap
   areas............... 51
   caulk... 336, 340-341, 343-344
   cement.......... 336, 342
   head.............. 111-112
   .... 234-235, 238, 249-250
   Schmid............. 274
   top. 45, 111-112, 197, 200, 208
   underlayment..... 43, 45, 51
Last strip width........ 150
Latex caulk........... 365
Lath strips.... 189, 200-201, 240
Lead
   apron.............. 217
   flange.......... 220, 223
   flashing........... 65, 70
   roofing............. 39
   saddle.............. 69
   shield............. 245

skirt........ 66, 215-216, 223
sleeve.............. 101
soaker......... 210-213
strips.............. 245
Lead-tin alloy......... 268
Leader......... 292, 407, 409
   head............ 409, 410
   strap.............. 409
Leaks
   asphalt shingles...... 105
   chimneys............ 382
   eaves.............. 383
   flashings. 61-62, 64, 69, 383-384
   metal roofing..... 281-284
   mid-roof........... 384
   repairing........ 384-386
   roll roofing......... 143
   tile roofing......... 208
   underlayment........ 35
   valleys............. 383
   vent............... 384
   wood shingles & shakes 381, 385
Life expectancy
   asphalt shingles...... 73
   built-up roofing...... 291
   butyl caulk......... 365
   CPE elastomeric...... 342
   EPDM elastomeric.... 338
   epoxy sealant....... 367
   fiberglass shingles... 40, 73
   hypalon elastomeric... 343
   metal roofing....... 255
   polysulfide caulk..... 366
   polyurethane caulk... 366
   roll roofing......... 131
   silicone caulk....... 366
   slate roofing...... 40, 231
   solvent acrylic caulk.. 366
   tile roofing...... 40, 197
   wood shingles & shakes... 159
Lightning............ 426
Lightweight insulating concrete
   ...... 297, 333, 338, 355
Limestone......... 73, 307
Limited service warranty... 327
Liquid
   elastomers......... 376
   emulsions........... 305
   water repellents..... 377
Live load............ 27-28
Loads
   dead...... 27-28, 315, 337
   live............... 27-28
   uneven............. 32
Lock seam........ 247, 269
Lock-down shingle.. 127-128
Locking tab........ 127-128
Long terne........... 279
Long-life roofs.... 37, 40, 131
Lookout rafters...... 23-24
Loose
   insulation.......... 357
   knots............... 27
   mortar............. 383
   shingles............ 384
Loosely-laid systems... 334, 338
   .... 339, 343
Louvers.......... 402, 404
Low-profile flange..... 321
Low-slope roofs...... 28, 35
   .... 159, 162, 303
Lugs............ 198-199, 206
Lumber
   asphalt treated...... 341
   concrete forms....... 26
   creosote-treated..... 341
   green.............. 26

kiln-dried............ 26
pressure-treated..... 202
scrap............... 26

M

Machine-grooved shakes. 163, 186
Malleted joints........ 265
Manhours, estimating labor
   costs............. 411-426
   asphalt shingles...... 130
   built-up roofs....... 326
   elastomeric roofing... 346
   insulation.......... 351
   metal roofing....... 289
   roll roofing......... 157
   slate roofing........ 254
   tile roofing......... 229
   wood shingles & shakes... 196
Mansard roofs. 171-172, 195, 256
Marble chips......... 307
Masonry
   asphalt shingles...... 99
   built-up roofing... 311, 316
   insulation.......... 350
   waterproofing... 371, 377-379
   wood shingles & shakes... 192
Mastic............... 308
   asphalt....... 305, 373, 378
Material waste......... 5
Maximum span for plywood... 28
Mean temperature..... 24
Measuring roofs...... 5-6
Mechanical fasteners
   built-up roof.... 293, 295-297
   elastomeric roofing.. 336, 338
   .... 340, 342-344
   insulation....... 359-360
Membrane
   built-up roof...... 300-303
   elastomeric......... 336
   vented............. 298
Membrane roofing..... 291
Metal
   clips............. 66-67
   closures........ 201, 206
   coatings........... 255
   decks.............. 293
   eaves drip....... 207, 211
   foil composites...... 377
   patches.......... 26, 392
   valley flashing... 64-67, 383
Metal roof decks.... 310, 359
   roofing repair.. 382, 391, 395
Metal roof overhangs... 257
Metal roofing..... 255-289
   aluminum....... 277-278
   batten seams....... 259
   copper.......... 286-287
   corrugated... 24, 29, 271-276
   decking requirements.. 256-257
   estimating labor costs... 289
   fasteners........... 263
   flashings........ 261-262
   galvanized....... 270-271
   lead............... 286
   life expectancy..... 255
   monel.............. 285
   ribbed...... 24, 29, 271-273
   ridges............. 260
   stainless steel... 279-280, 285
   standing seams...... 258
   terne........... 278-284
   valleys............. 261
Metal roofing, estimating quantities
   copper.......... 286-287

corrugated sheets.... 274-275
per square........ 269-272
terne metal....... 280-285
Metal roofing, waste factors....
   269-270
copper.............. 287
corrugated steel sheets 274-275
galvanized steel panels... 272
terne metal....... 280-285
Metal shingles... 24, 38, 288-289
Metal siding...... 269, 273, 278
   aluminum.......... 277
   estimating quantities 269-270, 274
   siding panels....... 263
   terne........... 278-279
Micrometer........... 335
Mid-span, rafters...... 33
Mildew......... 159, 362, 383
Mineral
   colloids............ 306
   fiber insulation..... 353
   fillers............. 309
   granules....... 73-74, 131
   .... 308-309, 384
   stabilizers.......... 73
Mineral spirits........ 309
Mineral wool......... 348
Mineral-surfaced roll roofing
   131-157
   See also Roll roofing
   built-up roofing.... 300, 308
   roofing repair...... 394
   slate roofing....... 242
   tile roofing.. 214, 220-221, 223
   vent flashing....... 105
   underlayment..... 38, 48-49
   .... 64, 67-68
   waste factors..... 152-156
Minimum live load....... 27
Minimum roof slope .. 31, 38-39
Minimum span, plywood.... 28
Minimum underlayment... 37
Miter............. 409-410
Mitered corner........ 187
Mitered hip slates.... 239-241
Mitered joints......... 169
Mitered tiles....... 210-211
Mobile home insulation.... 350
Mobile placers........ 423
Modification factor.... 419-420
Modified bitumen asphalt (MBA).
   .... 132-133, 299
Moisture content of aggregate 307
Mold................ 362
Molded gutter........ 409
Mopping, spot..... 296, 298
   .... 309, 331, 337
Mortar
   built-up roofing..... 316
   dampproofing....... 378
   flashing....... 98-100, 370
   roofing repairs.... 382, 384
   slate roofing....... 242
   tile roofing..... 198, 203-212
   .... 217, 224
   waterproofing...... 372
Mortar-set method........ 48
Moss............. 159-160
Muriatic acid......... 256

N

Nail hole......... 382, 384, 385
Nailable decks, built-up roofs...
   ...... 309, 311, 313, 318, 331

Nailing flange 318
Nailing, plywood 29
Nailing strips
   built-up roof 313
   sheathing 31, 35
   slate roofing 238, 244
   wood shingles & shakes . . . 171
Nails. *See* Fasteners
National Construction Estimator,
   labor costs 411-426
   asphalt shingles 130
   built-up roofs 326
   elastomeric roofing 346
   metal roofing 289
   production rates 411-412
   published costs 415-421
   roll roofing 157
   sheathing 34
   slate roofing 254
   tile roofing 229
   unit prices 411
   wood shingles & shakes . . . 196
Negative method 7-8
Neoprene 344, 376
   adhesive 306
   sealants 370
Neoprene-hypalon roofing . . . 336
Net roof area 17-19, 21
. . . . 52, 59, 155
Net vent area 402, 404-405
Nickel 285
No-tab shingles 93, 107
Non-metal valley flashing . . 67-68
Non-nailable decks
   built-up roof 296, 309, 313
. . . . 315, 331
   elastomeric roofing 339
Non-skid surface 335
Nose lugs 198-199
Nosing 40
Number of eaves units 111
Number of slates required . . . 250

O

Ogee gutter 408
Open cornice 30, 405
Open valley 213, 243, 394
Organic felts 37, 310
Organic hydrocarbon 303
Organic shingles 74
Organic solvents 305
Organosilane sealers 379
Outside design temperature . . . .
. . . . 24-25, 299
Overcut allowance 52
Overhang
   asphalt shingles . 81-82, 93, 114
   metal roofs 257
   roll roofing 133
   roof 21, 23, 30, 63
   shakes 164, 177
   slates 236
   tiles 200
   underlayment 44
   wood shingles 164, 177
Overhead 415
Overhead power lines 426
Overlaps 52
Over-walling 190
Oxidation 310, 329, 379
Ozone 342, 343, 376

P

Paint
   aluminum 382, 392, 393

asphaltic 377
bituminous 66, 192, 245
plaster bond 378
vapor-retarder 364
Painting galvanized metal . . . 256
Pan flashing, tile roofing . 206, 208
. . . . 210, 217-219, 222
Panel clips 258
Panels
   concealed clip 278
   corrugated metal 29, 257
. . . . 274, 276, 295
   job-fabricated seams . . 265-266
   pre-cut metal 259
   uncoated metal 276
   vented metal 338
Paper, building 188-191
   asphalt-laminated 362
   rosin . . . 39, 297, 300, 313, 360
   rosin-sized 300, 308-309
Paper-faced insulation 349
Paper-pulp board 371
Parapet wall flashing
   built-up roofs . 309-312, 315-318
   elastomeric roofing 341
   metal roofing 262
   underlayment 35
Parge coat 378-379
Particleboard, compressed . . . 353
Paver blocks 339
Pavers 337, 339
Penetrating oils 306
Percentage-of-increase factor . . . .
. . . . 225, 397
Perforated pipe 372
Perforated soffit 403, 404
Perforated steel deck 295
Perimeter of roof 8-9, 17, 42
Perimeter insulation 355
Periphery 8
Perlite board . 298, 313, 354, 360
Phasing, built-up roof 310
Phenol formaldehyde insulation 339
Pier-and-beam foundation . . . 364
Pine shingles 160
Pipe
   bituminous fiber 372
   perforated 372
Pipe flashing 318-319
Pipe penetrations 300, 369
Pitch 11, 77, 236
   pan 319-320
   starter 289
Pitch bitumen 292, 297
. . . . 10, 333, 342, 355
Pits in shingles 388
Plan area 7, 19
Plan length 21
Planks, precast 297
Plaster bond paint 378
Plaster lath 238, 241
Plastic binders 36
Plastic corrugated sheets 275
Plastic films . . . 343, 362, 376-377
Plastic foam insulation 354
Plastic-bitumen composites . . 344
Plasticizers 314
Plate, bright 270
Plenum 299
Ply sheets 36-37
Plywood
   built-up roof . . . 292, 296, 300
   elastomeric roofing . . . 338-340
   grades 28
   grain direction 29
   rigid insulation 360
   sheathing 24, 27-29
   underlayment 43

Plywood panels
   clips 28
   spans 27
   thickness 28
   tongue and groove 29
Polyester fabric 303
Polyester polyurethane foam
   sealant 370
Polyester-reinforced fabrics . . . . .
. . . . 335, 342
Polyethylene
   film 355, 370
   rope 369
Polyisobutylene 336
Polyisocyanurate insulation . . . . .
. . . . 339, 353
Polysiloxane 335
Polystyrene beads 351
Polystyrene foam insulation . . 334
Polystyrene insulation 339
. . . . 343, 354, 360
Polysulfide caulk 365
Polyurethane caulk 365
Polyurethane insulation 335
. . . . 353-354
Polyvinyl chloride 336, 343
Ponded water
   built-up roof 292, 303
. . . . 310, 315, 328
   elastomeric roofing 334
Porous substrates 301
Positive method 7-8
Poured insulation 351
Power lines, overhead 426
Power tools 423, 424
Precast concrete decks 315
Precast planks 297
Precoating 245
Precut metal panels 259
Precut shingle 166-167
Prefabricated units
   apex tile 212
   hip 115
   ridge 115
Prefabrication 414
Preformed drip edge 258
Preformed pans 265-266
Pressure-cleaning 387
Pressure-treated lumber 202
Pressure-treated wood shingles 160
Prices, labor unit 414
Production rates 411-412
   maximizing 420, 422-424
   required 413
Profit 415
Protection board
   built-up roofs 296
   waterproofing . 373-375, 377-378
Protective gear 424
Protractor 12
Published hourly wages 414
Published labor unit prices . . . 414
Purified iron particles 379
Purlins . . . 25, 174, 256-259, 274
PVC flashing 317

Q

Quartz carbide sealer 379

R

R-values, insulation 349-362
   blown fiberglass 352
   blown rock wool 352
   calculating 361-362

   fiberglass 349, 352
   loose fill 350
   recommendations 356
   rigid 353
   rock wool 349, 352
Racking 86-87
Radiant heat 354
Rafter extension 405
Rafter spacing 27, 29
Rafters 13
   actual lengths 14-15
   common 13
   fly 24
   heel-cut 14
   hip 13-14, 21
   hip jack 13-14
   lookout 23-24
   mid-span 33
   plan length 14
   ridge 13
   run 14, 21
   tail-cut 14
   types 13-14
   valley 13-14
   valley jack 13-14
Raised fascia 203, 206
Rake fascia 24
Rake tiles . . . 208-210, 213, 226
Rakes 23, 24
   roll roofing 134-139, 146
   shingles 84, 164, 177
   underlayment 40, 43, 52
Random pattern 82-83, 87
Random slates 234, 237
Rebutted shingles 185
Recess 9
Recommended live loads 28
Recovery board 331
Rectangular gutter 408
Redwood 270
Redwood shingles 160
Reglet 341
Reinforced mineral fillers . . . 309
Reinforced polyester sheets . . 308
. . . . 335
Rejointed shingles 186
Rejuvenation 329
Relative humidity 29
Relief vents 295, 298, 319
Repairing
   asphalt shingles 384-385
   built-up roof 329-330
   shakes 385
   slates 386
   splits 390
   valleys 383
   wood shingles 385
Re-roofing 394-402
   asphalt shingles 394-399
   built-up roof 308
   estimating 401-402
   metal roofing 400
   roll roofing 400
   slate roofing 400-401
   tile roofing 401
   wood shingles & shakes . . . 400
Resaturant 306, 329
Resawn shakes 162-163, 175
Residual moisture 333
Resin, epoxy 410
Re-walling 190
Ribbon courses 124-127
. . . . 190-191, 392
Ridge 13, 45, 84
   length, equation 22
   sagging 24
Ridge board 13, 25

Ridge caps, metal roofing . 259-260
Ridge covers
   metal roofing 269
   roll roofing 136-137
   . . . . 139-140, 142
Ridge flashing 170, 210
Ridge rafter 13
Ridge roll 92
Ridge saddle 213
Ridge seam 260
Ridge shingles 382
Ridge slates 233
Ridge tiles 202-203, 212, 217, 226
Ridge units
   asphalt shingles 92-95
   . . . . 109, 111, 115
   estimating quantities . . . 114-115
   roll roofing 145
   roof area. 18
   underlayment. 51
   wood shingles & shakes. 169-170
   178
Ridge vents 260, 269, 403
Rigid fiberboard 313
Rigid insulation 354
   built-up roofs. 296
   elastomeric roofing 334, 337, 346
   roofing repair 398
   wood shingles & shakes . 173-174
Rise, roof 11-12
Rock wool insulation 349, 352, 361
Roll roofing 131-157
   See also Mineral-surfaced
      roll roofing
   built-up 300, 308
   concealed nail (blind nail). . . .
   . . . . 134, 137-139
   double-coverage 140-143
   estimating costs 157
   estimating quantities . . . 144-156
   exposed nail 134-137
   flashing 134, 142-143
   hip and ridge units 136
   . . . . 139-140, 142
   installation 133-144
   live expectancy 131
   modified bitumen asphalt (MBA)
   132-133
   pattern-edge 132
   re-roofing 400-401
   selvage. 132, 140
   sheathing under 24
   shed-roof 142
   single-coverage 134
   split-sheet 132, 140
   starter course. 80
   storage 132
   waste factors 144-156
Roll roofing, as flashing
   built-up roofs. 305
   roofing repair 393-394
   tile roofing 214, 220
   underlayment. 65, 68
Roll valley metal 65
Roll waterproofing 35
Roof area 17-19
Roof curb 269
Roof deck 23-24, 32
Roof drainage areas 408
Roof drains
   built-up roofs. 292, 315, 319-320
   elastomeric roofing 345
Roof edge flashing . . . 142, 315
   . . . . 317, 340
Roof edging 40
Roof frame 23
Roof inspection 381, 386

Roof insulation 358-360
   built-up roofs. . . . 292, 298-301
   . . . . 309-311, 315, 321
   elastomeric 337-338
   . . . . 340, 342-343
   wood shingles & shakes . 172-174
Roof juncture. 192
Roof load 27, 32
Roof maintenance . . 381, 386-388
Roof overhang . . . 21, 23, 30, 63
Roof penetrations
   asphalt shingles 89
   built-up roofs. . . 309, 315, 318
   elastomeric 340
   roofing repair . . . 383-384, 389
   underlayment. 49-50
Roof perimeter . . 8-9, 17, 339-340
Roof periphery 8
Roof pitch 11-12
Roof relief vent 295
Roof rise 11-12
Roof run 11-12
Roof sectioning. 20
Roof slope
   area. 17-21
   asphalt shingles 76-77
   measuring 6, 10-13, 15
   minimum 31, 38-39
   roll roofing 140
   underlayment. 38-40, 46, 52-53
   varying. 64
Roof-slope factors . . . 15, 17, 428
Roof span 11
Roof structure 32
Roof supports. 25
Roof types. 10
Roof walkway 314-315, 333
Roofer's hatchet 82, 422
Roofing cement
   Asphalt shingles 81, 88-89
   underlayment. 41, 43
   . . . . 46, 62, 67-69
Roofing equations 430-435
Roofing felts 26, 31
Roofing injuries 334
Roofing tape 382
Roofs
   level 7
   loading 32-34
   mansard 172, 194-195
   shed. 100
Rosin flux 268
Rosin paper 39, 297, 300, 313, 360
Rosin-sized paper . 300, 308-309
Rotten deck 29
Round gutter 408
Round valley 242-243
Rounding off 49
Rubber closure 206
Rubber vent flange . . . 102, 261
Rubber-vinyl composites. . . . 344
Rubberized asphalt . 132, 343, 377
Rubberized composites. 344
Rubberized membrane 376
Run
   rafter 21
   roof. 11-12

S

Saddle flashing . . 70, 96, 168-169
Saddle hip 240-241, 252
   strip. 240
Saddle ridge . . . 238, 240-241
   strip. 240
Sagging roof 24

Salvaged shingles 115
Saturant 73
Scaffolding 425
Scheduling. 423
Schmid lap 274
Scoria 307
Scrap lumber 26
Screen 409
Screened vent. . . . 402, 405, 406
Seal cap 382
Sealant 364
   epoxy 366
   foam 370
   neoprene 370
Sealant tape 369
Sealed underlayment system 46-48
   . . . . 202, 208, 217-220
Sealers
   organosilene 379
   quartz carbide 379
Seam solvent 343
Seams, metal roofing . . . 263-268
   batten. . . 266-267, 281-282, 284
   brazed 269-270
   cleats 265-266, 268
   cross 268, 271
   flat . 264-265, 271-273, 280-282
   integral standing 269
   job-fabricated. 263-272
   lock. 247, 269
   riveted 263-264
   soldered . . . 263-265, 268-269
   standing 247, 260, 262
   . . . . 265-266, 280-286
   welded 269
Sectioning a roof. 20
Seismic movement. 321
Seismic zones. 24
Selvage edge 132
Selvage strip 140
   starter strip 140-141
   waste 145
Separation sheet . . . 333, 337, 342
Settlement 99, 100
Shading. 87
Shadow coursing. 124-127
Shake
   bundles. . . . 163, 174, 175, 176
   spacing 165
   waste 176-180
Shakes 159-196
   Certi-guard 159
   Certi-sawn 163
   covering capacity 174-175
   estimating quantities . . 176-185
   exposures 165-166
   fire-retardant 159
   grading. 163
   hand-split 162-163, 175
   installation 164-174
   life expectancy 159
   machine-grooved . . . 163, 186
   resawn 162-163, 175
   repairing 386
   sheathing 24, 29-32
   sidewall 185
   staggered pattern 185
   starter-finish 166
   straight-split 162, 174-175
   tapersplit. 162-163, 175
   underlayment. 35
   valley flashing 65, 70
Shale 307
Sheathing 23-32
   deflection 292-293
   eaves 30-31, 164-165
   estimating quantities 34

   plywood 27-29
   skip 29
   solid 24-32, 76, 134
   . . . . . . 170-171, 200, 274, 400
   spaced 29-31, 39
   . . . . 57, 164-165, 172, 400
   support 29
   waferboard 29
Sheet lead 286
Sheet metal 235
   cold-rolled 235
   galvanized. 270-271
   hot-rolled 235
Sheet metal copper 287
Sheet metal gauge 255
Sheets
   plastic corrugated 275
   reinforced polyester. . . 308, 335
   V-beam 278
Shingle
   nails. 106
   patterning 87
   tabs 78
Shingle patterns
   asphalt 82-84
   Dutch weave 185
   gauged 185
Shingle siding. . 187-189, 194, 288
Shingle undercoursing . . 185, 188
Shingles
   Certi-guard 159
   Certi-last. 159
   clip-down 127-128
   damaged . 76, 385-386, 388, 395
   edge grain . . . 160, 162, 389-390
   enamel-coated 288
   fire-resistant 75, 159
   flatgrain . . . 160, 162, 389-390
   hexagonal . . . 79, 127-129, 395
   individual 395
   lock-down 127-128
   no-tab 93, 107
   pre-cut 166-167
   rebutted 185
   rejointed 186
   ridge 382
   salvaged 115
   slashgrain . . . 160, 162, 389-390
   T-lock 64, 79-80, 100
   . . . 111, 127, 395, 398-400, 422
   three-tab 74, 78-79, 422
   two-tab 106
   wind-resistant 185
Shiplap boards 26
Shoe. 408-409
Short terne 279
Side lap. 274, 293, 301
Side seam 268
Sidewall shingles
   bundles. 194
   coursing 187
   double-coursing 188-190
   exposure 189
   flashing 288
   installation 186-188
   ribbon coursing 190
   single-coursing . . . 188, 189-190
   spacing 188
   staggered coursing . . . 189-190
Siding
   flashing with 100
   metal 269, 273, 278
   shingle 187-189, 194, 288
Silicone caulk 365
Silicone rubber 335
Silicone sealant 226, 366, 379, 382
Single-coverage roll roofing
   . . . . 132, 142

442

Single-coverage underlayment . . .
. . . . 38-39, 43, 51, 202
Skylight flashing 315-316
Slab
   insulation 355
   topping. 375
Slag. 306-307, 315
Slashgrain wood shingles . . . 160
. . . . 162, 389-390
Slate. 231-254
   colors. 231
   combing 238-239
   commercial standard 232
   coursing 233
   damaged 386
   estimating costs 254
   estimating quantities . . . 249-254
   exposure 236
   fasteners 244-245
   flashing 245-249
   grades 232
   graduated . . . 232-234, 237, 242
   head lap 236-237
   hips 240-241
   nailing 239
   life expectancy 40, 231
   mitered-hip 239, 241
   random 234, 237
   repair 386
   re-roofing 401
   ridges 233, 238-240
   sizes 232
   standard 232-245
   textural 232-234, 242
   thickness 232
   tools 249
   trade names 232
   trapezoidal 240
   triangular 240
   under-eaves 236, 251
   underlayment 234
   unfading 232
   unpunched 234
   valleys 242-244
   weathering 232
   weight 234-235
Slate jointing 234, 237
Slate roofs . . . 35, 36, 38, 65, 286
Slate-surfaced felt 309
Slater's punch 249
Slating nails 245
Slip joint connectors 410
Slip sheet . . . 300, 333, 337, 343
Slippery surfaces 425-426
Slope 10, 12-13
   degrees 12
   variable 20
Sloping fascia 14
Slotted corrugated panels . . . 295
Smooth-surfaced roof 317
. . . . 338, 400
Snow loading . 256, 264, 292, 359
Snow precautions
   elastomeric roofing 334
   flashing 61, 96
   interlayment 57
   maintenance 387
   tile roofing 202
   wood shingles & shakes 164, 173
Soffit 195, 256
   perforated 403-404
   ventilation 61, 172, 402
Soft copper 245, 286
Solder 245, 247-248, 408
   acid-flux 249
   aluminum 268
   50-50 lead-tin 268

   high-tin 268
   rosin-flux 245
Solid sheathing . . . 24-32, 76, 134
. . . . 170, 200
Solvent
   organic 305
   seam 343
Solvent acrylic caulk 365
Sound control batts 349
Sound transmission 349
Spaced battens 171-172
Spaced sheathing 29-31
. . . . 57, 164, 172
   over solid sheathing . . . 32, 400
Spacer tube 408-409
Spacers 31, 170-171
Spacing, sheathing boards 30
Span 11, 278
Span rating 28
Spandrel flashing 370
Splash block 409-410
Splash diverter 64, 66, 410
Splice covers 321, 322
Sponge 212, 217
Spot mopping
   built-up roof 296, 298, 309, 331
   elastomeric roofing 337
Sprayed insulation 359, 360
Spud 391
Squangle® 12
Square 18
   factory 51
Square-edged boards 26, 30
Stacking 111
Stagger nailing 80, 106, 138
Staggered butts 78
Staggered joints, decks . . . 26, 29
Stainless steel composites . . . 343
Stainless steel roofing 39
Standing seams, metal roofing . . .
. . . . 265-266, 280-282, 284
Staples 107-108, 167, 401
Starter course
   asphalt shingles . . 88, 100, 111
   estimating 18
   metal roofing 289
   roofing repair 396, 399
   underlayment 57-59
   wood shingles & shakes . . . 164
. . . . 171, 177
Starter roll 80, 398-399
Starter strip 203
   allowance 145
Starter-finish shake 166
Statement of compliance 327
Steel
   deck 174
   door 357
Steel-core shingles 288
Steep asphalt
   built-up roof 293, 300, 311, 316
   insulation 360
   underlayment 48
Step flashing
   asphalt shingles 97, 99-100
   roofing repair 393
   wood shingles & shakes 168, 169
Stop, gravel 41, 299, 317-318, 340
Storm clips . . . 206-207, 210, 213
Storm door 357
Storm sash 357
Story pole 186-187
Straight flange 321, 322
Straight-bond method, tile . . . 208
Straight-split shakes . 162, 174, 175
Straight-up method, asphalt
   shingle 86

Strainers 410
Strapping 172-173, 297-298
Stretch, asphalt shingles 24
Strip copper 287
Strip saddle ridge 238
Strip-shingle roofs 64
Structural cement fiber deck . 300
Structural corrugated glass . . 277
Structural expansion joints . . . 321
Structural wood fiber deck . . 297
Struts 24-25
Stucco . . . 100, 192, 218-219, 372
Subcontractors 424
Sulfur dioxide 370
Sulfuric acid 370
Supervision 415
Support spacing 28-29
Supports, horizontal 25
Surface primer 366
Surface-coated tile 198
Sweat sheet 43
Swept eaves 171-172
Synthetic rubber 342, 376

T

T-bevel 12-13
T-lock shingles
   asphalt shingles 79-80, 100
. . . . 111, 127
   flashing 64
   roofing repair . . . 395, 398-400
Tab notches 78
Tabs, shingle 78
   locking 127-128
Tape measure 5
Tapered edge strips 313
Tapered insulation 320
Tapersplit shakes . . 162, 163, 175
Tar saturant 36
Tar strip 75
Tarred felt 36
Tedlar sheets 329
Tees 322
Temporary roof 313
Termite shield 279
Terne metal roofing & siding . . . .
. . . . 278-280, 283
   estimating quantities . . 283, 285
   seams 263-266, 268
   underlayment 39-40
   waste factors . 281-282, 284-285
Terne-coated stainless steel . . . .
. . . . 279-280, 283
Terne plate 270, 278
Textural slate 232-234, 242
Thermal conductivity . . . 361-362
Thermal expansion 264
Three-ply waterproofing . . 373-374
Three-tab shingles . 74, 78-79, 422
Through-wall flashing 370
Tie-in 420
Tile gaps 208
Tile roofing 197-229
   accessories 226
   barge 208
   barrel 207-208
   body-colored 198
   clay 197
   concrete 197
   direct-nail 204
   estimating quantities . . 225-226
   farm 372
   fasteners 204-206
   glazed 198
   installation 200-217

   life expectancy 40, 197
   loading 33-34
   mitered 210-211
   mortar finishing 210, 217
   mortar mix 226
   mortar-set 204, 206-207
   prefabricated apex 212
   rake 208-210, 213, 226
   replacement 224
   repairing 382
   re-roofing 401
   ridge . . 202-203, 212, 217, 226
   shapes 197
   surface coated 198
   straight-bond method 208
   trimming 211
   underlayment 35-38, 65
   under-eaves 203-204
   unglazed 198
   V-ridge 210-211
   weights 197-198
Tin caps 44, 360
Tin flashing 65, 245
Tin roofing 263
Tin tags 44, 47, 48, 221
Tinning 245
Titanium dioxide 342
Tongue-and-groove
   boards 26, 32
   deck 26
   plywood 29
Top lap
   asphalt shingles 111-112
   tile roofing 197, 200, 208
   underlayment 45
Topping slab 375
Torching 132
Traffic pads 314-315
Trap rock 73
Trapezoidal slates 240
Tree sap 27
Triangular slates 240
Trowel 206, 207, 217
Two-tab shingles 106

U

U-value 347, 361
Ultraviolet light
   built-up roof 310, 329
   caulking 365-366
   elastomeric roofing 334-335
. . . . 342-343
   insulation 359
   waterproofing 376
Umbrella 320
Uncoated felts 301
Uncoated metal panels 276
Under-eaves slates 236, 251
Under-floor ventilation 364
Under-shimming 390
Undercoursing shingles . . 185, 188
Underlayment 35-72
   buckling 62, 133
   coursing 43
   double-coverage 38-39, 43
. . . . 51, 62
   end laps 45-46
   estimating quantities 49-56
   installation 43-49
   laps 45-47, 51, 62, 68-69
   minimum 37
   non-sealed . . . 46-47, 199, 213
. . . . 215, 217-220
   overhang 44
   recommended 36-40

saturated felt 36
saturated fiberglass 36
sealed 46-48, 199-200
. . . . 202, 214, 216-221
single coverage 38-39, 43
. . . . 51, 202
unsealed system 66, 215
valley 43-45, 68-69
waste factors 50, 52-53, 58
weights 36-37
Uneven roof loads 32
Unfaced insulation 349
Unfading slate 232
Unglazed tile 198
Uniform live loads 28
Unpunched slate 234
Unsaturated felt 337
Unsealed underlayment system . . .
. . . . 66, 215
Upper roof edge flashing . . . 142
Uprights 24-25
Urethane 335, 353, 359, 366

V

V-beam sheets 278
V-ridge tile 210-212
Valley
blocks 242, 244
canoe 243-244
clip 215
closed 64-65, 242, 244
closed-cut 65, 67, 89-92
half-lace 65, 67, 89-92
length factors 429
open 64-65, 89, 242
radius 244
roll roofing 134, 137
round 242-243
underlayment 43-45, 49
woven 65, 67, 89-92
Valley debris 64, 66, 89, 382
Valley flashing 64-72
asphalt shingles 80
estimating 70-72
installing, metal 68-70
installing, non-metal . . . 67-68
metal 64-67, 214, 242
metal roofing 261
roll roofing 134
tile roofing 200, 213-216
underlayment . . . 35, 50, 64-72
wood shingles & shakes . . . 164
Valley jack rafter 13-14
Valley length factors 22, 429
Valley rafter 13-14
plan length 21
Valley-ridge juncture . . . 213, 215
Valley-slope factors 14-15
Vapor barrier (retarder) . . 362-364
aluminum foil 362
batt insulation 349-350
built-up roof 295, 298-300
. . . . 312-313, 363-364
crawl space, location 365
disadvantage 362

effectiveness 363
elastomeric roofing 337
floor location 365
foil 362
insulating doors & windows 357
materials 362
placement 362-363
rigid insulation 353-354
tile roofing 218-219, 221
wood shingles & shakes . 173-174
Vapor retarder paint 364
Varying roof slopes 64
Vaulted ceiling 360
Vent collar 101
Vent flange 101, 103, 105
. . . . 168, 220
flange, rubber 102, 261
Vent flashing
asphalt shingles 101-105
boot 344-345
elastomeric roofing . . . 344-345
roll roofing 142-143
slate roofing 248
tile roofing 220, 223-224
Vent jack 220, 223
Vent louvers 402, 404
Vent pipe
asphalt shingles 104-105
built-up roof 318
roll roofing 143
roofing repair 382
slate roofing 248
tile roofing 220
Vent screen 402
Vent sleeve 101
Vent space 63
Vent stack 101, 168, 344
Vented base sheet 312, 364
Vented membrane 298
Vented metal panels 338
Ventilation 29, 63, 202, 295
attic 402-407
underfloor 364
Ventilator 248
Vents
adjustable 406
dryer 362
gable-end 406
gross area 402
inlet 403-404
relief 295, 298, 319
ridge 260, 269, 403
screened 405-406
vinyl ridge 403
Vermiculite insulation 351
Vinyl acrylic caulk 365
Vinyl acrylic roofing 335
Vinyl ridge vent 403
Viscosity 305
Volcanic glass 355
Vulcanized rubber roofing . . 337

W

Waferboard 29
sheathing 24

Wage rates 415, 419
Walking in 48, 133
Walkways 333
Wall flashing 194, 370-371
Wall insulation 336, 353
Warped felts 26
Warped shingles 395
Warranty 5, 40, 327, 338
Waste 5
Waste factors 124, 269
Water course 199
Water cut-offs 311
Water dam 258, 383
Water diverter 213
Water guards
metal roofing 261
slate roofing 242
tile roofing . . 210, 213-214, 217
valley flashing 64, 66
Water jackets 78
Water lines 78
Water repellents 371
Water shield 61, 96
Water-retaining roofs . . . 306, 315
Waterproof wax 378
Waterproofing 371
bentonite 377
built-up 372-375
cold-applied 374
composite 377
elastomeric 334, 376-377
epoxy 379
estimating quantities . . 375-376
installation 373-375
integral 372
liquid plastic 377
phase method 374
roll 35
shingle method 374
Waterstops 371
Weather 424
Weather checks 198-199
Weathering slate 232
Weatherstripping 402
Woven corner 187
Weep holes 207, 211-212
Weight
built-up roof 27
clay tile 27
roof deck 27
roofing felt 27
shakes 162
Welding 285
Wellpoints 377
White cedar shingles 185
White lead 265
Wicking 37
Wind block 213-214
Wind damage 384
Wind-resistant roofing
asphalt shingles 781, 93
built-up roof 307
elastomeric roofing . . . 337, 340
roofing repair 384
shingles 75, 88
underlayment 48

Windows 62, 63, 247
casing 186
flashing 186, 279
Wood batten strips 264, 266
Wood ceiling 32
Wood closure strips 263
Wood nailers . . 297-298, 311, 341
Wood resins 35, 305, 344
Wood shingles 159-196
bell eaves 171, 172
bundles 161, 177
covering capacity 174, 176
decking, metal 173
Dutch weave 185
estimating costs 196
estimating quantities . . 176-185
exposure 175, 177, 179-180
fancy-butt 186
fasteners 167
fire-retardant 159
flashing 167-169, 192-194
grading 161-162
grain 160
hip units 169-170
installation 164-174
insulation, rigid 171-172
joints 166
life expectancy 159
low-slope applications . 170-171
mansard roof 171, 172
nails 167
panels 194-195
patterns 185
pressure-treated 160
redwood 160
repair 385
ridge units 169-170
roof junctures 192-194
sheathing under 24, 29-30
sidewall 185-192
staggered patterns 185
starter course 177
steep-slope applications . . . 171
underlayment 37, 39
valleys 65, 69-70, 166-167
waste factors 177-185
white cedar 185
Wood starter strip 202
Workers' compensation insurance .
. . . . 424
Workspace limitations 420
Woven fiberglass flashing . . . 370
Woven valley 90, 92
Wye tile 212, 213

XYZ

Z-bar flashing 218-219
Z-closure 260
Zinc flashing 65, 286
Zinc napthenate 160

Practical References for Builders

Basic Engineering for Builders

This book is for you if you've ever been stumped by an engineering problem on the job, yet wanted to avoid the expense of hiring a qualified engineer. Here you'll find engineering principles explained in non-technical language and practical methods for applying them on the job. With the help of this book you'll be able to understand engineering functions in the plans and how to meet the requirements, how to get permits issued without the help of an engineer, and anticipate requirements for concrete, steel, wood and masonry. See why you sometimes have to hire an engineer and what you can undertake yourself: surveying, concrete, lumber loads and stresses, steel, masonry, plumbing, and HVAC systems. This book is designed to help you, the builder, save money by understanding engineering principles that you can incorporate into the jobs you bid. **400 pages, 8½ x 11, $36.50**

Basic Plumbing with Illustrations, Revised

This completely-revised edition brings this comprehensive manual fully up-to-date with all the latest plumbing codes. It is the journeyman's and apprentice's guide to installing plumbing, piping, and fixtures in residential and light commercial buildings: how to select the right materials, lay out the job and do professional-quality plumbing work, use essential tools and materials, make repairs, maintain plumbing systems, install fixtures, and add to existing systems. Includes extensive study questions at the end of each chapter, and a section with all the correct answers.
384 pages, 8½ x 11, $33.00

National Concrete & Masonry Estimator

Since you don't get every concrete or masonry job you bid, why generate a detailed list of materials for each one? The data in this book will allow you to get a quick and accurate bid, and allow you to do a detailed material takeoff, only for the jobs you are the successful bidder on. Includes assembly prices for bricks, and labor and material prices for brick bonds, brick specialties, concrete blocks, CMU, concrete footings and foundations, concrete on grade, concrete specialties, concrete beams and columns, beams for elevated slabs, elevated slab costs, and more. Includes a CD-ROM with an electronic version of the book with *National Estimator*, a stand-alone *Windows*™ estimating program, plus an interactive multimedia video that shows how to use the disk to compile construction cost estimates.
672 pages, 8½ x 11, $54.00. Revised annually

Craftsman's Construction Installation Encyclopedia

Step-by-step installation instructions for just about any residential construction, remodeling or repair task, arranged alphabetically, from Acoustic tile to Wood flooring. Includes hundreds of illustrations that show how to build, install, or remodel each part of the job, as well as manhour tables for each work item so you can estimate and bid with confidence. Also includes a CD-ROM with all the material in the book, handy look-up features, and the ability to capture and print out for your crew the instructions and diagrams for any job.
792 pages, 8½ x 11, $65.00

Handbook of Construction Contracting, Volume 1

Everything you need to know to start and run your construction business: the pros and cons of each type of contracting, the records you'll need to keep, and how to read and understand house plans and specs so you find any problems before the actual work begins. All aspects of construction are covered in detail, including all-weather wood foundations, practical math for the job site, and elementary surveying. **416 pages, 8½ x 11, $32.75**

Handbook of Construction Contracting, Volume 2

Everything you need to know to keep your construction business profitable: different methods of estimating, keeping and controlling costs, estimating excavation, concrete, masonry, rough carpentry, roof covering, insulation, doors and windows, exterior finishes, specialty finishes, scheduling work flow, managing workers, advertising and sales, spec building and land development, and selecting the best legal structure for your business.
320 pages, 8½ x 11, $33.75

Basic Lumber Engineering for Builders

Beam and lumber requirements for many jobs aren't always clear, especially with changing building codes and lumber products. Most of the time you rely on your own "rules of thumb" when figuring spans or lumber engineering. This book can help you fill the gap between what you can find in the building code span tables and what you need to pay a certified engineer to do. With its large, clear illustrations and examples, this book shows you how to figure stresses for pre-engineered wood or wood structural members, how to calculate loads, and how to design your own girders, joists and beams. Included FREE with the book — an easy-to-use limited version of NorthBridge Software's *Wood Beam Sizing* program.
272 pages, 8½ x 11, $38.00

Finish Carpenter's Manual

Everything you need to know to be a finish carpenter: assessing a job before you begin, and tricks of the trade from a master finish carpenter. Easy-to-follow instructions for installing doors and windows, ceiling treatments (including fancy beams, corbels, cornices and moldings), wall treatments (including wainscoting and sheet paneling), and the finishing touches of chair, picture, and plate rails. Specialized interior work includes cabinetry and built-ins, stair finish work, and closets. Also covers exterior trims and porches. Includes manhour tables for finish work, and hundreds of illustrations and photos. **208 pages, 8½ x 11, $22.50**

National Home Improvement Estimator

Current labor and material prices for home improvement projects. Provides manhours for each job, recommended crew size, and the labor cost for the removal and installation work. Material prices are current, with location adjustment factors and free monthly updates on the Web. Gives step-by-step instructions for the work, with helpful diagrams, and home improvement shortcuts and tips from an expert. Includes a CD-ROM with an electronic version of the book, and *National Estimator*, a stand-alone *Windows*™ estimating program, plus an interactive multimedia tutorial that shows how to use the disk to compile home improvement cost estimates.
504 pages, 8½ x 11, $53.75. Revised annually

Construction Forms & Contracts

125 forms you can copy and use — or load into your computer (from the FREE disk enclosed). Then you can customize the forms to fit your company, fill them out, and print. Loads into *Word for Windows*, *Lotus 1-2-3*, *WordPerfect*, *Works*, or *Excel* programs. You'll find forms covering accounting, estimating, fieldwork, contracts, and general office. Each form comes with complete instructions on when to use it and how to fill it out. These forms were designed, tested and used by contractors, and will help keep your business organized, profitable and out of legal, accounting and collection troubles. Includes a CD-ROM for *Windows*™ and *Mac*™.
432 pages, 8½ x 11, $41.75

Residential Steel Framing Guide

Steel is stronger and lighter than wood — straight walls are guaranteed — steel framing will not warp, shrink, split, swell, bow, or rot. Here you'll find full page schematics and details that show how steel is connected in just about all residential framing work. You won't find lengthy explanations here on how to run your business, or even how to do the work. What you will find are over 150 easy-to-read full-page details on how to construct steel-framed floors, roofs, interior and exterior walls, bridging, blocking, and reinforcing for all residential construction. Also includes recommended fasteners and their applications, and fastening schedules for attaching every type of steel framing member to steel as well as wood.
170 pages, 8½ x 11, $38.80

Profits in Buying & Renovating Homes

Step-by-step instructions for selecting, repairing, improving, and selling highly profitable "fixer-uppers." Shows which price ranges offer the highest profit-to-investment ratios, which neighborhoods offer the best return, practical directions for repairs, and tips on dealing with buyers, sellers, and real estate agents. Shows you how to determine your profit before you buy, what "bargains" to avoid, and how to make simple, profitable, inexpensive upgrades. **304 pages, 8½ x 11, $24.75**

A Roof Cutter's Secrets to Custom Homes

A master framer spills his secrets to framing irregular roofs, jobsite solutions for rake walls, and curved and two-story walls. You'll also find step-by-step techniques for cutting bay roofs, gambrels, and shed, gable, and eyebrow dormers. You'll even find instructions on custom work like coffered ceilings, arches and barrel vaults; even round towers, hexagons, and other polygons. Includes instructions for figuring most of the equations in this book with the keypad of the Construction Master Pro calculator.
342 pages, 8½ x 5½, $32.50

Plumber's Handbook Revised

This new edition shows what will and won't pass inspection in drainage, vent, and waste piping, septic tanks, water supply, graywater recycling systems, pools and spas, fire protection, and gas piping systems. All tables, standards, and specifications are completely up-to-date with recent plumbing code changes. Covers common layouts for residential work, how to size piping, select and hang fixtures, practical recommendations, and trade tips. It's the approved reference for the plumbing contractor's exam in many states. Includes an extensive set of multiple-choice questions after each chapter, with answers and explanations in the back of the book, along with a complete sample plumber's exam.
352 pages, 8½ x 11, $36.50

Stair Builders Handbook

If you know the floor-to-floor rise, this handbook gives you everything else: number and dimension of treads and risers, total run, correct well hole opening, angle of incline, and quantity of materials and settings for your framing square for over 3,500 code-approved rise and run combinations — several for every 1/8-inch interval from a 3 foot to a 12 foot floor-to-floor rise. **416 pages, 5½ x 8½, $19.50**

Contractor's Guide to *QuickBooks Pro* 2007

This user-friendly manual walks you through *QuickBooks Pro*'s detailed setup procedure and explains step-by-step how to create a first-rate accounting system. You'll learn in days, rather than weeks, how to use *QuickBooks Pro* to get your contracting business organized, with simple, fast accounting procedures. On the CD included with the book you'll find a *QuickBooks Pro* file for a construction company. (Open it, enter your own company's data, and add info on your suppliers and subs.) You also get a complete estimating program, including a database, and a job costing program that lets you export your estimates to *QuickBooks Pro*. It even includes many useful construction forms to use in your business.
352 pages, 8½ x 11, $53.00

Also available: **Contractor's Guide to *QuickBooks Pro* 1999, $42.00**
Contractor's Guide to *QuickBooks Pro* 2001, $45.25
Contractor's Guide to *QuickBooks Pro* 2002, $46.50
Contractor's Guide to *QuickBooks Pro* 2003, $47.75
Contractor's Guide to *QuickBooks Pro* 2004, $48.50
Contractor's Guide to *QuickBooks Pro* 2005, $49.75
Contractor's Guide to *QuickBooks Pro* 2006, $51.50

Markup & Profit: A Contractor's Guide

In order to succeed in a construction business, you have to be able to price your jobs to cover all labor, material and overhead expenses, and make a decent profit. The problem is knowing what markup to use. You don't want to lose jobs because you charge too much, and you don't want to work for free because you've charged too little. If you know how to calculate markup, you can apply it to your job costs to find the right sales price for your work. This book gives you tried and tested formulas, with step-by-step instructions and easy-to-follow examples, so you can easily figure the markup that's right for your business. Includes a CD-ROM with forms and checklists for your use. **320 pages, 8½ x 11, $32.50**

Rough Framing Carpentry

If you'd like to make good money working outdoors as a framer, this is the book for you. Here you'll find shortcuts to laying out studs; speed cutting blocks, trimmers and plates by eye; quickly building and blocking rake walls; installing ceiling backing, ceiling joists, and truss joists; cutting and assembling hip trusses and California fills; arches and drop ceilings — all with production line procedures that save you time and help you make more money. Over 100 on-the-job photos of how to do it right and what can go wrong. **304 pages, 8½ x 11, $26.50**

Electrician's Exam Preparation Guide

Need help in passing the apprentice, journeyman, or master electrician's exam? This is a book of questions and answers based on actual electrician's exams over the last few years. Almost a thousand multiple-choice questions — exactly the type you'll find on the exam — cover every area of electrical installation: electrical drawings, services and systems, transformers, capacitors, distribution equipment, branch circuits, feeders, calculations, measuring and testing, and more. It gives you the correct answer, an explanation, and where to find it in the latest NEC. Also tells how to apply for the test, how best to study, and what to expect on examination day. Includes a FREE CD-ROM with all the questions in the book in interactive test-yourself software that makes studying for the exam almost fun! Updated to the 2005 NEC. **352 pages, 8½ x 11, $39.50**

Estimating Electrical Construction

Like taking a class in how to estimate materials and labor for residential and commercial electrical construction. Written by an A.S.P.E. National Estimator of the Year, it teaches you how to use labor units, the plan take-off, and the bid summary to make an accurate estimate, how to deal with suppliers, use pricing sheets, and modify labor units. Provides extensive labor unit tables and blank forms for your next electrical job. **272 pages, 8½ x 11, $35.00**

Contractor's Survival Manual Revised

The "real skinny" on the down-and-dirty survival skills that no one like to talk about – unique, unconventional ways to get through a debt crisis: what to do when the bills can't be paid, finding money and buying time, conserving income, transferring debt, setting payment priorities, cash float techniques, dealing with judgments and liens, and laying the foundation for recovery. Here you'll find out how to survive a downturn and the key things you can do to pave the road to success. Have this book as your insurance policy; when hard times come to your business it will be your guide. **336 pages, 8½ x 11, $38.00**

National Construction Estimator

Current building costs for residential, commercial, and industrial construction. Estimated prices for every common building material. Provides manhours, recommended crew, and gives the labor cost for installation. Includes a CD-ROM with an electronic version of the book with *National Estimator*, a stand-alone *Windows*™ estimating program, plus an interactive multimedia video that shows how to use the disk to compile construction cost estimates. **656 pages, 8½ x 11, $52.50. Revised annually**

Estimating With Microsoft Excel

Most builders estimate with *Excel* because it's easy to learn, quick to use, and can be customized to your style of estimating. Here you'll find step-by-step instructions on how to create your own customized automated spreadsheet estimating program for use with *Excel*. You'll learn how to use the magic of *Excel* to create detail sheets, cost breakdown summaries, and links. You'll put this all to use in estimating concrete, rebar, permit fees, and roofing. You can even create your own macros. Includes a CD-ROM that illustrates examples in the book and provides you with templates you can use to set up your own estimating system. **148 pages, 7 x 9, $39.95**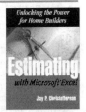

Electrical Blueprint Reading Revised

Shows how to read and interpret electrical drawings, wiring diagrams, and specifications for constructing electrical systems. Shows how a typical lighting and power layout would appear on a plan, and explains what to do to execute the plan. Describes how to use a panelboard or heating schedule, and includes typical electrical specifications.
208 pages, 8½ x 11, $18.00

Construction Estimating Reference Data

Provides the 300 most useful manhour tables for practically every item of construction. Labor requirements are listed for sitework, concrete work, masonry, steel, carpentry, thermal and moisture protection, doors and windows, finishes, mechanical and electrical. Each section details the work being estimated and gives appropriate crew size and equipment needed. Includes a CD-ROM with an electronic version of the book with *National Estimator*, a stand-alone *Windows*™ estimating program, plus an interactive multimedia video that shows how to use the disk to compile construction cost estimates. **432 pages, 11 x 8½, $39.50**

National Repair & Remodeling Estimator

The complete pricing guide for dwelling reconstruction costs. Reliable, specific data you can apply on every repair and remodeling job. Up-to-date material costs and labor figures based on thousands of jobs across the country. Provides recommended crew sizes; average production rates; exact material, equipment, and labor costs; a total unit cost and a total price including overhead and profit. Separate listings for high- and low-volume builders, so prices shown are specific for any size business. Estimating tips specific to repair and remodeling work to make your bids complete, realistic, and profitable. Includes a CD-ROM with an electronic version of the book with *National Estimator*, a stand-alone *Windows*™ estimating program, plus an interactive multimedia video that shows how to use the disk to compile construction cost estimates.
296 pages, 8½ x 11, $53.50. Revised annually

Wood-Frame House Construction

Step-by-step construction details, from the layout of the outer walls, excavation and formwork, to finish carpentry and painting. Packed with clear illustrations and explanations updated for modern construction methods. Everything you need to know about framing, roofing, siding, interior finishings, floor covering and stairs — your complete book of wood-frame homebuilding. **320 pages, 8½ x 11, $25.50. Revised edition**

The Contractor's Legal Kit

Stop "eating" the costs of bad designs, hidden conditions, and job surprises. Set ground rules that assign those costs to the rightful party ahead of time. And it's all in plain English, not "legalese." For less than the cost of an hour with a lawyer you'll learn the exclusions to put in your agreements, why your insurance company may pay for your legal defense, how to avoid liability for injuries to your sub and his employees or damages they cause, how to collect on lawsuits you win, and how to protect yourself from claims related to mold. It also includes a FREE computer disk with contracts and forms you can customize for your own use.
376 pages, 8½ x 11, $69.95

CD Estimator

If your computer has *Windows*™ and a CD-ROM drive, CD Estimator puts at your fingertips over 135,000 construction costs for new construction, remodeling, renovation & insurance repair, home improvement, framing & finish carpentry, electrical, concrete & masonry, painting, and plumbing & HVAC. Monthly cost updates are available at no charge on the Internet. You'll also have the *National Estimator* program — a stand-alone estimating program for *Windows*™ that *Remodeling* magazine called a "computer wiz," and *Job Cost Wizard*, a program that lets you export your estimates to *QuickBooks Pro* for actual job costing. A 60-minute interactive video teaches you how to use this CD-ROM to estimate construction costs. And to top it off, to help you create professional-looking estimates, the disk includes over 40 construction estimating and bidding forms in a format that's perfect for nearly any *Windows*™ word processing or spreadsheet program.
CD Estimator is $78.50

Renovating & Restyling Older Homes

Any builder can turn a run-down old house into a showcase of perfection — if the customer has unlimited funds to spend. Unfortunately, most customers are on a tight budget. They usually want more improvements than they can afford — and they expect you to deliver. This book shows how to add economical improvements that can increase the property value by two, five or even ten times the cost of the remodel. Sound impossible? Here you'll find the secrets of a builder who has been putting these techniques to work on Victorian and Craftsman-style houses for twenty years. You'll see what to repair, what to replace and what to leave, so you can remodel or restyle older homes for the least amount of money and the greatest increase in value.
416 pages, 8½ x 11, $33.50

Roof Framer's Bible

68 different pitch combinations of "bastard" hip roofs at your fingertips. Don't curse the architect — let this book make you an accomplished master of irregular pitched roof systems. You'll be the envy of your crew, and irregular or "bastard" roofs will be under your command. This rare pocket-sized book comes hardbound with a cloth marker like a true bible.
216 pages, 3¾ x 7½, $24.00

Contractor's Plain-English Legal Guide

For today's contractors, legal problems are like snakes in the swamp — you might not see them, but you know they're there. This book tells you where the snakes are hiding and directs you to the safe path. With the directions in this easy-to-read handbook you're less likely to need a $200-an-hour lawyer. Includes simple directions for starting your business, writing contracts that cover just about any eventuality, collecting what's owed you, filing liens, protecting yourself from unethical subcontractors, and more. For about the price of 15 minutes in a lawyer's office, you'll have a guide that will make many of those visits unnecessary. Includes a CD-ROM with blank copies of all the forms and contracts in the book.
272 pages, 8½ x 11, $49.50

Roof Framing

Shows how to frame any type of roof in common use today, even if you've never framed a roof before. Includes using a pocket calculator to figure any common, hip, valley, or jack rafter length in seconds. Over 400 illustrations cover every measurement and every cut on each type of roof: gable, hip, Dutch, Tudor, gambrel, shed, gazebo, and more.
480 pages, 5½ x 8½, $24.50

How to Succeed With Your Own Construction Business

Everything you need to start your own construction business: setting up the paperwork, finding the work, advertising, using contracts, dealing with lenders, estimating, scheduling, finding and keeping good employees, keeping the books, and coping with success. If you're considering starting your own construction business, all the knowledge, tips, and blank forms you need are here. **336 pages, 8½ x 11, $28.50**

National Renovation & Insurance Repair Estimator

Current prices in dollars and cents for hard-to-find items needed on most insurance, repair, remodeling, and renovation jobs. All price items include labor, material, and equipment breakouts, plus special charts that tell you exactly how these costs are calculated. Includes a CD-ROM with an electronic version of the book with *National Estimator*, a stand-alone *Windows*™ estimating program, plus an interactive multimedia video that shows how to use the disk to compile construction cost estimates.
568 pages, 8½ x 11, $54.50. Revised annually

Steel-Frame House Construction

Framing with steel has obvious advantages over wood, yet building with steel requires new skills that can present challenges to the wood builder. This book explains the secrets of steel framing techniques for building homes, whether pre-engineered or built stick by stick. It shows you the techniques, the tools, the materials, and how you can make it happen. Includes hundreds of photos and illustrations, plus a FREE download with steel framing details, and a database of steel materials and manhours, with an estimating program. **320 pages, 8½ x 11, $39.75**

Moving to Commercial Construction

In commercial work, a single job can keep you and your crews busy for a year or more. The profit percentages are higher, but so is the risk involved. This book takes you step-by-step through the process of setting up a successful commercial business; finding work, estimating and bidding, value engineering, getting through the submittal and shop drawing process, keeping a stable work force, controlling costs, and promoting your business. Explains the design/build and partnering business concepts and their advantage over the competitive bid process. Includes sample letters, contracts, checklists and forms that you can use in your business, plus a CD-ROM with blank copies in several word-processing formats for both Mac and PC computers. **256 pages, 8½ x 11, $42.00**

Painter's Handbook

Loaded with "how-to" information you'll use every day to get professional results on any job: the best way to prepare a surface for painting or repainting; selecting and using the right materials and tools (including airless spray); tips for repainting kitchens, bathrooms, cabinets, eaves and porches; how to match and blend colors; why coatings fail and what to do about it. Lists 30 profitable specialties in the painting business.
320 pages, 8½ x 11, $33.00

Builder's Guide to Accounting Revised

Step-by-step, easy-to-follow guidelines for setting up and maintaining records for your building business. This practical guide to all accounting methods shows how to meet state and federal accounting requirements, explains the new depreciation rules, and describes how the Tax Reform Act can affect the way you keep records. Full of charts, diagrams, simple directions and examples, to help you keep track of where your money is going. Recommended reading for many state contractor's exams. Each chapter ends with a set of test questions, and a CD-ROM included FREE has all the questions in interactive self-test software. Use the Study Mode to make studying for the exam much easier, and Exam Mode to practice your skills. **360 pages, 8½ x 11, $35.50**

2006 International Residential Code

Replacing the *CABO One-* and *Two-Family Dwelling Code*, this book has the latest technological advances in building design and construction. Among the changes are provisions for steel framing and energy savings. Also contains mechanical, fuel gas and plumbing provisions that coordinate with the *International Mechanical Code* and *International Plumbing Code*.
578 pages, 8½ x 11, $68.00

Also available: **2003 International Residential Code, $62.00**
2000 International Residential Code, $59.00
2000 on interactive CD-ROM, $48.00

Residential Structure & Framing

With this easy-to-understand guide you'll learn how to calculate loads, size joists and beams, and tackle many common structural problems facing residential contractors. It covers cantilevered floors, complex roof structures, tall window walls, and seismic and wind bracing. Plus, you'll learn field-proven production techniques for advanced wall, floor, and roof framing with both dimensional and engineered lumber. You'll find information on sizing joists and beams, framing with wood I-joists, supporting oversized dormers, unequal-pitched roofs, coffered ceilings, and more. Fully illustrated with lots of photos. **272 pages, 8½ x 11, $34.95**

Contractor's Guide to the Building Code Revised

This new edition was written in collaboration with the International Code Council, writers of the code. It explains in plain English exactly what the latest edition of the *Uniform Building Code* requires. Based on the 1997 code, it explains the changes and what they mean for the builder. Also covers the *Uniform Mechanical Code* and the *Uniform Plumbing Code*. Shows how to design and construct residential and light commercial buildings that'll pass inspection the first time. Suggests how to work with an inspector to minimize construction costs, what common building shortcuts are likely to be cited, and where exceptions may be granted.
320 pages, 8½ x 11, $39.00

Craftsman Book Company
6058 Corte del Cedro
P.O. Box 6500
Carlsbad, CA 92018

☎ **24 hour order line**
1-800-829-8123
Fax (760) 438-0398

In A Hurry?
We accept phone orders charged to your
○ Visa, ○ MasterCard, ○ Discover or ○ American Express

Card#_____
Exp. date_____Initials_____

Order online http://www.craftsman-book.com
Free on the Internet! Download any of Craftsman's estimating database for a 30-day free trial!
www.craftsman-book.com/downloads

Name_____
e-mail address (for order tracking and special offers)
Company_____
Address_____
City/State/Zip ○ This is a residence
Total enclosed_____(In California add 7.25% tax)
We pay shipping when your check covers your order in full.

Tax Deductible: Treasury regulations make these references tax deductible when used in your work. Save the canceled check or charge card statement as your receipt.

Download all of Craftsman's most popular costbooks for one low price with the Craftsman Site License. http://www.craftsmansitelicense.com

10-Day Money Back Guarantee

- ○ 36.50 Basic Engineering for Builders
- ○ 38.00 Basic Lumber Engineering for Builders
- ○ 33.00 Basic Plumbing with Illustrations
- ○ 35.50 Builder's Guide to Accounting Revised
- ○ 78.50 CD Estimator
- ○ 39.50 Construction Estimating Reference Data with FREE *National Estimator* on a CD-ROM
- ○ 41.75 Construction Forms & Contracts with a CD-ROM for *Windows*™ and *Macintosh*™
- ○ 53.00 Contractor's Guide to *QuickBooks Pro* 2007
- ○ 51.50 Contractor's Guide to *QuickBooks Pro* 2006
- ○ 49.75 Contractor's Guide to *QuickBooks Pro* 2005
- ○ 48.50 Contractor's Guide to *QuickBooks Pro* 2004
- ○ 47.75 Contractor's Guide to *QuickBooks Pro* 2003
- ○ 46.50 Contractor's Guide to *QuickBooks Pro* 2002
- ○ 45.25 Contractor's Guide to *QuickBooks Pro* 2001
- ○ 42.00 Contractor's Guide to *QuickBooks Pro* 1999
- ○ 39.00 Contractor's Guide to the Building Code Revised
- ○ 69.95 Contractor's Legal Kit
- ○ 49.50 Contractor's Plain-English Legal Guide
- ○ 38.00 Contractor's Survival Manual Revised
- ○ 65.00 Craftsman's Construction Installation Encyclopedia
- ○ 18.00 Electrical Blueprint Reading Revised
- ○ 39.50 Electrician's Exam Preparation Guide
- ○ 35.00 Estimating Electrical Construction
- ○ 39.95 Estimating with Microsoft Excel
- ○ 22.50 Finish Carpenter's Manual
- ○ 32.75 Handbook of Construction Contracting Volume 1
- ○ 33.75 Handbook of Construction Contracting Volume 2
- ○ 28.50 How to Succeed w/Your Own Construction Business
- ○ 68.00 2006 *International Residential Code*

- ○ 62.00 2003 *International Residential Code*
- ○ 59.00 2000 *International Residential Code*
- ○ 48.00 2000 *International Residential Code* on CD-ROM
- ○ 32.50 Markup & Profit: A Contractor's Guide
- ○ 42.00 Moving to Commercial Construction
- ○ 54.00 National Concrete & Masonry Estimator with FREE *National Estimator* on a CD-ROM
- ○ 52.50 National Construction Estimator with FREE *National Estimator* on a CD-ROM
- ○ 53.75 National Home Improvement Estimator with FREE *National Estimator* on a CD-ROM
- ○ 54.50 National Renovation & Insurance Repair Estimator with FREE *National Estimator* on a CD-ROM
- ○ 53.50 National Repair & Remodeling Estimator with FREE *National Estimator* on a CD-ROM
- ○ 33.00 Painter's Handbook
- ○ 36.50 Plumber's Handbook Revised
- ○ 24.75 Profits in Buying & Renovating Homes
- ○ 33.50 Renovating & Restyling Older Homes
- ○ 38.80 Residential Steel Framing Guide
- ○ 34.95 Residential Structure & Framing
- ○ 32.50 A Roof Cutter's Secrets to Custom Homes
- ○ 24.50 Roof Framing
- ○ 24.00 Roof Framer's Bible
- ○ 26.50 Rough Framing Carpentry
- ○ 19.50 Stair Builder's Handbook
- ○ 39.75 Steel-Frame House Construction
- ○ 25.50 Wood-Frame House Construction
- ○ 38.00 Roofing Construction & Estimating
- ○ FREE Full Color Catalog

Prices subject to change without notice

Craftsman Book Company
6058 Corte del Cedro
P.O. Box 6500
Carlsbad, CA 92018

☎ 24 hour order line
1-800-829-8123
Fax (760) 438-0398

In A Hurry?
We accept phone orders charged to your
○ Visa, ○ MasterCard, ○ Discover or ○ American Express

Card#_____
Exp. date_____Initials_____

Tax Deductible: Treasury regulations make these references tax deductible when used in your work. Save the canceled check or charge card statement as your receipt.

*We pay shipping when your **check** covers your order in full.*

Name_____
e-mail address (for order tracking and special offers)_____
Company_____
Address_____
City/State/Zip_____ ○ This is a residence
Total enclosed_____(In California add 7.25% tax)

Order online http://www.craftsman-book.com
Free on the Internet! Download any of Craftsman's estimating database for a 30-day free trial! www.craftsman-book.com/downloads

- ○ 36.50 Basic Engineering for Builders
- ○ 38.00 Basic Lumber Engineering for Builders
- ○ 33.00 Basic Plumbing with Illustrations
- ○ 35.50 Builder's Guide to Accounting Revised
- ○ 78.50 CD Estimator
- ○ 39.50 Construction Estimating Reference Data with FREE *National Estimator* on a CD-ROM
- ○ 41.75 Construction Forms & Contracts with a CD-ROM for Windows™ and Macintosh™
- ○ 53.00 Contractor's Guide to *QuickBooks Pro* 2007
- ○ 51.50 Contractor's Guide to *QuickBooks Pro* 2006
- ○ 49.75 Contractor's Guide to *QuickBooks Pro* 2005
- ○ 48.50 Contractor's Guide to *QuickBooks Pro* 2004
- ○ 47.75 Contractor's Guide to *QuickBooks Pro* 2003
- ○ 46.50 Contractor's Guide to *QuickBooks Pro* 2002
- ○ 45.25 Contractor's Guide to *QuickBooks Pro* 2001
- ○ 42.00 Contractor's Guide to *QuickBooks Pro* 1999
- ○ 39.00 Contractor's Guide to the Building Code Revised
- ○ 69.95 Contractor's Legal Kit
- ○ 49.50 Contractor's Plain-English Legal Guide
- ○ 38.00 Contractor's Survival Manual Revised
- ○ 65.00 Craftsman's Construction Installation Encyclopedia
- ○ 18.00 Electrical Blueprint Reading Revised
- ○ 39.50 Electrician's Exam Preparation Guide
- ○ 35.00 Estimating Electrical Construction
- ○ 39.95 Estimating with Microsoft Excel
- ○ 22.50 Finish Carpenter's Manual
- ○ 32.75 Handbook of Construction Contracting Volume 1
- ○ 33.75 Handbook of Construction Contracting Volume 2
- ○ 28.50 How to Succeed w/Your Own Construction Business
- ○ 68.00 2006 *International Residential Code*
- ○ 62.00 2003 *International Residential Code*
- ○ 59.00 2000 *International Residential Code*
- ○ 48.00 2000 *International Residential Code* on CD-ROM
- ○ 32.50 Markup & Profit: A Contractor's Guide
- ○ 42.00 Moving to Commercial Construction
- ○ 54.00 National Concrete & Masonry Estimator with FREE *National Estimator* on a CD-ROM
- ○ 52.50 National Construction Estimator with FREE *National Estimator* on a CD-ROM
- ○ 53.75 National Home Improvement Estimator with FREE *National Estimator* on a CD-ROM
- ○ 54.50 National Renovation & Insurance Repair Estimator with FREE *National Estimator* on a CD-ROM
- ○ 53.50 National Repair & Remodeling Estimator with FREE *National Estimator* on a CD-ROM
- ○ 33.00 Painter's Handbook
- ○ 36.50 Plumber's Handbook Revised
- ○ 24.75 Profits in Buying & Renovating Homes
- ○ 33.50 Renovating & Restyling Older Homes
- ○ 38.80 Residential Steel Framing Guide
- ○ 34.95 Residential Structure & Framing
- ○ 32.50 A Roof Cutter's Secrets to Custom Homes
- ○ 24.50 Roof Framing
- ○ 24.00 Roof Framer's Bible
- ○ 26.50 Rough Framing Carpentry
- ○ 19.50 Stair Builder's Handbook
- ○ 39.75 Steel-Frame House Construction
- ○ 25.50 Wood-Frame House Construction
- ○ 38.00 Roofing Construction & Estimating
- ○ FREE Full Color Catalog

10-Day Money Back Guarantee
Prices subject to change without notice

Craftsman Book Company
6058 Corte del Cedro
P.O. Box 6500
Carlsbad, CA 92018

☎ 24 hour order line
1-800-829-8123
Fax (760) 438-0398

In A Hurry?
We accept phone orders charged to your
○ Visa, ○ MasterCard, ○ Discover or ○ American Express

Card#_____
Exp. date_____Initials_____

Tax Deductible: Treasury regulations make these references tax deductible when used in your work. Save the canceled check or charge card statement as your receipt.

*We pay shipping when your **check** covers your order in full.*

Name_____
e-mail address (for order tracking and special offers)_____
Company_____
Address_____
City/State/Zip_____ ○ This is a residence
Total enclosed_____(In California add 7.25% tax)

Order online http://www.craftsman-book.com
Free on the Internet! Download any of Craftsman's estimating database for a 30-day free trial! www.craftsman-book.com/downloads

- ○ 36.50 Basic Engineering for Builders
- ○ 38.00 Basic Lumber Engineering for Builders
- ○ 33.00 Basic Plumbing with Illustrations
- ○ 35.50 Builder's Guide to Accounting Revised
- ○ 78.50 CD Estimator
- ○ 39.50 Construction Estimating Reference Data with FREE *National Estimator* on a CD-ROM
- ○ 41.75 Construction Forms & Contracts with a CD-ROM for Windows™ and Macintosh™
- ○ 53.00 Contractor's Guide to *QuickBooks Pro* 2007
- ○ 51.50 Contractor's Guide to *QuickBooks Pro* 2006
- ○ 49.75 Contractor's Guide to *QuickBooks Pro* 2005
- ○ 48.50 Contractor's Guide to *QuickBooks Pro* 2004
- ○ 47.75 Contractor's Guide to *QuickBooks Pro* 2003
- ○ 46.50 Contractor's Guide to *QuickBooks Pro* 2002
- ○ 45.25 Contractor's Guide to *QuickBooks Pro* 2001
- ○ 42.00 Contractor's Guide to *QuickBooks Pro* 1999
- ○ 39.00 Contractor's Guide to the Building Code Revised
- ○ 69.95 Contractor's Legal Kit
- ○ 49.50 Contractor's Plain-English Legal Guide
- ○ 38.00 Contractor's Survival Manual Revised
- ○ 65.00 Craftsman's Construction Installation Encyclopedia
- ○ 18.00 Electrical Blueprint Reading Revised
- ○ 39.50 Electrician's Exam Preparation Guide
- ○ 35.00 Estimating Electrical Construction
- ○ 39.95 Estimating with Microsoft Excel
- ○ 22.50 Finish Carpenter's Manual
- ○ 32.75 Handbook of Construction Contracting Volume 1
- ○ 33.75 Handbook of Construction Contracting Volume 2
- ○ 28.50 How to Succeed w/Your Own Construction Business
- ○ 68.00 2006 *International Residential Code*
- ○ 62.00 2003 *International Residential Code*
- ○ 59.00 2000 *International Residential Code*
- ○ 48.00 2000 *International Residential Code* on CD-ROM
- ○ 32.50 Markup & Profit: A Contractor's Guide
- ○ 42.00 Moving to Commercial Construction
- ○ 54.00 National Concrete & Masonry Estimator with FREE *National Estimator* on a CD-ROM
- ○ 52.50 National Construction Estimator with FREE *National Estimator* on a CD-ROM
- ○ 53.75 National Home Improvement Estimator with FREE *National Estimator* on a CD-ROM
- ○ 54.50 National Renovation & Insurance Repair Estimator with FREE *National Estimator* on a CD-ROM
- ○ 53.50 National Repair & Remodeling Estimator with FREE *National Estimator* on a CD-ROM
- ○ 33.00 Painter's Handbook
- ○ 36.50 Plumber's Handbook Revised
- ○ 24.75 Profits in Buying & Renovating Homes
- ○ 33.50 Renovating & Restyling Older Homes
- ○ 38.80 Residential Steel Framing Guide
- ○ 34.95 Residential Structure & Framing
- ○ 32.50 A Roof Cutter's Secrets to Custom Homes
- ○ 24.50 Roof Framing
- ○ 24.00 Roof Framer's Bible
- ○ 26.50 Rough Framing Carpentry
- ○ 19.50 Stair Builder's Handbook
- ○ 39.75 Steel-Frame House Construction
- ○ 25.50 Wood-Frame House Construction
- ○ 38.00 Roofing Construction & Estimating
- ○ FREE Full Color Catalog

10-Day Money Back Guarantee
Prices subject to change without notice

Mail This Card Today
For a Free Full Color Catalog

Over 100 books, annual cost guides and estimating software packages at your fingertips with information that can save you time and money. Here you'll find information on carpentry, contracting, estimating, remodeling, electrical work, and plumbing.

All items come with an unconditional 10-day money-back guarantee.
If they don't save you money, mail them back for a full refund.

Name_____
e-mail address (for special offers)_____
Company_____
Address_____
City/State/Zip_____

Craftsman Book Company / 6058 Corte del Cedro / P.O. Box 6500 / Carlsbad, CA 92018

Download all of Craftsman's most popular costbooks for one low price with the Craftsman Site License. http://www.craftsmansitelicense.com

BUSINESS REPLY MAIL
FIRST CLASS MAIL PERMIT NO. 271 CARLSBAD, CA

POSTAGE WILL BE PAID BY ADDRESSEE

Craftsman Book Company
6058 Corte del Cedro
P.O. Box 6500
Carlsbad, CA 92018-9974

NO POSTAGE
NECESSARY
IF MAILED
IN THE
UNITED STATES

Download all of Craftsman's most popular costbooks for one low price with the Craftsman Site License. http://www.craftsmansitelicense.com

BUSINESS REPLY MAIL
FIRST CLASS MAIL PERMIT NO. 271 CARLSBAD, CA

POSTAGE WILL BE PAID BY ADDRESSEE

Craftsman Book Company
6058 Corte del Cedro
P.O. Box 6500
Carlsbad, CA 92018-9974

NO POSTAGE
NECESSARY
IF MAILED
IN THE
UNITED STATES

Download all of Craftsman's most popular costbooks for one low price with the Craftsman Site License. http://www.craftsmansitelicense.com

BUSINESS REPLY MAIL
FIRST CLASS MAIL PERMIT NO. 271 CARLSBAD, CA

POSTAGE WILL BE PAID BY ADDRESSEE

Craftsman Book Company
6058 Corte del Cedro
P.O. Box 6500
Carlsbad, CA 92018-9974

NO POSTAGE
NECESSARY
IF MAILED
IN THE
UNITED STATES